MINE SURVEYING AND MINING AREA PROTECTION

The conference on Mine Surveying and Mining Area Protection is devoted to exploring both theoretical and practical aspects of surface and mine surveying, with a special emphasis on new measurement techniques and technologies as well as mining area protection.

This proceedings compilation contains cutting-edge research and innovative insights into various aspects of surface deformation, including description methods, measurement techniques, and accuracy of forecasts. It also looks at the development of post-mining areas and hazard monitoring and management. Through diverse case studies and practical examples, this book serves as an essential resource for both industry professionals and research and development teams.

This book is intended for a diverse audience including engineering students, academic researchers, industry practitioners, regulatory specialists, mining consultants, surveying professionals, government authorities, and individuals interested in mining operations and oversight.

PROCEEDINGS OF THE 18TH CONFERENCE ON MINE SURVEYING AND PROTECTION OF MINING AREAS, USTROŃ, POLAND, OCTOBER 22ND–24TH, 2025

Mine Surveying and Mining Area Protection

Edited by

Violetta Sokoła-Szewioła and Patrycja Jarczyk

Silesian University of Technology, Poland

CRC Press
Taylor & Francis Group
Boca Raton London New York Leiden

CRC Press is an imprint of the
Taylor & Francis Group, an **informa** business

First edition published 2025
by CRC Press
4 Park Square, Milton Park, Abingdon, Oxon, OX14 4RN

and by CRC Press
2385 NW Executive Center Drive, Suite 320, Boca Raton FL 33431

CRC Press is an imprint of Informa UK Limited

British Library Cataloguing-in-Publication Data
A catalogue record for this book is available from the British Library

ISBN: 978-1-041-10531-2 (hbk)
ISBN: 978-1-041-10533-6 (pbk)
ISBN: 978-1-003-65555-8 (ebk)

DOI: 10.1201/9781003655558

Typeset in Times New Roman
by MPS Limited, Chennai, India

 ISBN: 978-1-041-10531-2

Table of contents

Preface ix
Editor biographies xi
Organizing committee xiii
Scientific committee xv
Partners xvii
Partners of the conference xix

Long-term ground subsidence caused by historical coal mining: Case study of Eastern Canada 1
B. Apanowicz & A. Szostak-Chrzanowski

POMHAZ project – spatial decision support system for support of the development of towns in post-mining areas in Poland 12
B. Apanowicz, M. Wysocka, K. Niedbalska, P. Gruchlik, K. Kura, M. Bonczyk & A. Kowalski

Multi-stage reconstruction and long-term maintenance of the road viaduct over DK81 in relation to mining exploitation and economic transformation in Poland 20
P. Bętkowski

Analysis of ground movements in sinkhole prone post-mining area: Case study of the Babina mine in SW Poland 30
J. Blachowski, N. Walerysiak & P. Modlińska

Comprehensive evaluation of mining-induced deformations and their consequences for a residential structure 40
I. Bryt-Nitarska, M. Pomykoł & K. Szafulera-Bałecka

Natural and anthropogenic rock mass deformations as threats to sustainable development 50
L. Florkowska & I. Bryt-Nitarska

Analysis of building reinforcement against discontinuous deformation in mining terrain 60
K. Gromysz

Measurement of water surface slope using tachymetric observations 69
M.F. Gruszczyński, J. Błotnicki, W. Sowa, A. Pałęcki, K. Stasch & M. Popczyk

Assessment of mining impacts at the location of a building situated in the area of a former mine – A case study 79
P. Jarczyk & K. Szafulera-Bałecka

Integrated diagnostics of a Heritage Palace: A multi-modal sensor fusion approach with machine learning for structural assessment 89
K. Karsznia & K. Książek

An integrated approach to assessing mine shaft deviation: Surveying measurements versus theoretical calculations 101
M. Kruczkowski & G. Dyduch

Uncertainty of land subsidence model parameter values and its impact on the risk assessment of buildings 111
A. Kwinta, R. Hejmanowski, A. Malinowska & Hua-yang Dai

Multivariate analysis of the accuracy of the basic geodetic control network for guiding the axis of the tunnel built using a TBM machine on the LK104 railway line 127
T. Lipecki & W. Jaśkowski

A multi-criteria framework for assessing groundwater retention potential 136
S. Łabojko, A. Malinowska & B. Stypiński

Geographical information systems in terms of available data sources on mining areas 145
A. Mierzejowska

Accuracy analysis of the digital elevation model made on the basis of a low-altitude photogrammetry 155
J. Paszek & J. Mazur

Delineation of the potentially flooded area in the rock mass 164
M. Poniewiera, M. Pomykoł & P. Pierzyna

Correction of errors due to earth's curvature using selected photogrammetric software 173
M. Poniewiera

Sustainable post-mining transition in Europe: Integrating environmental engineering and urban regeneration 181
R. Rodríguez-Galindo Díez, M. Menéndez Fernández, L.A. Menéndez-García & V. Sokoła-Szewioła

Remote sensing and low-altitude photogrammetry in monitoring the condition of mining waste disposal sites 187
Z. Różański, A.P. Niewiadomski, G. Pach, P. Wrona, L. Balcarczyk, J. Pawliński & A. Wala

Numerical assessment of building safety under mining-induced ground curvature: Verification of ITB instruction 364/2007 197
L. Słowik, L. Chomacki & J. Rusek

Forecasting changes in seismicity induced by mining operations based on variations in mining depth 206
V. Sokoła-Szewioła & P. Jarczyk

Towards the use of low-cost sensors in automated systems for monitoring ground surface vibrations 217
L. Stolecki & K. Fuławka

Applications and selection criteria of fly ash from hard coal fired power plants for post-mining activities in Upper Silesia Coal Basin 226
G. Strozik

Factors affecting structure damage in a post-mining area: Case study 238
P. Strzałkowski & L. Szojda

Factors influencing the accuracy of recording resources extracted from a deposit exploited beneath the water table 248
A. Szafarczyk

Assessment of drainage condition of structures using UAVs- case studies 255
M. Wróblewska & M. Grygierek

Analysis of the impact of Earth's curvature on the transformation of boundary point coordinates using the Method of Independent Models 263
M. Żogała, V. Sokoła-Szewioła & M. Poniewiera

Author index 273

Preface

Mine Surveying and Protection of Mining Areas has been organized in Poland every two years since 1991. The first Surveying Days were organized by the University of Science and Technology in Krakow, the second in 1993 by the Central Mining Institute in Katowice, and the third in 1995 by the Silesian University of Technology in Gliwice. The conferences are organized in this cycle until 2025. The current 18th Days of Mine Surveying and Protection of Mining Areas are organized by the Silesian University of Technology, Faculty of Mining, Safety Engineering and Industrial Automation.

Conference participants include practitioners from the mining industry, academics, researchers, consultants, including mining surveyors, specialists in the field of mining area protection, geomechanics, geophysicists, specialists in construction in mining areas, representatives of mining supervision and geodetic and cartographic supervision authorities, engineering students, representatives of mining municipalities and interested individuals and groups from the wider community. The presentations are given by people working in very diverse environments, with extensive knowledge, who present their unique problems and common solutions, including speakers from abroad.

The conference can be considered the most prestigious industry event due to the importance of the issues addressed for the mining sector. It provides a forum for the exchange of experiences between science and industry, enables the discovery of new theoretical solutions and innovative technical solutions, and facilitates discussion of challenges and the joint creation of development directions for safe mining operations while minimising their harmful impact on the surface, taking into account environmental and climate protection requirements. In connection with the implementation in Poland of the European Union's policy to achieve climate neutrality by 2050, this series of conferences also addressed issues related to post-mining areas, in particular in terms of the safe development of such areas.

Therefore, the theme of the current *Conference* is *'Energy Transition Challenges'*, and the conference programme covers issues related to surface and mining surveying, in particular new measurement techniques and technologies, including measurements of surface and building deformation, protection of mining areas, including new methods of describing surface deformation, deformation of discontinuous surfaces, accuracy of continuous deformation forecasting, and hazards and development of post-mining areas and monitoring of these hazards.

In the current edition of the Conference, the following range of detailed themes has been proposed:

- legal, social, and economic aspects of mining area protection,
- modern technologies in mine surveying,
- prediction of surface deformation and its consequences,
- monitoring of the effects of mining operations on the rock mass, surface, and structures,
- resistance of building structures in mining and post-mining areas, prevention, and restoration of damage,
- mining maps and the integration of spatial data in mining and post-mining areas,
- hazards and management of post-mining areas during the transition period,
- transfer of surveying and geological documentation from closed mining facilities to the State Mining Authority.

We hope that the Conference will be a source of inspiration for you and will provide many fruitful opportunities for future cooperation between Participants.

We wish you a very productive *Conference,* a lively exchange of knowledge and many interesting discussions.

Gliwice, October 22–24, 2025
On behalf of the Scientific and Organizing Committee of the Conference

The Chairman of Scientific Committee
Prof. SUT, Violetta Sokoła-Szewioła, PhD, DSc, Eng.

Mine Surveying and Mining Area Protection – Sokoła-Szewioła and Jarczyk (Eds)
© 2025 The Editor(s), ISBN: 978-1-041-10531-2

Editor biographies

Violetta Sokoła-Szewioła, PhD, DSc, Eng., prof. SUT, is Deputy Dean for Science and Cooperation at the Faculty of Mining, Safety Engineering and Industrial Automation, Silesian University of Technology. Her research focusses, among others, on issues related to underground mining activities, such as its impact on rock masses and mining terrain in seismically threatened areas. She has developed a new method to forecast seismic threats based on continuous observations of subsidence in mining areas. She also deals with monitoring of deformation and seismic activity in post-mining areas. She has been part of many research projects, including those financed by the European Union. She has authored over 130 publications and received various awards for her scientific research. She is a member of the Commission for the Protection of Mining Areas of the Polish Academy of Sciences. Since 2012, she has been assigned as the expert in the Technical Group on Coal operating at the European Commission.

Patrycja Jarczyk is a researcher and academic teacher at the Faculty of Mining, Safety Engineering and Industrial Automation, Silesian University of Technology in Gliwice. She is the author of numerous scientific publications and conference presentations. Her research interests include underground construction, rock and rock mass mechanics, as well as mining-induced seismicity. She is also actively involved in the field of geodesy and cartography, being the author and co-author of many studies. She has participated in several international projects concerning seismicity in mining and post-mining areas.

 ISBN: 978-1-041-10531-2

Organizing committee

Professor VIOLETTA SOKOŁA-SZEWIOŁA, Vice Dean for Science and Cooperation of the Faculty of Mining, Safety Engineering, and Industrial Automation at the Silesian University of Technology
Professor RYSZARD MIELIMĄKA, Silesian University of Technology, Poland
Professor GRZEGORZ STROZIK, Silesian University of Technology, Poland
Doctor GRZEGORZ DYDUCH, Silesian University of Technology, Poland
Doctor JOANNA HERCZAKOWSKA, Silesian University of Technology, Poland
Doctor PATRYCJA JARCZYK, Silesian University of Technology, Poland
Doctor ALEKSANDRA MIERZEJOWSKA, Silesian University of Technology, Poland
Doctor ADAM NIEWIADOMSKI, Silesian University of Technology, Poland
Doctor JUSTYNA PASZEK, Silesian University of Technology, Poland
Doctor MACIEJ POMYKOŁ, Silesian University of Technology, Poland
Doctor MARIAN PONIEWIERA, Silesian University of Technology, Poland
Doctor MARCIN POPCZYK, President of the Management Board of the Foundation for the Faculty of Mining, Safety Engineering, and Industrial Automation at the Silesian University of Technology
Doctor MONIKA ŻOGAŁA, Silesian University of Technology, Poland
MSc Eng SZYMON SZCZERBA, Silesian University of Technology, Poland

© 2025 The Editor(s), ISBN: 978-1-041-10531-2

Scientific committee

Mine Surveying and Mining Area Protection – Sokoła-Szewioła and Jarczyk (Eds)
© 2025 The Editor(s), ISBN: 978-1-041-10531-2

Partners

Main Partner

Partners of the conference

tpi

- when it has to be right Leica Geosystems

HyperView

GEOMATIX®

GEOTRONICS DYSTRYBUCJA

JSW SA

Mine Surveying and Mining Area Protection – Sokoła-Szewioła and Jarczyk (Eds)
© 2025 The Author(s), ISBN: 978-1-041-10531-2

Long-term ground subsidence caused by historical coal mining: Case study of Eastern Canada

B. Apanowicz
Central Mining Institute – National Research Institute, Katowice, Poland

A. Szostak-Chrzanowski
University of New Brunswick, Fredericton, New Brunswick, Canada

ABSTRACT: Extraction of underground solids causes changes of the initial state of stress in the rock mass and results in ground surface deformation, known as subsidence. The magnitude of the subsidence depends on the geometry of the underground excavation, depth of the mining operation, mining method, mechanical characteristics of the rock mass. The magnitude of subsidence may be determined either from surface monitoring surveys or using empirical or deterministic prediction theories. The goal of the study was to determine direct and residual ground surface subsidence caused by historical coal mining in the Coal Field in Nova Scotia Canada. Direct surface subsidence was calculated using Knothe's empirical model. Long-term subsidence was determined using satellite radar interferometry (SAR). The results of InSAR analysis gave long-term (recent 8 years) subsidence reaching in some places 56 mm. The study showed that long-term subsidence may last even more than 60 years after mining activities.

1 INTRODUCTION

Extraction of underground solids causes changes of the initial state of stress in the rock mass and results in ground surface deformation, known as subsidence. The magnitude of the subsidence depends on the geometry of the underground excavation, depth of the mining operation, mining method, mechanical characteristics of the rock mass that include the type of the rock and geomechanical parameters, and initial stress (Szostak-Chrzanowski and Chrzanowski 2011).

Time dependent subsidence can be divided into two types: direct and long-term (residual). Direct subsidence occurs on the surface during the extraction and shortly after it is completed. The subsidence vector at any point located on the surface has two components: vertical and horizontal. The following parameters describing a subsidence may be identified: slope, horizontal strain, and curvature. In case of coal mining long-term residual deformations appear long after extraction was completed (Kowalski *et al.* 2000).

Long-term residual ground subsidence caused by historical coal extraction create a significant challenge to the sustainable land management. Even dozens of years after mining is finished, the development of horizontal and vertical ground displacements may impact the environment, and consequently the stability of infrastructure and the safety of land residents. The land management is particularly complex in the case of historical mines, where is the lack of geological data and mining documentation is limited. The lack of appropriate documentation typically was caused by insufficient legal regulations and administrative changes in areas of mining. In the past, mining activity was conducted when most countries were not bound by rigorous laws concerning requirements of conducting surface surveys and keep precise geological and mining documentation. Furthermore, administrative changes in some geographical areas resulted in loss or dispersion of documentation related to mining

DOI: 10.1201/9781003655558-1

activities. Examples of long-term surface changes caused by underground coal mining include mining in Atlantic Canada (Bazanowski and Szostak-Chrzanowski 2016), the Upper Silesian Coal Basin (USCB) (Apanowicz *et al.* 2024) or the brown coal mining "Babina" in Poland (Buczyńska and Blachowski 2023), coal mining in Belgium and France (Declercq *et al.* 2023, Gueguen *et al.* 2009).

The subsidence may be determined using results of surface surveying, empirical and deterministic prediction theories. Surveying methods such as levelling, Global Navigation Satellite System (GNSS), and lately also Satellite Radar Interferometry (SAR) may be used in determination of subsidence. Traditional precise levelling enables performing measurements with sub-millimetre accuracy. However, it can only be applied for measuring the vertical component of mining-induced deformations. GNSS allows measuring the deformations in three dimensions (X, Y and Z coordinates), but achieving accuracy comparable to levelling in this case requires very long measurement sessions that span many hours. Satellite radar interferometry measures the ground displacement towards the satellite's line of sight (LOS), that is at a set angle. Afterwards, by decomposing the two displacement vectors towards the LOS, it is possible to calculate the vertical and horizontal displacements running east-west and north-south (Hanssen 2001).

Both levelling and GNSS surveys gives information at discrete points. SAR gives the continuous field of displacements. Use of SAR is limited with respect to the quality of signal reflection. SAR should not be use in wooded areas, areas covered by high vegetation or snow-covered areas (Ferretti *et al.* 2007). Another benefit of SAR is the availability of archive satellite images taken by satellite radar interferometry, which are kept by the European Space Agency (ESA) as part of the Copernicus program. This makes it possible to perform back-analysis of subsidence.

The subsidence prediction theories are based on empirical (stochastic) or deterministic modeling. The empirical models are formed based on functions describing a relationship between the geometry of the subsidence trough and dimensions and depth of the underground openings with a use of empirically determined parameters. The deterministic models are derived from the a priori known mechanical properties of rocks, from mechanical relations between the loads and deformations, and from the physical laws governing the stress-strain relation.

In simple geological and mining conditions, either empirical or deterministic methods may be used in modeling and predicting ground deformations. The use of empirical models requires long history of surveys in order to determine the empirical parameters of the mining area. However, in cases of complex geology and mining conditions, deterministic methods must be used. The deterministic methods give an advantage of supplying information on the stress redistribution within the rock mass. Application of deterministic methods requires very careful scaling of the rock mass parameters and systematic investigation of the model sensitivity to changes of the parameters. Physical interpretation of ground deformations must be based on an integration of modeling results with the results of 3-D monitoring of displacements.

Most of the empirical theories for predicting subsidence have been developed in the coal mining industry in central Europe and in the United Kingdom where systematic monitoring surveys have been conducted in hundreds of mining areas for several decades. Among the empirical theories, there are theories based on the so called 'influence functions' and 'profile functions' (Kratzsch 1983). Within the theories based on influence functions, Knothe's theory (Knothe 1984), developed in Poland, has gained the most popularity and has been used (with some modifications) in many countries. Among the 'profile function' theories, the British theory (NCB 1975) has found many applications in coal mines with long wall mining method. The NCB theory is based on a graphical presentation of the relationship between width and depth of the long walls. In both, the 'influence function' and 'profile function' models, some coefficients of the functions must be calibrated empirically through a comparison with many years of the observed subsidence. Since other parameters, such as rock geomechanical properties and tectonic stresses, are not taken into account, the prediction theories are applicable only to the areas where the mining, geological, and tectonic conditions are the same or very similar to the area where the empirical data for the theory had been collected. The empirical theories are generally not reliable in cases of complicated geometry of mined deposits and in areas of previous extensive mining operations.

Deterministic modeling of mining subsidence is performed using principles of continuum mechanics. Rock mass in its natural geological state is largely discontinuous, anisotropic, and inhomogeneous. Rock mass may be presented as an assemblage of intact rock blocks, called rock material, separated by different types of geological discontinuities. The strength of rock mass depends on the strength of rock material and the strength of geological discontinuities. Since it is almost impossible to define in practice all the joint blocks, one can assume that rock mass may behave, in a global sense, like a continuous body. Therefore, a continuum mechanics deterministic model can be used with the effect of discontinuities adequately considered in the model. This is achieved by a large-scale approach, where equivalent geomechanical properties of the rock mass are derived based on the geometry of the contained fracture system and physical properties of the intact rock mass and the fractures (Szostak-Chrzanowski and Chrzanowski 2021).

Deterministic method used to calculate subsidence caused by underground coal mining was first used by Sałustowicz (1953). He used closed form solution with the assumption that the overburden was behaving as an elastic beam. Presently numerical methods such as the finite element method (FEM) (Bazanowski and Szostak-Chrzanowski 2016, Szostak-Chrzanowski and Chrzanowski 2021, Warchala *et al.* 2017, Warchala and Szostak-Chrzanowski 2020), the boundary element method (BEM) and the finite difference method (FDM) (Parmar *et al.* 2019, Sikora and Wesołowski 2021) are being used to calculate subsidence.

The goal of the presented studies was to determine a direct and residual mining-induced ground subsidence in the area of the historical coal mining in Nova Scotia, Canada. Because of the historical character of mining there was very limited information relating to mining method and geology. The direct subsidence was calculated using Knothe's influence function using extensive assumptions relating to mining method and values of empirical coefficients. The residual subsidence was determined using InSAR results obtained for the period of the last 8 years using satellite images taken by the Sentinel-1 satellite constellation. The image processing and mining-induced deformation determination was performed by the advanced SBAS (small baseline) time series method (Berardino *et al.* 2002).

2 SATELLITE RADAR INTERFEROMETRY

Satellite radar interferometry (SAR) is a known remote sensing technique using synthetic aperture radar (InSAR) mounted on satellites circling low Earth orbits (LEO). The principle of SAR operation consists of sampling a given fragment of land from various satellite positions and at various angles. The radars are active sensors that allow operation regardless of the time of day or weather conditions (Zebker and Goldstein 1986).

InSAR technique is based on analysing the phase differences of radar signals reflected from the ground and registered by the satellite at various points in time or from various positions. The principle of operation consists in calculating the difference between two satellite images (SAR images) that contain information on the reflected signal phase and amplitude for the same area. The result of this process is a matrix (interferogram) that contains the phase difference, that is the interferometric phase, described by (Ferretti *et al.* 2007):

$$\Delta\varphi = \frac{4\pi}{\lambda} \times \Delta r \tag{1}$$

where: λ is the radar wavelength; Δr is the difference in distance between the ground surface and one or another satellite, or the same satellite but at various points in time. By using precise satellite orbit parameters, it is possible to obtain an interferometric phase representing the topography of the surveyed area, what applies to the case when no ground movement is recorded between the times when the two SAR images are taken by the satellite. In practice the InSAR measurement is influenced by numerous factors. To obtain the relative ground displacement towards the satellite line of sight (LOS), the wrapped phase requires unwrapping, which is typically done using two-dimensional statistical methods (Chen and Zebker 2002).

InSAR was used to determine mining-induced deformations in Australia (Hay Man Ng *et al.* 2010), China (Chen *et al.* 2020) and Poland (Apanowicz *et al.* 2024).

3 SMALL BASELINE METHODS

The small baseline method (SBAS) is one of the methods for processing SAR images taken as part of InSAR measurements (Berardino *et al.* 2002). It belongs to the group of time series analysis methods. The method involves ground displacement determination based on series of interferograms generated from SAR image pairs with small spatial and temporal baselines. A spatial baseline is the perpendicular distance between two orbital passes of two satellites or the same satellite but at various points in time. A temporal baseline corresponds to the time interval between the taking of two SAR images that are afterwards used to calculate the interferograms. The interferometric phase in the SBAS method is expressed per relationship:

$$\Delta\phi(r,A) = \frac{4\pi}{\lambda}[d(t_B,r,A) - d(t_A,r,A)] + \frac{4\pi B_n \Delta z}{\lambda r \sin\theta} \\ + [\phi_{atm}(t_B,r,A) - \phi_{atm}(t_A,r,A)] + \Delta n, \quad \forall = 1, ..., M \tag{2}$$

where: r, A are the coordinates of a given pixel within range r and at azimuth A; t_B and t_A is the time when a given image was taken; ϕ is the signal phase value; d is the total displacement towards the satellite LOS; Δz is the topographic error; ϕ_{atm} is the phase component determined by atmospheric delays; Δn is the phase component determined by phase noise.

The SBAS method also allows reducing atmospheric delays, which usually exhibit spatial correlation and may otherwise affect the accuracy of ground displacement results.

4 KNOTHE'S EMPIRICAL MODEL OF SUBSIDENCE

Knothe's influence function method was developed in the 1950s and uses the Gaussian function to describe the profile of a subsidence trough (Knothe 1984). Two horizontal Cartesian coordinate systems are assumed for the three-dimensional description of the ground displacement area, one of which is connected to the mined deposit ($\xi\eta$ system), and the other to any calculation point P located on the surface (x, y). In such a coordinate system, the influence function approximated by the Gaussian is given as:

$$w(x,y) = \frac{w_{\max}}{r^2}\iint_S e^{\left[-\frac{\pi}{r^2}\left[(\xi-x)^2+(\eta-y)^2\right]\right]} d\xi d\eta \tag{3}$$

where: w_{max} is the maximum ground subsidence; r is the critical influence radius, S is the mined area; dS is the surface area of an element. Differentiating (3) yields relationships that determine deformation parameters, such as: slope (T), horizontal strain (ε), curvature and radius of curvature (R).

Knothe's theory formulation includes two empirical parameters a and r, where *a* is the extraction coefficient dependent on the mining method, and is given:

$$a = \frac{w_{\max}}{g} \tag{4}$$

where: w_{max} is the maximum ground subsidence; g is the mined deposit thickness. The second parameter is critical radius related to the tangent of influence angle (tgβ) what expressed as:

$$tg\beta = \frac{H}{r} \quad (5)$$

where: H is the mined deposit depth; r is the critical radius.

5 MINING IN NOVA SCOTIA

The analysed historical mining area is located in the northern part of the Nova Scotia province in Canada. Four coal seams are located in the analysed mining area with two coal seams being extracted. Extracted seams Acadia and Scott had an average depth of about 70 and 130 m respectively. The average coal seams thickness ranged between 4.4m - 4.6 m. The average coal seam dip in the area was 20°. The coal seams were extracted in the years 1866-1963, using a pillar and room method (Table 1). Location of the extraction area is shown in Figure 1.

Table 1. Geological and mining data.

Coal seam	Depth [m]	Thickness [m]	Dip [°]	Azimuth of dip [°]	Commencement date	Conclusion date
Acadia	70	4.4	20	58	1866	1963
Scott	130	4.6	21	49	1890	1953

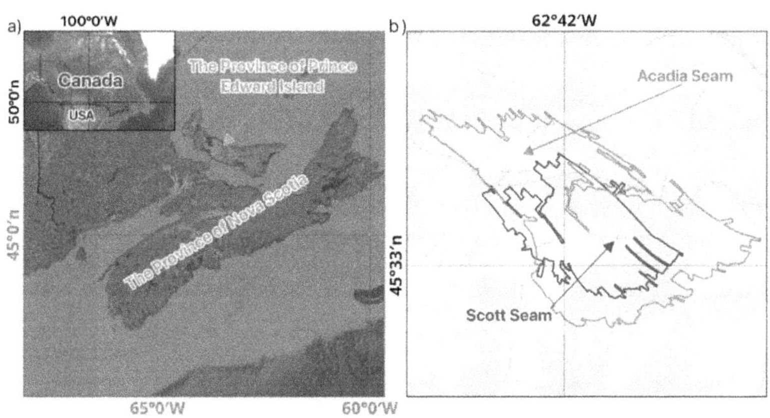

Figure 1. a) Location of the Nova Scotia province; b) Location of the area of study: boundaries of the mined coal seams Acadia (red line) and Scott (blue line).

6 DETERMINATION OF DIRECT GROUND SUBSIDENCE USING KNOTHE'S METHOD

The direct subsidence caused by mining activity was determined using Knothe's method. The empirical Knothes coefficients had to be assumed using data from other locations because of the lack of geodetic surveys data. The extraction coefficient was defined as the value commonly characterised room and pillar mining method and was accepted as 0.6. The range of primary influence was determined based on relationships between the measured average mining-induced deformation range (L_r), defined as 1,22H (Apanowicz 2023), and the theoretical mining-induced deformation range (L_t), adopted in Knothe's model as 1,5r, where: H is the mined deposit depth; r is critical radius. By using formulas (5), above assumptions (L_r and L_t), and assuming that in the studied case the actual range is equal to the theoretical range, the primary influence was calculated as:

$$tg\beta = 1,23 \tag{6}$$

The model data were used as described earlier. The the mined deposit thickness, which was based on the technical capabilities available for the years 1866-1963 and a local field study conducted at the location of mining was assumed to be 2 m (Table 2).

Table 2. Model parameters.

Coal seam / Parameter	Extraction coefficient, a	Rock mass parameter, tgβ	Depth, H [m]	Thickness, g [m]	Dip, α [°]	Azimuth of dip, Az [°]
Acadia	0.6	1.23	70	2.0	20	58
Scott	0.6	1.23	130	2.0	21	49

The custom software SZKODY 5.0 developed at the Central Mining Institute – National Research Institute (GIG-PIB) in Poland which includes Knothe's prediction model (Jędrzejec 2008) was used to calculate the deformation parameters.

Calculated subsidence had irregular trough with a maximum subsidence reaching 2.5 m, and a slope and horizontal strain reaching 15 and 16 mm/m respectively (Figure 2). This indicated mining terrain category V. The subsidence trough had a steplike distribution and while the subsidence within a single mined coal seam (Acadia) reached 1.0 m, a stepwise increase of the subsidence to a level of 2.5 m occurred within the area of two coal seams (Acadia and Scott) overlaying. The subsidence trough was characterised by very high slope and horizontal strain increments, with maximum values at a distance of 100 m - 120 m. The maximum subsidence was located in the central part of the subsidence trough, while the slope and horizontal strain overlapped the boundary of the mined coal seam. There is no classic asymmetry of influence associated with the dip of mined strata, which is also related to the shallow depth of the mined coal seams.

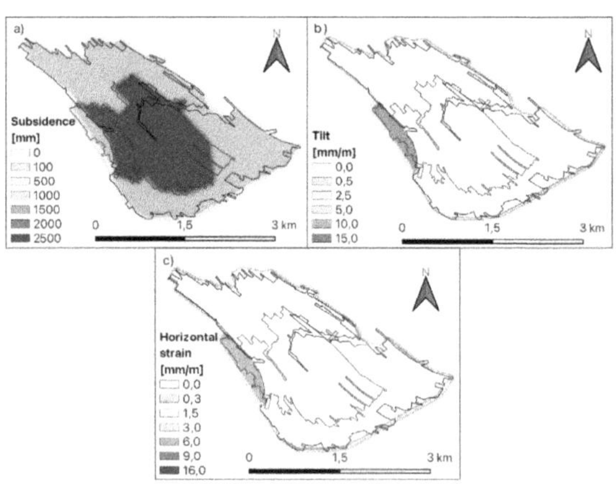

Figure 2. Surface deformation, Acadia (black line) and Scott (blue line); a) subsidence [mm]; b) slope [mm/m]; c) strain [mm/m].

7 DETERMINATION OF RESIDUAL GROUND SUBSI DENCE BY SBAS METHOD

The calculations used SAR images taken by the Sentinel-1 satellite constellation were processed by SBAS. The calculations were performed using the GMTSAR software (Sandwell *et al.* 2011). The image time series encompassed 231 images taken from October 2, 2016 to August 9, 2024. The spatial and temporal baseline criteria were adopted at 200 m and 60 days respectively. The adopted values made it possible to optimise the calculation process and avoid network continuity loss, while minimising the occurrence of signal decorrelation resulting from excess baselines. The phase was unwrapped by two-dimensional phase unwrapping implemented in the SNAPHU application (Chen and Zebker 2002), and the topography-dependent phase component was offset by means of SRTM-1 data (Farr *et al.* 2007). The first obtained results exhibited a problem of significant noise, which was a consequence of the difficulty posed to InSAR technology by the area of study. The vicinity of test site exhibits a high level of vegetation due to the presence of woodland areas. In order to eliminate this problem, signal filtration was applied by fitting the unwrapped phase to a low-degree polynomial trend by the method of least squares using an algorithm implemented in the GMT software (Wessel *et al.* 2019). This process enabled the elimination of phase noise from the obtained results to a degree acceptable in the further analysis. The calculation diagram is presented in Figure 3.

The calculation process yielded long-term ground displacements towards the LOS occurring from October 2016 to August 2024 in the analysed area. The total slope and horizontal strain on the surface were calculated using the relationships that define the deformation parameters, which made it possible to determine the mining land category 61 years after the mine closure for last 8 years. In order to obtain the maximum values, the parameters were calculated at the locations where the maximum values were found, the slope was calculated between the maximum subsidence and half the maximum subsidence, and the horizontal strain was calculated between the maximum subsidence and one quarter of the maximum subsidence.

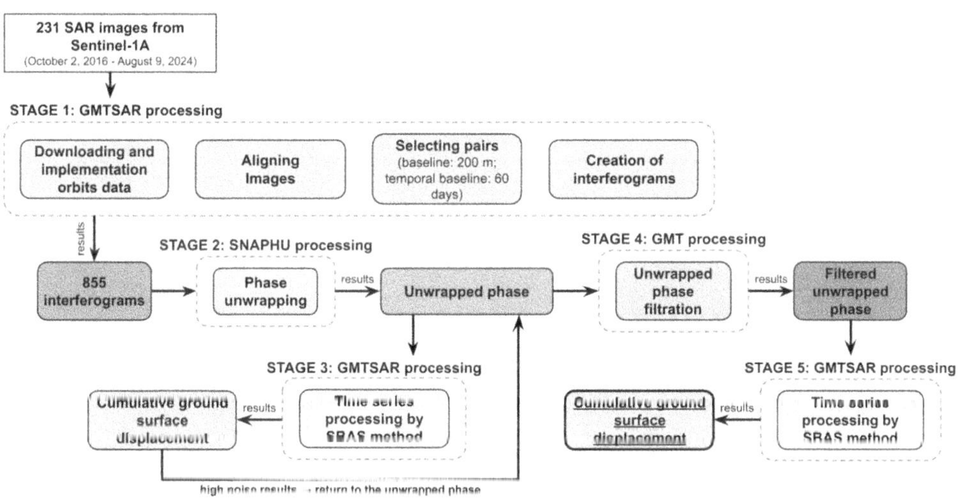

Figure 3. Diagram of SAR image processing by SBAS.

8 RESIDUAL GROUND DEFORMATION DETERMINED BY SBAS METHOD

The identified surface subsidence towards the LOS over the entire analysed time period ranged within +70 mm to -60 mm. In the central part, that is where the extraction was conducted in two coal seams, a zone of subsidence reached 56 mm. It was characterised by maximum subsidence of 50 mm (Figure 4). Given the temporal baseline and the level of subsidence, it can be concluded that the ground in this area underwent subsidence at an average rate of up to about 7 mm/year. The topographic changes as a function of time are presented in graphs prepared for selected scattered points numbered 1 - 4 (Figure 6) and along profiles A-A’, B-B’, C-A’ and D-D’ (Figure 5). The largest subsidence was found at the 208th m of profile C-A’, that is in point 3, and at the 208th m of profile D-D’, that is in point 4 (Figures 5–6). In the first case, the subsidence reached 56 mm and occurred at the end of the analysed period, in August 2024. In the second case, the largest subsidence occurred in October 2022 and April 2023 and reached 59 mm. Afterwards, during the next year, a slow uplift of the ground by 10 mm can be observed, as well as a stabilisation of the subsidence at a level of about 50 mm. By analysing the ground subsidence as a function of time in half-year intervals (Figure 6), The largest changes in subsidence were between April 2019 and October 2022. A zone of acceleration can be seen in the charts during this period.

The slope and horizontal strain were calculated along the established profiles. They ranged within 0.15 - 0.39 mm/m and 0.022 - 0.093 mm/m respectively. The greatest slope of 0.39 mm/m occurred along the southwestern part of profile C-A’, while the greatest horizontal strain of 0.093 mm/m was found along the northeastern part of profile C-A’. These are very low values which indicated that the area constitutes mining land of category 0. The values obtained along the profiles are shown in Table 3.

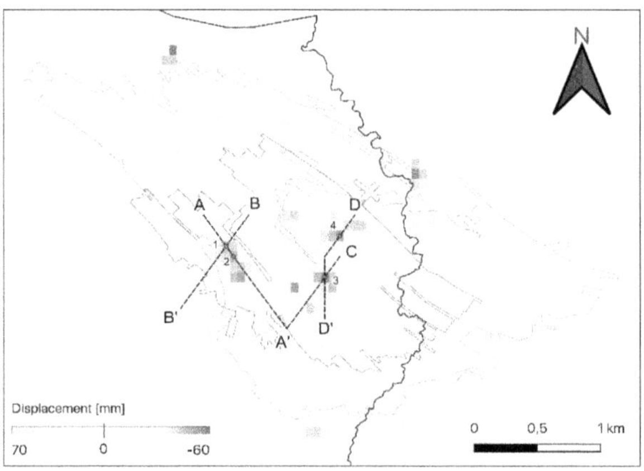

Figure 4. Subsidence - October 2016 to August 2024 (coal seam Acadia - red line; Scott - blue line).

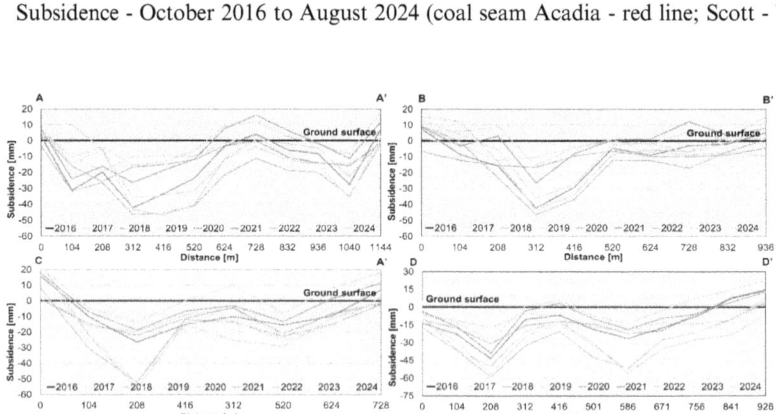

Figure 5. Subsidence from October 2016 to August 2024.

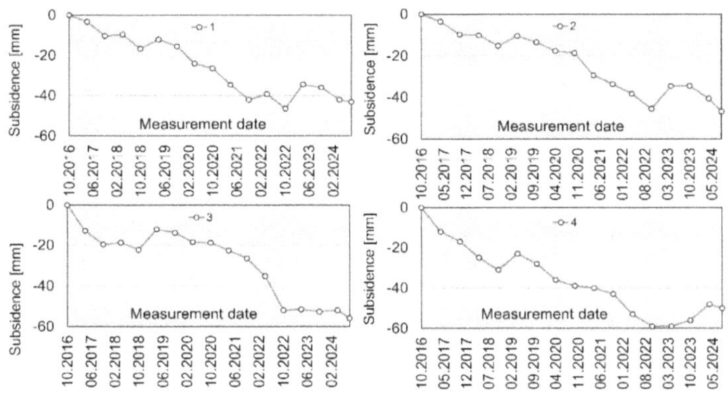

Figure 6. Subsidence as a function of time (October 2016 to August 2024).

Table 3. Maximum slopes and horizontal strains along the profiles, (October 2016 to August 2024).

Profile	Profile side	T [mm/m]	T_{max} [mm/m]	ε [mm/m]	ε_{max} [mm/m]	Area category
A-A’	NW	0.15	0,15	0.022	0,022	0
	SE	0.14		0.008		
B-B’	NE	0.29	0,29	0.028	0,028	0
	SW	0.14		0.001		
C-A’	NE	0.33	0,39	0.093	0,093	0
	SW	0.39		0.007		
D-D’	SW	0.28	0,28	0.004	0,036	0
	NE	0.17		0.036		

9 DISCUSSION

The results of this study confirm that ground deformations caused by underground coal mining can continue for decades after the end of mining activity. The maximum residual subsidence of 56 mm, observed during the eight-year InSAR monitoring period, corresponds to an average rate of about 7 mm/year. These values are comparable with results obtained in other post-mining areas in the world. For example, Guéguen *et al.* (2009) reported residual subsidence rates of 5–10 mm/year in the Nord/Pas-de-Calais basin more than 30 years after the closure of the mines, while Declercq *et al.* (2023) showed that displacements in the Campine basin continued for more than two decades. Similar displacements were also found in the Upper Silesian Coal Basin (Apanowicz *et al.* 2024), where SBAS analysis showed subsidence of several tens of millimetres.

The concentrated character of the subsidence trough in Nova Scotia and the short-term uplift observed in 2023 may indicate slow pillar failure in the rock mass or delayed roof collapses in the room and pillar system. Similar processes were described by Kowalski (2007), who connected stepwise increases of subsidence with stress release in the rock mass and delayed roof falls.

The direct subsidence calculated with Knothe’s method showed maximum values of 2.5 m, which classifies the area as mining category V during the exploitation period. Such high values are typical for room-and-pillar mining at shallow depth (Kratzsch 1983), although the stepwise subsidence profile indicates overlapping influence of several coal seams – in this case two. Despite the limited historical mining documentation, the adopted assumptions on the extraction coefficient and range of influence gave results consistent with the expected subsidence values for these geological conditions.

In summary, the results from Nova Scotia are consistent with previously published observations from other coal basins, confirming that residual ground deformations of a few centimetres are not a phenomenon present everywhere in post-mining areas. The applied methodology, combining the empirical model and SBAS-InSAR time series, proved effective in reconstructing both historical and present-day ground movements and can be recommended for similar post-mining regions where direct geodetic data are missing.

10 CONCLUSIONS AND RECOMENDATIONS

Based on the modelling results and InSAR time-series analysis, several important conclusions can be drawn regarding both historical and current ground deformation in the study area such as:

(1) Using Knothe's model for direct deformations determination and InSAR technology determination of the residual deformations proved to be effective, even when detailed mining documentation was not sufficient.
(2) After the coal mining in Nova Scotia was closed, the maximum subsidence, slope and strain were 2.5 m, 15 mm/m and 16 mm/m respectively. These values correspond to mining terrain category V, which indicates a significant impact of the mining on the surface. The deformations exhibited a steplike character, with subsidence reaching up to 1 m at the edge of Acadia coal seam and up to 2.5 m at the edge of Scott seam.
(3) The determination of residual deformations using InSAR showed that ground subsidence continues to occur in the area even 61 years after extraction was ended. The maximum subsidence was 56 mm, and the average ground subsidence rate was about 7 mm/year. The subsidence exhibited the largest changes from April 2019 to October 2022. The subsidence exhibited a concentrated character, which can be the result of relaxation processes occurring in the rock mass, or the gradual failure of the pillars left in the rock mass during extraction.
(4) Currently, the analysed area is characterised by very low slopes (0.15 - 0.39 mm/m) and horizontal strains (0.022 - 0.093 mm/m), which gives the mining terrain category 0. The subsidence has no impact on the infrastructure.
(5) The study results indicate that residual deformations can persist for dozens of years after extraction is finished, much longer than previously was accepted.
(6) It is recommended to conduct long-term post-mining surface deformation monitoring and analysis using InSAR methods at least every half year. It is particularly important as regards the safety of the infrastructure and residents, as well as the spatial planning in such regions.

REFERENCES

Apanowicz, B. (2023). Analiza obniżeń powierzchni na terenach górniczych i pogórniczych w aspekcie zasięgu i czasu z wykorzystaniem interferometrii satelitarnej. PhD thesis. Katowice: Central Mining Institute – National Research Institute.

Apanowicz, B., Milczarek, W. and Kowalski A. (2024). Novel method for determining the time coefficient c in Knothe's function and disappearance time of deformation increase using SAR data. *Measurement* 235: 114898.

Berardino, P., Fornaro, G., Lanari, R. and Sansosti, E. (2002). A new algorithm for surface deformation monitoring based on small baseline differential SAR interferograms. *IEEE Transaction on Geoscience and Remote Sensing* 40: 2375–2383.

Bazanowski, M. and Szostak-Chrzanowski, A. (2016). *Salt rock deformations caused by mining activity.* Wroclaw: Wroclaw University of Science and Technology

Buczyńska, A. and Blachowski, J. (2023). Proposal of a new Mining and Geology Impact Factor (MaGIF) index for the study of the post-mining environment. *CATENA* 232: 107463.

Chen, Y., Tong, Y. and Tan, K. (2020). Coal mining deformation monitoring using SBAS-InSAR and offset tracking: A case study of Yu County, China. *IEEE Journal of Selected Topics in Applied Earth Observations and Remote Sensing* 13: 6077–6087.

Chen, C. and Zebker, H. (2002). Phase unwrapping for large SAR interferograms: Statistical segmentation and generalized network models. *IEEE Transactions on Geoscience and Remote Sensing* 40(8): 1709–1719.

Declercq, P.Y., Dusar, M., Pirard, E., Verbeurgt, J., Choopani, A. and Devleeschouwer, X. (2023). Post mining ground deformations transition related to coal mines closure in the Campine Coal Basin, Belgium, evidenced by three decades of MT-InSAR data. *Remote Sensing* 15(3).

Farr, T., Rosen, P., Caro, E., Crippen, R., Duren, R., Hensley, S., Kobrick, M., Paller, M., Rodriquez, E., Roth, L., Seal, D., Shaffer, S., Shimada, J., Umland, J., Werner, M., Oskin, M., Burbank, D. and Alsdorf, D. (2007). The shuttle radar topography mission. *Reviews of Geophysics* 45: 1–33.

Ferretti, A., Monto-Guarnieri, A., Prati, C. and Rocca, F. (2007). *InSAR Principles: Guidelines for SAR Interferometry Processing and Interpretation*. European Space Agency.

Great Britain. National Coal Board. Mining Department. (1975). *Subsidence Engineers Handbook*. N.C.B.

Guéguen, Y., Deffontaines, B., Fruneau, B., Al Heib, M., De Michele, M., Raucoules, D., Guise, Y. and Planchenault, J. (2009). Monitoring residual mining subsidence of Nord/Pas-de-Calais coal basin from differential and persistent scatterer interferometry (northern France). *Journal of Applied. Geophysics* 69 (1): 24–34.

Hanssen, R. (2001). Radar interferometry: Data interpretation and error analysis. PhD thesis. Delf: Technische Universiteit Delf.

Hay Man Ng, A., Ge, L., Yan, Y., Li, X., Chang, H.C., Zhang, K. and Rizos, C. (2010). Mapping accumulated mine subsidence using small stack of sar differential interferograms in the southern coalfield of new south wales, australia. *Engineering Geology* 115: 1–15.

Jędrzejec E. (2008). Eksperymentalna wersja 6.0 systemu Szkody do prognozowania deformacji nieustalonych spowodowanych eksploatacją o zmiennych dobowych postępach. *Scientific Works of Central Mining Institute – National Research Institute, Mining and Environment*: 115–118.

Knothe, S. (1984). *Prognozowanie wpływów eksploatacji górniczej*. Katowice: Śląsk

Kowalski, A. (2000). *Eksploatacja górnicza a ochrona powierzchni. Doświadczenia z wałbrzyskich kopalń*. Katowice: Central Mining Institute – National Research Institute.

Kowalski, A. (2007). *Nieustalone górnicze deformacje powierzchni w aspekcie dokładności prognoz*. Katowice: Central Mining Institute – National Research Institute.

Kratzsch, H. (1983). *Mining Subsidence Engineering*. Berlin, Heidelberg, New York: Springer-Verlag.

Parmar, H., Bafghi, A.Y. and Najafi, M. (2019). Impact of Ground Surface Subsidence Due to Underground Mining on Surface Infrastructure: the Case of the Anomaly No. 12 Sechahun, Iran. *Environmental Earth Science* 78: 1–14.

Sałustowicz, A. (1953). Profil niecki osiadania jako ugięcie warstwy na sprężystym podłożu. *Archiwum Górnictwa i Hutnictwa* 1(1).

Sandwell, D., Mellors, R., Tong, X., Wei, M. and Wessel, P. (2011). Open radar interferometry software for mapping surface deformation. *EOS, Transactions American Geophysical Union* 92(28): 233–240.

Sikora, P. and Wesołowski M. (2021). Numerical assessment of the influence of former mining activities and plasticity of rock mass on deformations of terrain surface. *International Journal of Mining Science and Technology* 31(2): 209–214.

Szostak-Chrzanowski, A. and Chrzanowski, A. (2021). *INTEGRAN: integrated analysis of rock mass deformation due to underground mining*. Wroclaw: Wroclaw University of Science and Technology.

Szostak-Chrzanowski, A. and Chrzanowski, A. (2011). Challenges of Deterministic Modeling of Ground Subsidence in Complex Geological and Mining Conditions. Schriftenreihe des Institutes fur Markscheidewesen und Geodasie an der Technischen Universitat Bergakademie Freiberg, 11-th Geomatischer Tag. Freiberg.

Warchala, E., Szostak-Chrzanowski, A. and Stefanek, P. (2017). Wykorzystanie MES do prognozowania przemieszczeń terenu wywołanych eksploatacją górniczą. *The Bulletin of The Mineral and Energy Economy Research Institute of the Polish Academy of Sciences* 99: 171–184.

Warchala, E. and Szostak-Chrzanowski, A. (2020). An Analysis on the Effect of Crosscuts within Shaft Protective Pillars on Deformations of the Surrounding Rock Mass Deformations. *Mining Science* 27: 253–265.

Wessel, P., Luis, J., Uieda, L., Scharroo, R., Wobbe, F., Smith, W. and Tian, D. (2019). The Generic Mapping Tools version 6. *Geochemistry, Geophysics, Geosystems* 20: 5556–5564.

Zebker, H. and Goldstein, R. (1986). Topographic mapping from interferometric SAR observations. *Journal of Geophysical Research* 91: 4993–4999.

Mine Surveying and Mining Area Protection – Sokoła-Szewioła and Jarczyk (Eds)
© 2025 The Author(s), ISBN: 978-1-041-10531-2

POMHAZ project – spatial decision support system for support of the development of towns in post-mining areas in Poland

B. Apanowicz, M. Wysocka, K. Niedbalska, P. Gruchlik, K. Kura, M. Bonczyk & A. Kowalski
Central Mining Institute-National Research Institute, Katowice, Poland

ABSTRACT: Project POMHAZ (Post-mining Multihazard Evaluation for Land Planning) is co-financed by the Coal and Steel Research Fund and the Polish Ministry of Science and Education. Since September 2022, it has been implemented by an international consortium led by the French National Institute of Industrial Environment and Risks (INERIS). Project focuses on improving the hazard assessment and the risk management of abandoned coalmines. The main objective is to improve the methodological knowledge for practical realization of multi-hazards analyses, at the scale of a mining basin, in correlation with the main post-mining hazards. To achieve this objective, the hazards of the abandoned coalmines, their interactions and consequences on the urban development, were identified and analyzed based on multicriteria approaches and Decision Support System. An operational methodology and GIS tools was developed to provide a global representation of the hazards and have been applied to real case studies.

1 INTRODUCTION

The POMHAZ project, in line with the European idea of mining transformation, industrial decarbonization and the closure of hard coal and lignite mines, focuses on improving the methodology of hazard analysis and risk assessment. The systematic reduction of coal mining, the closure of mines and consequently the cessation of mining activities is resulting in the transformation of post-mining areas (Grozdanovic *et al.* 2015, Smith & Underwood 2000). As a result of centuries of mining activity, such areas are exposed to many different hazards so called 'multi-hazards' (Chang *et al.* 2022, Kappes *et al.* 2012, Kim *et al.* 2016, Krzemień *et al.* 2016). Sometimes, individual hazards occur simultaneously in the same area. Sometimes they interact, amplifying the development of adverse phenomena and their effects (Merad *et al.* 2004). The primary objective of the project was to identify and assess the conditions for the occurrence of multi-hazards in chosen post-mining areas. The project aims to estimate the total risk, referred to as 'multi-risk', by integrating the analysis of multiple individual hazards (Al Heib *et al.* 2023). It was agreed that the risk (multi-risk) would be calculated according to the Equation (1), taking into account factors such as hazard, vulnerability and exposure (Deck *et al.* 2009, Dalezios 2017, Haske *et al.* 2025, Pomhaz Consortium 2024):

$$Risk = Hazard \times Vulnerability \times Exposure \tag{1}$$

where: *Hazard* can be post-mining, natural, and technical; *Vulnerability* plays a key role in multi-hazard analysis, representing the degree to which specific elements are sensitive to risk. It includes various factors such as demographics, infrastructure, socio-economic conditions, and community resilience. Vulnerability plays a crucial role in multi-hazard analysis,

DOI: 10.1201/9781003655558-2

representing the degree to which specific elements (persons, structures and infrastructures) are sensitive to risk. The calculation of Social Vulnerability Index (SoVI) is the basis for assessing vulnerability. Index SoVI, developed by Cutter (Cutter *et al.* 2014), evaluates vulnerability through a complex index that includes a range of socio-economic and demographic variables, such as income levels, age distribution, education, and housing quality, factors that can affect a community's ability to cope with and recover from hazards (Haske *et al.* 2024). *Exposure* (Exposed Elements at Risk) understood as an indication of which elements are located in areas of risk and which of them are subject to destruction.

The challenge of the project was to develop a comprehensive multi-hazard risk assessment framework and decision support system (DSS) for post-mining areas. By using the developed decision support system (DSS) in real locations in Europe, it will be possible to improve management and spatial planning in post-mining areas during their transformation. Each of the project partners identified locations where the system was tested. In Poland, the system's capabilities were analysed on the example of three post-mining municipalities: Piekary Śląskie, Wałbrzych and Sosnowiec.

2 TESTING AND VERIFICATION OF THE DEVELOPED POST-MINING RISK INFORMATION SYSTEM ON FULLY KNOWN HISTORICAL LOCATIONS – POLISH EXPERIENCES

2.1 *Selected locations – case studies*

In Poland, the system's capabilities were analysed on the example of three post-mining municipalities: Piekary Śląskie, Wałbrzych and Sosnowiec. The first two cities are located in southern Poland in the Upper Silesian Coal Basin. Wałbrzych is located in the western part of the country. The cities have a long history of mining of mineral resources. However, in each of the cities, there are significant differences related to different geological conditions, the presence or absence of historical ore mining. A very important factor that differentiates the selected locations is the time when the mines were completely liquidated. Coal mines in Sosnowiec had been in operation since the 19th century. Last one was closed in 2021. In Piekary Śląskie mining activity has been carried out for more than 800 years. The effects of historical shallow exploitation of metal ores and deep coal mining are overlapping. The last coal mine was closed in 2020. The last ton of hard coal within the borders of Wałbrzych and its surroundings was exploited in 1998. The mining areas of mines operating in cities at various times occupied a significant part of their area: 49% in Wałbrzych, over 90% in Sosnowiec, and 52% in Piekary Śląskie.

2.2 *Work stages*

The first stage of the work consisted in identifying hazards occurring in chosen cities - the areas of interest of the project. Their nature has been specified as natural, mining-related, or artificial. Interactions between the individual hazards have been analyzed, with a focus on preliminary and secondary hazards. In Polish coal mining areas, the most significant hazards causing degradation of large areas and significant material losses are land subsidence (Apanowicz 2022, Apanowicz *et al.* 2024, Kowalski 2001) and hydrological disturbances, mining induced floods (Bukowski 2024, Krogulec *et al.* 2023). In recent years, the effects of exploitation have become increasingly apparent in the form of discontinuous deformations such as sinkholes and cervices (Kotyrba *et al.* 2024). During certain stages of activity and subsequently during the liquidation of mines, induced seismicity is observed (Mutke *et al.*2020). In Poland, there are still unliquidated mining heaps, which in turn are a source of environmental pollution. In addition, in some of them the phenomena of self combustion and fires are still observed (Gogola *et al.* 2020, Smoliński *et al.* 2021). In addition to the obvious hazards present in post-mining areas, specific hazards were identified – radiation emission, elevated due to mining activity (Chmielewska *et al.* 2024, Kies *et al.* 2006, Nowak *et al.* 2024,

Wysocka 2016, Wysocka *et al.* 2019). For each of the cities, an analysis of the significance/importance of each identified risk was conducted. Due to the fact that the geological, mining, and technical conditions vary in each city, discussions were held with specialists from the cities and mining company. An important stage in the analysis of multi-hazards was establishing a hierarchy of their importance for each location. A ranking of the significance of individual hazards in each city was prepared at the initial stage of the project. Changes were made based on the detailed data obtained and the results of research conducted in the cities. This stage of work was concluded with the presentation of the matrix that illustrates the results of the interaction analysis conducted for the Upper Silesia region (Poland). The analysis takes into account specific local geological and technical conditions. The matrix below illustrates the results of the interaction analysis conducted for the Upper Silesia region (Poland). The analysis takes into account specific local geological and technical conditions (Figure 1).

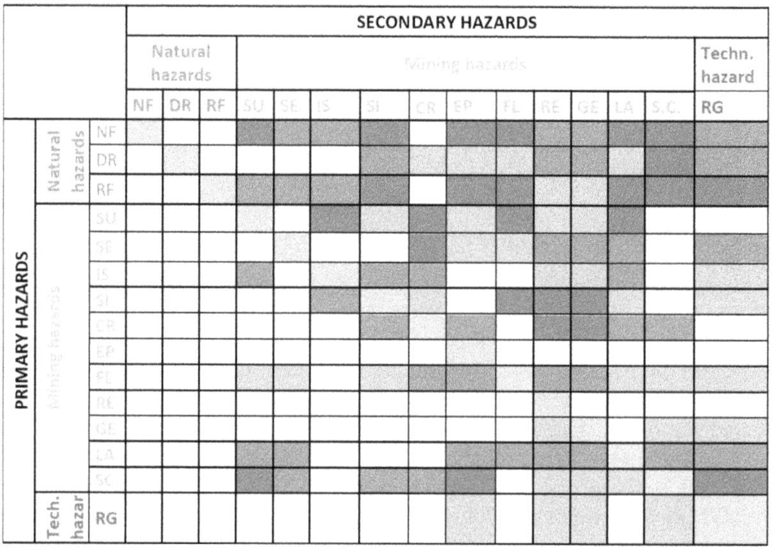

Figure 1. The risk matrix, Upper Silesian Coal Basin (Poland).

Levels of interaction: high interaction; intermediate interaction; low interaction; no interaction (white).

Hazard: natural hazards: flood (NF), droughts (DR), rainfall (RF);

mining hazards: subsidence (SU), settlement (SE), induced seismicty (IS), sinkholes (SI), cervices (CR), environmental pollutions from spoils (EP), hydrological disturbances, mining induced floods (FL), radiation emission (RE), gas emission linked to mining (GE), landslide (LA), self-combustion (SC);

Technical hazard: gases emission form spoils, heaps (RG).

The DSS system was developed by the German project partner Technische Hochschule Georg Agricola (DMT-THGA) In order for the system to be created, it had to be 'fed' with information characterizing individual testing locations. It was necessary to create a database of available data determining hazards in selected cities in mining areas. Polish project's partner GIG-PIB was responsible for collecting data and creating databases, according to the guidelines and common agreements with the partner THGA. The general information about each location was collected, such as: geology, topography, hydrology because it has a direct impact on hazards and risks. An important factor shaping the development of hazards is the described history of exploitation – past mining methods, methods of mine closure, and the time that has passed since the closure of the mining facilities. The actions that have been taken systematically, regulated by legal regulations or recommended instructions, aimed at monitoring adverse phenomena was analysed as well.

The data collected and prepared according to the requirements were presented in form of different maps and visualisations. The data was prepared in the form of a QGIS project, containing vector data layers in *.shp format. The databases covered the following issues:

- Seam maps
- Shafts, shafts and headings connected to the Surface
- Shallow mining
- Protective pillars
- Waste dumps and Settling ponds,
- Deformation and Faults
- Adits
- Occurences/outcrops of various geological formations,
- Depressions from the beginning to 1996
- Subsidences 1946 to 1996
- Age-related subsidences
- Reservoirs - extent of underground water bodies
- Overflow lands - areas of projected potential flooding
- Study areas - area of possible anomalous radon and other gases concentration.

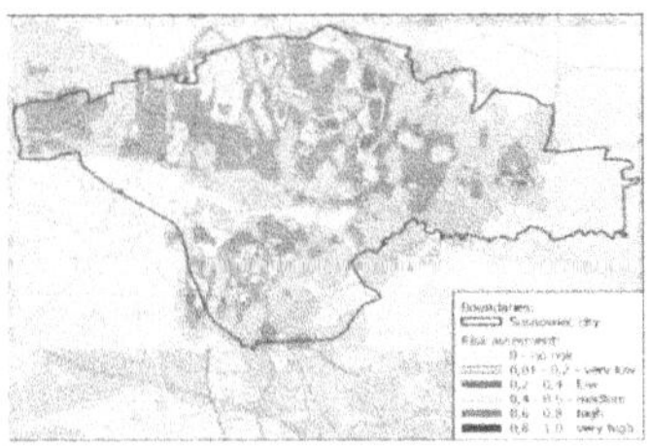

Figure 2. The risk map of subsidence in Sosnowiec.

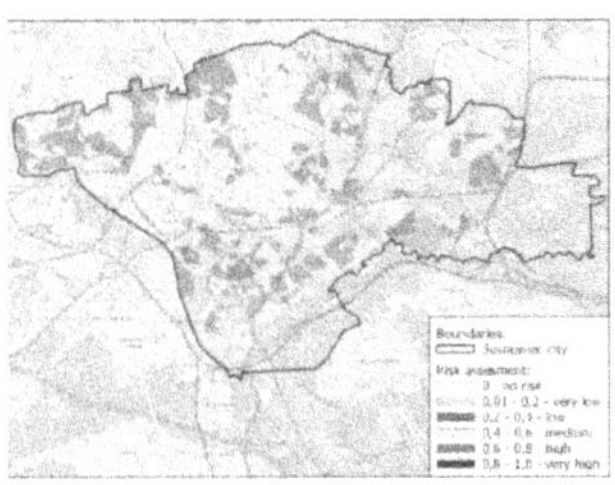

Figure 3. The risk map of sinkholes in Sosnowiec.

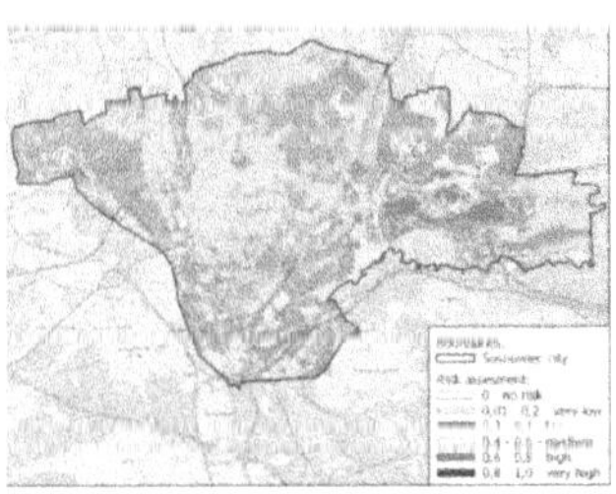

Figure 4. The risk map of floodplains in Sosnowiec.

The collected and formatted data and information concerning the hazards identified in the cities enabled the creation of risk maps for individual hazards. Figures (Figures 2–4) show, as an example, risk maps for sinkholes, ground deformation and floodplains in Sosnowiec.

3 APPLICATION OF DSS-GISS TOOL IN THE CASE STUDY

3.1 *Calculation of factors influencing multi-risk*

The main objective of the project was to develop a DSS-GIS tool that would allow for a better understanding of the interaction of many different hazards in post-mining areas, enable their visualisation, and allow for the selection of different assumptions (scenarios). The method of calculating the multi-risk base on pattern presented by Delazios (Delazios 2017). He identified three factors that are most important for post-mining areas: post-mining multi-hazard interaction, exposed elements at risk, and vulnerability. The DSS-GIS tool developed by project's partners form Technische Hochschule Georg Agricola (THGA) from Germany for the POMHAZ project enables the integration and visualisation of all multi-risk factors on maps (. The tool provides experts and authorities with the flexibility to adopt different risk management approaches based on specific needs, such as environmental impact assessment, structural damage mitigation, or other relevant concerns (e.g. groundwater contamination, land subsidence). The result of successive stages of calculations, scenario selection and weighting adjustments for individual elements is the Multi Hazard Index (MHI) (Al Haib *et al.* 2024, Haske *et al.* 2024, 2025).

The first step in calculation of MHI is selection of hazards occurring in specific areas. The user is able to choose hazards, level and number of interactions. The figure below shows examples of multi-hazard visualisations in the Sosnowiec area. It shows that the developed DSS tool allows for the analysis of hazards according to different scenarios, for example, assuming that we take into account all hazards (top part of the Figure 5), or only selected ones (bottom part of the Figure 5). In presented case in one scenario subsidence, sinkhole environmental pollution, hydrological disturbances, ionizing radiation emission and combustion/overheating of mine waste were analysed. In the second scenario subsidence, hydrological disturbances and sinkhole were taken into account.

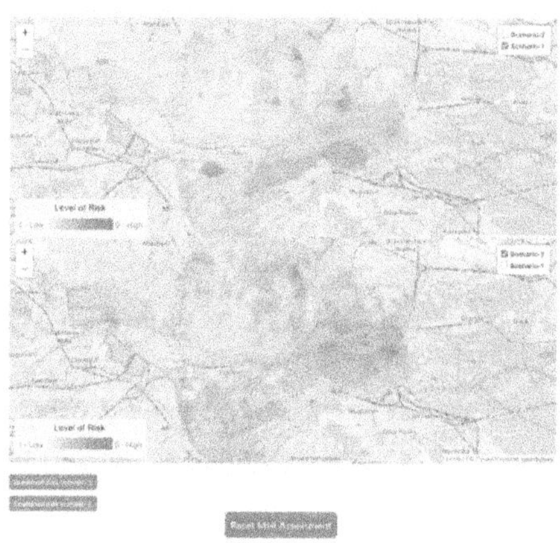

Figure 5. MHI risk map for Sosnowiec case. (Top) Scenario-1 with all involving hazards in the area. (Bottom) Scenario-2 with only Sinkhole, Subsidence and mining-induced flooding. After; Final version as of 2.4.2025 of deliverable D15: D4.4. of project PoMHaz.

Important risk factor that should be assessed is the Exposed Elements at Risk (Haske and Inojosa 2025). For this reason, proposed factors of Land Use and Land Cover (LULC) must be chosen by experts. In the case of selected cities, built-up areas are of greatest importance, followed by rangeland, areas designated for cultivation and forested areas. During the analysis of exposed elements, experts can also accurately assess and locate different classes. As with MHI results, the data can be visualised.

Another parameter assessed is Vulnerability Index (VI) for each location calculated in one of the previous stages of work. Once again, experts can then assess the importance of each category on the accepted scale and also view the result in the form of a VI map of the area in question.

3.2 *Multi-risk map*

The DSS GISS tool collects all calculations related to the factors described above and their expert analysis. The end result of the work is a map displaying various layers, depending on the user's choices and decisions. An important feature of the DSS tool is the ability to smoothly change the multi-risk value from low to medium to high. Of course, such actions require experience and expert knowledge, but this tool module will not be made publicly available. The map of Sosnowiec (Figure 6) shows the spatial distribution of multi-risk for two declared scenarios: one taking into account all hazards defined for this area and one taking into account two selected hazards. In addition, the tool suggests recommendations for actions to minimise the effects of adverse impacts for each risk level.

Figure 6. Multi-risk map of Sosnowiec study case. (Top) Scenario-1 with all hazards involved. (Bottom) Scenario-2 Risk map with Sinkhole, Subsidence and induced mining flooding. After: Final version as of 2.4.2025 of deliverable D15: D4.4. of project PoMHaz.

4 CONCLUSIONS

In Europe coal mines have been gradually closing for several decades. In Poland, about 20 mines are still in operation. In the middle of the last century, there were almost 60 mines operating. Therefore, the impact of mining, the risks for the population and the environment are caused by the historical effects of mineral extraction as well as the contemporary effects of the closure of further mines.

The selected case studies, i.e. the cities of Wałbrzych in Lower Silesia and Sosnowiec and Piekary Śląskie in Upper Silesia, are good testing grounds for observing the development of hazards, risks and their interaction. The collected hazard-specific databases provided good material for verification, validation and testing of the GIS -Decision Support System tool.

Each of the case studies has its own individual characteristics and conditions, determining the development of hazards and the intensity of their impacts. It can therefore be assumed that the correct implementation of the GIS-DSS tool in these cities confirms the system developers' assumption that the tool can be used throughout Europe.

The developed in frame of POMHAZ project system analyses and visualises the information prepared according to the imposed standard. This makes it possible to assume that multi-risk analyses carried out with the support of a GIS-DSS tool anywhere in Europe will be comparable. This will certainly allow the development of coherent and standardised instructions and guidelines for the administration of post-mining municipalities throughout Europe in the future. Stakeholders such as developers, industrial partners and others will be able to use a single, universal tool to assess the advantages and disadvantages of planned investments.

The results of the hazard interaction study presented in the document showed that theoretical analyses depend on the quality of the data collected.

It is also important to know that different post-mining areas have different vulnerabilities, different elements are exposed at risk, depending on many specific factors.

The problems presented in the paper show that expert knowledge and the ability to critically assess both the scope and quality of the data to be analysed and the resulting visualisations, models etc. are always necessary.

ACKNOWLEDGEMENTS

This work has received funding from the European Union's Research Fund for Coal and Steel (RFCS) under the European project "POMHAZ" grant agreement No 101057326 and from the Polish Ministry of Science and Higher Education under the Agreement No. 5270/FBWiS/2022/2.

The authors would like to thank everyone else involved in the research of project POMHAZ, specifically the international partners from INERIS, THGA, CERTH, PPC and SRK.

Additionally, we would like to thank the specialists from cities Sosnowiec, Piekary Śląskie and Wałbrzych as well as from Coal Mining Authority for their support and creative discussions.

REFERENCES

Al Heib, M.M., Franck, C., Djizanne, H. and Degas, M. (2023). Post-Mining Multi-Hazard Assessment for Sustainable Development. *Sustainability*. 15, 8139. https://doi.org/10.3390/su15108139

Apanowicz, B. (2022). Use of InSAR in Linear Discontinous Ground Deformation Generation Analysis: Case Study of a Mine in Poland. *Journal of Sustainable Mining* 21(1): 4

Apanowicz, B., Milczarek, W. and Kowalski A. (2024). Novel method for determining the time coefficient c in Knothe's function and disappearance time of deformation increase using SAR data. *Measurement* 235: 114898.

Bukowski, P. (2024). The impact of flooding of abandoned hard coal mines in the Upper Silesian Coal Basin on the state of general safety of post-mining areas (in Polish). *Przegląd Geologiczny* (72)5: 225–240.

Chang, M., Dou, X.,Tang, L. and Hengzhi X. (2022). Risk assessment of multi-disaster in Mining Area of Guizhou, China. *International Journal of Disaster Risk Reduction* 78:103128.

Chmielewska, I., Czerwiński, S. and Chmiela, A. (2024). Radionuclides in waters from underground coal mines. *Systemy Wspomagania w Inżynierii Produkcji*, 13(2): 145–153. (In Polish, abstract in English).

Cutter, S.L., Ash, K.D. and Emrich, C.T. (2014). The geographies of community disaster resilience. *Global environmental change*. 29:65–77.

Dalezios, N.R. (2017). Environmental Hazards Methodologies for Risk Assessment and Management. *Water Intell.* 16, 9781780407135.

Deck, O., Verdel, T., and Salmon, R. (2009). Vulnerability Assessment of Mining Subsidence Hazards. *Risk Analysis* 29(10):13811394.

Gogola, K., Rogala, T., Magdziarczyk, M. and Smoliński, A. (2020). The mechanisms of endogenous fires occurring in extractive waste dumping facilities. *Sustainability* 12:2856. doi:10.3390/su12072856
Grozdanovic, M., Savic, S. and Marjanovic, D. (2015). Assessment of the key factors for ergonomic design of management information systems in coal mines. *International Journal of Mining, Reclamation and Environment* 29(2):96–111.
Haske, B.; Al Heib, M.M.; Inojosa, V. and Bouaziz, M. (2024). Development of a spatial decision support system for multi risk assessment of interacting post-mining and natural hazards. *Sustainability*. https://doi: 10.20944/preprints202412.0155.v1
Haske, B., Al Heib, M.M., Inojosa, V. and Bouaziz, M. (2025). Spatial Decision Support System for Multi-Risk Assessment of Post-Mining Hazards. *Mining*. 5(1), 17; https://doi.org/10.3390/mining5010017.
Haske, B. and Inojosa, V. (2025). Development and evaluation of an interactive tool for multi-hazard analysis in European post-mining regions. *Mining report Glückauf*, August 2025, Germany.
Kappes, M.S., Keiler, M., von Elverfeldt, K. and Glade T.(2012). Challenges of analysing multi-hazard risk: a review. *Nat. Hazards* 64: 1925–1958.
Kies, A., Storoni, A. and Tosheva, Z. (2006). Radon measurements as a monitoring possibility for mining subsidence occurrence. *Journal of Sustainable Mining* 42: 518–522
Kim, S.M., Suh, J., Oh, S., Son, J., Hyun, C.U., Park, H.D., Shin, S.H. and Choi, Y. (2016). Assessing and prioritising environmental hazards associated with abandoned mines in Gangwon-do, South Korea: the total mine hazards index. *Environ. Earth Sci.* 75:1–14.
Kotyrba A., Kortas, Ł. and Siwek, S. (2024). Spatio-Temporal Changes in the Gravity And Post-Mining Seismic Activity in Abandoned Deep Coal Mine Kazimierz-Juliusz During Flooding. *Journal of Sustainable Mining* 23(4):357–372
Kowalski, A. (2001). Surface subsidence and rate of its increaments based on measurements and theory. *Archives of Mining Sciences* (4): 391–406.
Krogulec, E., Bukowski, P., Niedbalska, K., Trzeciak, J. and Zabłocki, S. (2023). Groundwater vulnerability to pollution in degraded coal mining areas: modifying the DRASTIC method using the factor of exploitation impact on land surface. *Geology, Geophysics and Environment* 49(4): 313–334. https://doi.org/10.7494/geol.2023.49.4.313
Krzemień, A., Suárez Sánchez, A., Riesgo Fernández, P., Zimmermann K. and González Coto, F. (2016). Towards sustainability in underground coal mine closure contexts: A methodology proposal for environmental risk management. *Journal of Cleaner Production* 139:1044–1056.
Merad, M., Verdel, T. and Kouniali, S. (2004). Use of multi-criteria decision-aids for risk zoning and management of large area subjected to mining-induced hazards. *Tunnelling and Underground SpaceTechnology* 19:125–138.
Mutke, G.Z., Lurka A.I. and Zembaty Z. (2020). Prediction of rotational ground motion for mining-induced seismicity – Case study from Upper Silesian Coal Basin, Poland. *Engineering Geology* 276(6):105767
Nowak, S., Wysocka, M., Czerwiński, S. and Chmiela, A. (2024). Multi-criteria analysis of factors forming radon risk in mining areas. *Systemy Wspomagania w Inżynierii Produkcji*, 13(2):135–144. (In Polish, abstract in English).
POMHAZ Consortium. *Technical Report (Part B): RFCS Project Number: 101057326*; POMHAZ Consortium: Brussels, Belgium, 2024.
Smith, F.W. and Underwood, B. (2000). Mine closure: the environmental challenge. *Mining Technology* 109 (3):202–209
Smolinski A., Dombek V., Pertile E., Drobek L., Gogola K., Zechowska S. and Magdziarczyk M. (2021). An analysis of self-ignition of mine waste dumps in terms of environmental protection in industrial areas in Poland. *Scientific Reports* 11:8851 https://doi.org/10.1038/s41598-021-88470-7
Wysocka, M., Niedbalska, K., Kura K., Apanowicz B., Gruchlik, P., Haske, B. and Inojosa, V. (2025). Coupled GIS-DSS module with an intuitive interface and guide document, documented in a verification case. *Final version as of 2.4.2025 of deliverable D15: D4.4. of project PoMHaz.*
Wysocka, M. (2016). *Radon problems in mining and post-mining areas in Upper Silesia region*, Poland. Nukleonika, 61(3). DOI: 10.1515/nuka-2016–0051.
Wysocka, M., Skubacz, K., Chmielewska, I., Urban, P. and Bonczyk, M. (2019). Radon migration in the area around the coal mine during closing process. *International Journal of Coal Geology*, 212. DOI: 10.1016/j.coal.2019.103253

Mine Surveying and Mining Area Protection – Sokoła-Szewioła and Jarczyk (Eds)
© 2025 The Author(s), ISBN: 978-1-041-10531-2

Multi-stage reconstruction and long-term maintenance of the road viaduct over DK81 in relation to mining exploitation and economic transformation in Poland

P. Bętkowski
Silesian University of Technology, Gliwice, Poland

ABSTRACT: The article describes the interesting technical history of a road viaduct over busy dual carriageway national road. Economic transformation in Poland caused that this bridge changed owners several times. The lack of maintenance by previous owner and his agreement to active mining exploitation despite the inadequate technical condition of bridge object, led to the situation threatening a serious construction failure/disaster. It was necessary to take unusual measures that would have prevented a construction disaster and allowed mining exploitation to continue, because suspending mining operations would result in millions losses for the local mine every day. Due to lack of time to prepare the documentation for rebuilding and obtain formal-legal permits, immediate action was taken to prevent spans from sliding off the pier, including: constructing an auxiliary frame to stabilize the bridge bearings and anchors to connect adjacent spans. Then, technical documentation was prepared and the viaduct was rebuilt, adapting construction to predicted mining area deformations. Original, individual solutions are worth presenting, they can be applied in engineering practice. The article also describes the experience of over 12 years of observing this road viaduct and the cooperation between new owner, local mine and local municipality.

1 INTRODUCTION

The viaduct has two freely supported spans made of prefabricated WBS bridge beams (Bętkowski 2017a) (Figure 1). This viaduct was not adapted to transfer the impacts of mining exploitation. Mining influences (Kwiatek 2007, Kawulok 2010, Ren *et al.* 2022) have a significant (unfavourable) impact on bridge constructions. A significant (unfavourable) factor is the elevation of rigid spans above the ground and the separation and distance between span support points (e.g.: bridge bearings on abutments and piers) (Bętkowski 2017b, Rosikoń 1979, Salamak 2013, Wytyczne/Guidelines 1977, Zhang *et al.* 2025). At the time of building of the viaduct in the 1970s, there was no mining activity in this area, so the viaduct was not protected against effects of mining activities. Due to its inadequate (bad) technical condition (Bień 2010), this bridge was not able to transfer the mining impacts revealed on the surface. The poor technical condition of the viaduct was result of its complicated ownership situation, a consequence of the economic transformation in Poland in the 1990s, the transition from a centrally planned communist economy to a free-market economy with a significant role for local governments in maintaining road infrastructure. This period of political and economic transformation was characterized by a shortage of money for road infrastructure maintenance – generally a deterioration of the technical condition of road infrastructure. Example of such economic changes was the liquidation of former manager/owner of the viaduct, it was Generalna Dyrekcja Dróg Publicznych (GDDP) (eng. General Directorate of Public Roads). GDDP was centrally financed institution managing all national roads in Poland. Then the

DOI: 10.1201/9781003655558-3

viaduct and the road came under the management of Generalna Dyrekcja Dróg Krajowych i Autostrad (GDDKiA) (eng. General Directorate for National Roads and Motorways), strictly in its Katowice division. GDDKiA was established on April 1, 2002.

The very busy dual carriageway national road DK81 runs under the viaduct. The viaduct is located along the municipal road and connects two parts of the municipality, which are cut by this national road, so it is very important for the local community. The local municipality had been trying for many years, since 2002, to take over the bridge from the national road administrator, and it was successful. However, the viaduct meant not only freedom to create communication within the municipality but also necessitated responsibilities (e.g., regular technical inspections) and maintenance costs, as the next elected (new) local authorities found out – no repairs were carried out, what quickly led to a significant deterioration of the viaduct's technical condition. After several more years, local municipality managed to return the viaduct to GDDKiA. After taking over the viaduct, the national road administrator was unable to carry out renovation immediately, because GDDKiA did not have money reserved for it in the budget. The technical condition of the construction suddenly deteriorated after the frosty winter and rainy spring of 2011 – the cover of reinforced concrete bars on the bearing bench on the pillar was destroyed, some bridge bearings were hanging on the reinforcing bars (there was no proper support for bearings). Other serious damage included the lack of movement control on sliding bearings due to advanced corrosion of the sliding plates (displacement blockage) and cracked (broken) concrete mortar under bearings. The spans were in danger of slipping off the pier. Danger was real, as mining exploitation were approaching the viaduct, and ground creep/stretching began to appear. Due to the impossibility of preparing and agreeing on technical documentation within a very short period of time (including the required formal and legal permits), it was decided to carry out only the necessary urgent repairs in the first stage, so that the bridge could withstand the emerging mining impacts.

In the first stage, two independent spans were connected by tie bars above the pier using crossbeams as support for anchors to prevent longitudinal displacement above selected bearings (pier had im-properly supported fixed bearings). A non-standard method was also used to repair the bearing bench, using additional steel frame – due to the lack of support for the bearings, low strength and advanced corrosion of the concrete. These are non-standard, proprietary solutions, created individually for the repair of this viaduct.

The second stage of repair work was a complete renovation of the viaduct, works consisted of removing existing mining damage and adapting the construction to take over the impact of predicted mining impacts. This stage required the preparation and formal-legal approval of technical documentation, including, for example, agreeing the project with GDDKiA and the local municipality, obtaining a building permit, preparing and agreeing temporary traffic management project.

The article presents the renovation process and discusses the elements of mining prevention. The article presents conclusions from over 12 years of observations of the viaduct after the rebuilding.

2 TECHNICAL CONDITION OF THE VIADUCT IN JUNE 2011, BEFORE RENOVATION

This road viaduct is two-span construction, with two spans "freely supported" (Figure 1). Each span contains five prefabricated WBS-19 post-tensioned concrete beam girders (Bętkowski 2017a). Monolithic crossbeams are located at the end of each span and in the middle of the span. Bridge slab is made of monolithic reinforced concrete with small pedestrian bracket outside the girders. The intermediate support (pillar) and abutments are made of reinforced concrete. The bridge bearings was made of steel sheets and are placed directly on the bearing benches (no bearing pads) (Figures 2–3). Gaps between the span and the abutment were narrow and irregular, approximately 30-80 mm wide (Figures 2–3). These gaps were covered with bituminous "expansion joints" (Figure 4). On the northern sidewalk there is municipal water supply in 400 mm casing pipe.

The viaduct has a single-carriageway road with an asphalt surface. The sidewalks have a total width of 66 cm, including the cornice (Figure 4).

The national road DK81, a dual carriageway, runs under the viaduct, with two lanes in each direction, with a central reservation. The angle of intersection between the spans and the road is 90°. The viaduct spans are approximately 18.8 m long each, the total length of bridge (including abutment wings) is 50.75 m, the minimum vertical clearance is 5.30 m, the total width of the deck (including cornices) is 7.42 m, the height of the prefabricated girders is 1.00 m, the thickness of the bridge deck slab is at least 12 cm.

The viaduct had visible numerous damages (in 2011), resulting from mining impacts and insufficient maintenance.

Figure 1. View of the viaduct from the north side in September 2011, before rebuilding.

Bridge bearings made of tangential steel sheets were contaminated and heavily corroded and this corrosion blocked the ability of the upper part of the bearing to move relative to the lower part. Some of the lower bearing plates were detached from the bearing bench due to concrete corrosion – in such a situation it was impossible to control the displacements (Figures 2-3). Bearings under the external girders on the pier were in the worst condition – some of these bearings were not properly supported, e.g. the external bearing of the western span on the pier was partially hanging in the air (Figure 2). The technical condition of the bearings was unsatisfactory, especially in view of the expected future stretching (creeping) of mining area (e. g. Bętkowski 2018b).

Numerous stains, corrosion and concrete defects were visible near gaps between the span and the abutment. As a result of mining-related ground expansion (stretching) numerous cracks and fissures appeared in the area of bituminous expansion joint covers, these bituminous covers were irreversibly damaged. Water leaked through the expansion joint covers. This water caused damage to the pier cap (intermediate support) and created unsightly water stains on the abutments. The technical condition of the pillar cap (bearing bench) was pre-failure (Figure 2). Chemical tests indicated significant saturation of the pillar concrete with

Figure 2. Bridge bearings on the pillar –south side (before renovation).

Figure 3. Bridge bearing on the abutment – south-east side (before renovation).

chlorides and low pH. The concrete parameters were incorrect at a significant depth of i.e. over 8 cm.

Horizontal cracks were visible on the abutments, these cracks correspond to technological breaks in the concreting process. The lack of a sealing, bonding layer during concreting, was probably the main cause of leaks and cracks in places of old technological breaks. Corroded rods were visible on the front walls of the abutments – the result of a lack of concrete cover, moreover in many places the cover was detached from the structure (so-called "deaf" concrete).

On the lower surface of the bridge slab, numerous traces of leaks and stains were visible – which indicated leaky (damaged) insulation of the upper surface of the bridge slab.

The condition of the road surface on the bridge was insufficient – bituminous surface was cracked and scratched. Fault (threshold) was visible at entrance to viaduct on both sides (Figure 4) and was strongly felt while driving. In the area of this fault were visible traces of numerous surface repairs (repairs were carried out many times in the past), without the repairs the height of the fault would had been about 30 cm. The fault was created after each mining operation. Rusty sliding bearings blocked the kinematic freedom of the spans on the abutments, the viaduct with fixed (non-sliding) bearings on the pillar began to "work" like a frame structure (Bętkowski 2017b). The ground spreading (scratched) (viaduct with blocked bearings by corrosion) led not only to the formation of fault on the roadway (Figure 4), but also to the lowering of the ground line on the wings (Figure 5), the soil medium moved towards the basin, i.e. away from the viaduct. The condition of the prefabricated girders was satisfactory, the concrete strength, determined on the basis of Schmidt hammer tests, corresponded to class C30/37 (B35). Only the underside of the girders has visible signs of corrosion, which overlapped the beam steel stirrups. The nature of the corrosion indicated insufficient concrete cover thickness. The condition of the cornices on the spans was poor. There were numerous leaks, efflorescence, concrete defects, corrosion of the concrete and steel.

Figure 4. View from the east side on the roadway on the viaduct and fault/threshold at the roadway.

Figure 5. Lowering the ground line on the wings of the bridge abutments.

3 TEMPORARY VIADUCT REPAIR, AUTUMN 2011

Due to the poor technical condition of bridge bearings (lack of adequate bearing support due to concrete damage and corrosion, blockage of displacement due to corrosion of sliding steel plates) and mining impacts related to the approach of coal wall to the viaduct, there was a real danger that the spans could slide off the pier in 2011. In stage I, completed in October 2011, solutions were used to prepare the structure to absorb mining influences from approaching one mining wall. Type and scope of works, taking into account the poor condition of the intermediate sup-port cap and bridge bearings, were part of maintenance works restoring the proper functioning of bearings (freedom of displacement and appropriate support), such works did not require formal-legal permits to conduct repairs and could be performed almost immediately in emergency mode.

The crossbeams of spans above the pillar were connected with steel anchor (Figure 6). The use of steel tie bars made it impossible to displace the span on the pillar, i.e. it forced the original bearing scheme (fixed bearings on the pillar). Tie rods (anchors) were made of steel bar with a diameter of 40 mm, made of St3SY steel (Figure 6). Four tie rods were made in the crossbeam fields between the girders (one steel bar in each field). Use of steel tie bars (anchors) force bearings to have the appropriate kinematic freedom was an innovative and non-standard solution.

Analysis of the predicted mining deformations of the area showed that displacements would mainly occur on the direction of the viaduct longitudinal axis and only horizontal forces would appear in the tie bars. It was designed to make four anchors through the crossbeams above the pillar. Calculations assumed that a slight rotation of the spans relative to the viaduct axis could occur, hence two tie bars could first act as tension elements. It was therefore assumed that the two tie rods should transfer the entire horizontal force N coming from the resistance of the bearings on the abutments (1), where the resistance depends on the sum of the support reaction forces R from the bearings on the abutment. The friction coefficient was assumed μ=0.5, as the minimum static steel-steel friction for corroded and contaminated surfaces. Next tension in the tie bar was checked and the concrete crossbeam was numerical tested for punching shear; all ULS were met.

$$N = \mu \cdot R \qquad (1)$$

Predicted area deformation indices, related to the mining exploitation of longwall approaching viaduct were: ε=1.2mm/m, R=-125km, w=0.133m. Displacements on bearings at abutments (Wytyczne/Guidelines 1977) were determined as: $\Delta l \approx 30$mm. Length of bearing plates ensured implementation of predicted displacements, so immediate bearing replacement was not necessary.

Figure 6. View of the crossbeam and tie/anchor bar near the center girder above the pillar.

Figure 7. Outer bearing on pillar and steel frame (during temporary renovation).

Additionally, to protect the pier cap, weakened by corrosion of steel and concrete and concrete loss – steel frame made of sheet metal and angle iron was designed on its upper side. This frame strengthens the pillar cap by absorbing the tension, and at the same time prevents fragments of the new concrete from breaking away from the upper part of the cap (Figure 7). Concrete defects in the upper part of the cap were filled, which ensured stability and proper support of bearings on the pillar (Figure 7). The freedom of movement of the one-directional and multi-directional bridge bearings was restored – concrete defects were filled, bearings were cleaned and lubrication was applied to ensure the sliding of the contact plates (each bearing was made of two sliding plates).

4 FINAL REBUILDING OF THE VIADUCT, SUMMER AND AUTUMN 2013

In 2011, the local mine held concession for exploitation coal in the viaduct area only until 2020 (current mining concession runs until 2051). The final repair/rebuilding was designed to bring bridge object to such condition that would safely absorb the impacts of mining exploitation planned until 2020. At the viaduct location, surface deformations, effective curvatures and terrain slopes will not exceed the permissible/limit values for second category of mining area: w=0,971 m (subsidence of the area), T_{II} =-2,1 mm/m (inclination in direction of viaduct axis), $T\perp$=-2,2 mm/m (inclination in direction perpendicular to viaduct axis), ε_{II}= 1,0 mm/m (creeping/extension along the longitudinal axis of viaduct); $\varepsilon\perp$= 1,8 mm/m (creeping/extension in direction perpendicular to longitudinal axis of viaduct), R_{max} = -76,0 km (radius of mining area curvature). In viaduct area, only insignificant (safe for the structure) mining shocks with acceleration a $\leq$ 200 mm/s^2 were predicted, omitted in further analyses.

To calculate range of displacement, impacts corresponding to the full second category of mining area were assumed – it allowing for changes in length of mining walls and possibility of not exploitation some mining walls. Applied in calculation area deformation indicators correspond to the previously legally approved category of mining area – within this category, there are possible changes to long-term mining plans. Maximum displacements on the abutments related to influence of mining exploitation in the direction of viaduct axis were determined according to (2) (mining wall perpendicular to bridge) (Wytyczne/Guidelines 1977), where: ΔL - displacement caused by area deformations for the mining front perpendicular to bridge axis on bearings on abutments [mm]; *H*=7,5 m - distance from the base of the foundation to the upper surface of the roadway (abutment height) [m]; *L*=18,8 - total length of the span [m].

$$\Delta L = \pm\left(\varepsilon + \frac{H}{R}\right)\cdot L \qquad (2)$$

The maximum displacement on the abutments due to mining and thermal influences is estimated at ΔL≈80 mm. It is designed on the abutment to use pot bearings and block expansion joint devices with working range ±80 mm.

The existing pier was in poor condition. Concrete was class C12/15 (B15), concrete cover had lost its protective properties (pH <10). The chloride concentration was very high approx. 2% (chloride concentrations of Cl > 0.4% already led to chloride corrosion of the reinforcing steel).

Figure 8. View of the viaduct from the south side after rebuilding in December 2013.

In stage II, the pier was rebuilt using the existing foundation (demolition and rebuilding of the pier shaft and cap) (Figure 8). The bridge deck structure (road surface, sidewalks, cornices, handrails) had to be replaced – horizontal insulation of the bridge deck was leaking, which posed a risk of rusting through the reinforcing bars and damaging the deck. Surface repairs to the concrete were also carried out (filling in the gaps, hydrophobization, applying a paint coating) (Figure 8).

Width of gaps between each span and the abutment was small, and the back walls were too narrow for the installation of modern expansion joints. It was necessary to dismantle (remove) and reconstruct the back walls in a new location and with a new geometry.

The fault was visible at the entrance to the viaduct (Figure 4). As part of the rebuilding in 2013, the abutments were excavated from the inside to the ground level on the slope, and new horizontal insulation was made from the inside of the abutments to eliminate leaks through the front walls and wings. The anchoring of the spans was retained as an additional safety element in the event of lateral inclination of the viaduct pillar and "breaking" of the fixed bearings on this pillar.

Road substructure is constructed in an unusual way – approaching slabs above the bridge abutments were omitted. Solution with variable stiffness of the sub-base was designed: the thickness of the auxiliary sub-base (crushed aggregate mechanically stabilized) was gradually increased from 20 cm in front of the abutment to 40 cm in front of the rear wall, a geogrid was placed under the layer of the main/bituminous sub-base (10 cm, asphalt).

The "deaf" concrete from the external walls of the span and supports was removed, and the gaps were filled with mineral mortars (PCC). Additional reinforcement was made in places of old technological breaks in concreting. Slopes surface reinforcement was dismantled, the soil losses were filled and a new reinforcement was made (trylinka concrete slabs) (Figure 8).

The designed protection against mining impacts involves replacing old corroded tangential bearings with new pot bearings and replacing the bituminous expansion joints covering the expansion gaps with new block expansion joints with an appropriate range of displacements.

5 OBSERVATIONS OF THE VIADUCT AFTER REBUILDING (12.2013 ÷ 06.2025)

There are six benchmarks for height measurements on the viaduct (benchmarks 1-6, Figure 9), one on each side of the pier and each abutment. This enables geodetic measurement of non-uniform subsidence, e.g. the viaduct's inclination both parallel and perpendicular to the road axis on the viaduct. In addition, permanent measurement points are established in the vicinity of the expansion joints on both sides of the viaduct (12 points, four at each joint) in order to measure changes in the width of the expansion gaps (Figure 9). These measurements allow for assessment of torsion of the spans relative to the supports (widths of expansion gaps are then unequal).

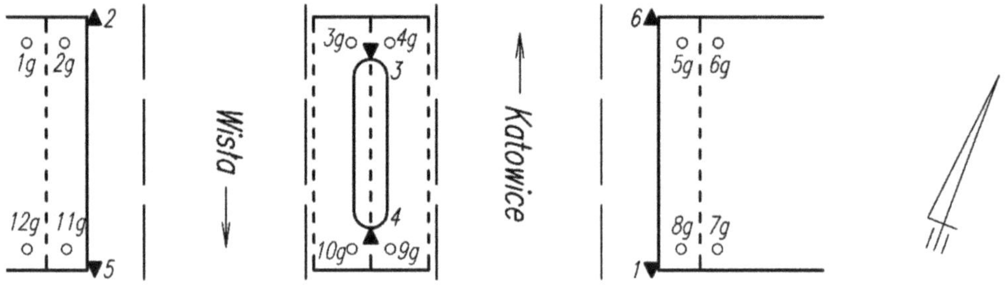

Figure 9. Location of measurement points/geodetic benchmarks on the viaduct.

Due to the lack of sudden threats to the structure, it was not decided to conduct remote SHM monitoring (e.g. Bętkowski *et al.* 2014) or other advanced monitoring methods (e.g.: InSAR, UAV/drone, terrestrial laser scanning TLS) (e.g.: Bętkowski 2018a, Yinfei *et al.* 2023).

Geodetic measurements are performed regularly three times a year (February or March, August, November or December) and specialized inspection of all bridge elements is performed monthly.

Settlements (subsidence) from December 2013 (i.e. from the renovation of the viaduct) to March 2025 are given below: point no. 1 w_1=887 mm, point no. 2 w_2=770 mm, point no. 3 w_3=818 mm, point no. 4 w_4=819 mm, point no. 5 w_5=771 mm, point no. 6 w_6=897 mm. In the direction of the viaduct axis, there is a slight inclination of the roadway to the east, approximately 3 mm/m (0.3%) (practically insignificant for bridge), i.e. in the direction of mining exploitation on the eastern side (several walls), consistent with the existing longitudinal slope of the road. In the direction perpendicular to the viaduct axis, the road inclination practically is the same (no significant change).

Benchmarks (measuring points) for measuring changes in support spacing are located at the bottom of the pier and at the bottom of the abutments, close to ground level. The lack of benchmarks at the top of the supports prevents determination of support inclination and complicates analysis of the impact of mining area curvature on the bridge. Horizontal displacements measure from December 2013 to March 2025 (i.e. change in the distance of abutments): distance of points 5 ÷ 1: increase in distance by ΔL=+64 mm, distance of points 2 ÷ 6: increase in distance by ΔL=+56 mm ($\varepsilon \approx 1.7$ mm/m) (Figure 9). Larger displacements occur on the eastern side of the viaduct; displacements of approximately 50 mm are measured on the bridge bearings and the expansion gaps on the eastern abutment (i.e. from the side of exploited mining walls). On the western side of the viaduct, a maximum displacement of approximately 20 mm is measured in the bridge bearings in the expansion joint on the eastern abutment. Such an uneven distribution of displacements is often encountered at viaducts located perpendicular to mining walls. In the author's article opinion, such distribution of displacements should be taken into account in analysis of resistance to mining impacts of multi-span viaducts – this problem is not sufficiently described in the technical literature and design guidelines (e.g. Wytyczne/Guidelines 1977) and requires individual analyses.

Results of 12-years measurements indicate ground spreading and abutments moving away. Viaduct has two freely-supported spans. Operating range of bearings and expansion joints after the rebuilding was ±80 mm, which indicates that measured maximum displacements (of approximately +60 mm) are safe for this construction, bridge still has kinematic reserves.

Although the viaduct was protected against mining impacts forecast until 2020, it is still in use in 2025 and does not require urgent preventative work or repairs. However, it is worth noting that two types of damage have recently appeared on the viaduct. One of the failures is dislodgement of typical (system) connector railing elements (Figure 10), standard (system) connector lengths turn out to be too short for mining displacements; situation is different for the individually designed railings (Figure 10), their guides/connectors are designed for the entire kinematic range of projected mining displacements. Second defect is related to the ineffective sealing (connecting) of cracks on wings and end walls of

Figure 10. Mining displacement visible on railing and barrier connectors.

Figure 11. Damage to repair mortar applied to the surface of old concrete.

abutments. Weak, porous concrete, despite reinforcement and new insulation inside abutments, still allowed water to permeate, which led to damage to repair mortar applied to old concrete (Figure 11). This damage is insignificant from perspective of construction safety, but deteriorates aesthetics. It was probably necessary to install under-plaster drains along the entire length of repaired cracks or use of specialized mineral mortars dedicated for waterproof.

6 CONCLUSIONS AND ADDITIONAL COMMENTS

For reasons beyond the control of local mine, viaduct's technical condition suddenly deteriorated in 2011. At the time, mining exploitation was underway near the viaduct, and one of mining walls was approaching the viaduct. Numerous construction repairs and rebuilding of bearings were necessary, but such actions require time to prepare design documentation and numerous formal-legal approvals and permits. Suspending mining exploitation would generate huge financial losses for the mine. Article shows how to divide repair actions into two stages in such an emergency situation: first quickly protect construction against predicted mining impacts from currently exploited coal wall, then carrying out comprehensive repairs and protections to prepare viaduct to absorb mining impacts cause by mine activity in the next years. Due to good cooperation local mine with viaduct owner (GDDKiA), several solutions were innovative, e.g. connecting spans with steel tie bars (anchors) to force appropriate kinematic freedom of bearings and abandoning approaching slabs on abutments in favor of mineral road substructure with variable stiffness in order to eliminate regularly occurring fault before and after the viaduct. Repairs are effective. Measured settlements and horizontal displacements are consistent with mining predictions. It is worth appreciating the good cooperation between owner of bridge (GDDKIA), local mine and local municipality.

Viaduct is under constant observation, and local mine has commissioned specialized supervision. This is due, among other things, to large permanent displacement of the bearings from their natural position, by up to 80% of their original operating range. Visual inspections of this bridge are conducted monthly, and semi-annual reports are prepared, analyzing the impact of mining operations on the viaduct, taking into account current permanent changes in the bearings‘ position. Current analyses extend to the end of 2026, during that period of time the bearings' operating range will be sufficient. It is likely (according to mining prediction) that in 2028, bearing rectification will be necessary to increase the range of displacement in the context of further ground spreading. The bearings will be rectified and the expansion joints replaced so that the viaduct can transfer mining impacts until the end of the mining concession (i.e., 2051).

REFERENCES

Będkowski, P., Bednarski, Ł. and Sieńko R. (2014). Structural health monitoring of a rail bridge structure impacted by mining operation. Kraków: Wydawnictwo Politechniki Krakowskiej. *Technical Transactions Civil Engineering* 6-B/2014 (i.21): 15–27.

Będkowski, P. (2017a). Dwuetapowa przebudowa wiaduktu drogowego nad DK81 z tytułu szkód górniczych. *Przegląd Górniczy*, T. 73, No. 12: 57–62.

Będkowski, P. (2017b). Repair and protection of small railway viaduct with jammed span at the mining. *World Multidisciplinary Civil Symposium WMCAUS. IOP Conference Series: Materials Science and Engineering*, No. 245, 2017, 022038: 1–11. (doi: 10.1088/1757-899X/245/2/022038)

Będkowski, P. (2018a). Modeling of information on the impact of mining exploitation on bridge objects in BIM. *BIG 2018 - 4th Nationwide Scientific Conference on Engineering-Infrastructure-Mining, E3S Web of Conferences*, No. 36, 01002: 1–8. (doi: https://doi.org/10.1051/e3sconf/20183601002)

Będkowski, P. (2018b). Selected bearing problems of old railway bridges located on mining areas in Poland. *IOP Conference Series: Materials Science and Engineering*, No. 471, 1757–8981: 1–10.

Bień, J. (2010). *Uszkodzenia i diagnostyka obiektów mostowych.* Warszawa: WKŁ.
Kawulok, M. (2010). *Szkody górnicze w budownictwie.* Warszawa: Wydawnictwo ITB.
Kwiatek, J. (2007). *Obiekty budowlane na terenach górniczych.* Katowice: GIG.
Ren, J., Kang, X., Tang, M., Gao, L., Hu, J. and Zhou, C. (2022). Coal Mining Surface Damage Characteristics and Restoration Technology. *Sustainability* 14(15), 9745. (MDPI: Special Issue Sustainability in Mining) (https://doi.org/10.3390/su14159745)
Rosikoń, A. (1979). *Budownictwo komunikacyjne na terenach objętych szkodami górniczymi.* Warszawa: WKŁ.
Salamak, M. (2013). *Obiekty mostowe na terenach z deformującym się podłożem w świetle kinematyki brył.* Gliwice: Wydawnictwo Politechniki Śląskiej.
Wytyczne (Design Guidelines). (1977). *Wytyczne techniczno-budowlane projektowania i wykonywania obiektów mostowych na terenach eksploatacji górniczej.* Warszawa: Ministerstwo Komunikacji.
Yinfei, C., Yutian, J., Zuoyang, W. and *et al.* (2023). A review of monitoring, calculation, and simulation methods for ground subsidence induced by coal mining. *International Journal of Coal Science & Technology*, 10(1):32. (https://doi.org/10.1007/s40789-023-00595-4)
Zhang, Ch., Wu, K., Huang, S., Li L., and Rao, X. (2025). Study on the simulation of bridge deformation in a mining subsidence area. Springer: *Scientific Reports*, 15(1):529. (doi: 10.1038/s41598-024-84220-7)

Mine Surveying and Mining Area Protection – Sokoła-Szewioła and Jarczyk (Eds)
© 2025 The Author(s), ISBN: 978-1-041-10531-2

Analysis of ground movements in sinkhole prone post-mining area: Case study of the Babina mine in SW Poland

J. Blachowski, N. Walerysiak & P. Modlińska
Wrocław University of Science and Technology, Wroclaw, Poland

ABSTRACT: Sinkholes can continue to develop in post-mining areas decades after mining activities have ended, posing ongoing hazard. This study aimed to assess ground movements in potentially sinkhole-prone areas in the former "Babina" brown coal mine (West Poland), closed over 50 years ago. Five research levelling networks were set-up and 6 levelling campaigns were conducted between September 2020 and September 2024. The geodetic surveys confirmed the presence of continuing ground deformations in selected parts of the study area. The greatest registered subsidence amounted to -24.0 mm. Based on a digital terrain model differentiation approach and in-situ verification 230 sinkholes were identified with diameters ranging from 2.7 to 11.2 m. Next, a geodatabase integrating sinkhole data, digitized mining maps, gravity surveys, and geodetic monitoring results was developed and used to determine spatial congruence between underground voids and sinkholes. The existence of subsurface disturbance (underground corridors) in the analysed sinkhole locations was confirmed by the observed gravimetric anomalies. These findings highlight that the risk of sinkhole formation can persist for decades after mine closure. The integrated monitoring approach proved effective in identifying subsidence risk zones, particularly in vegetated areas, where remote sensing methods are less applicable.

1 INTRODUCTION

A major challenge in managing the safety of post-mining grounds is the appearance of secondary ground movements. These can be classified into continuous and discontinuous ground deformations (Guzy and Malinowska 2020). The first type refers to residual subsidence following the settlement of the rock mass above exploited voids or elevation of the ground following the reconstruction of groundwater table and changes in hydrostatic pressure (Cuenca *et al.* 2013). The second group includes forms such as sinkholes, cracks and fissures that can develop suddenly on the ground surface, usually due to the destruction of underground voids and rapid propagation of post-mining voids to the surface (Kratzsch 1983). The unexpected character of sinkholes that can develop from weeks to decades after end of mining (Strzałkowski 2017, Wódka *et al.* 2024, Walczak *et al.* 2025) is one of the reasons why the mechanism of sinkhole formation is still insufficiently studied as compared to other forms mining induced ground deformations (Malinowska *et al.* 2020). Relatively few studies have explored the implications of the process of mine closures on sinkhole formation (Walczak, *et al* 2025) and taking precautionary measures. The abandonment of geodetic and hydrogeological monitoring after the end of mining can lead to situations posing threat to human safety and infrastructure integrity such as the observed development of sinkholes in the area of influence of the former "Siersza" Hard Coal Mine, Poland (Wódka *et al.* 2024). With many more mines expected to be closed in Europe and globally, due to economic and environmental constraints (Leowen 2022) this problem is likely to intensify and require more systematic monitoring and mitigation strategies. Detailed analysis of factors contributing to the risk of sinkhole formation

DOI: 10.1201/9781003655558-4

was conducted by Strzałkowski (2021) who investigated the presence of several contributing factors. The reported factors driving development of sinkholes include: lithology, shallow depth and inclination of underground voids, restoration of groundwater level following the end of mine dewatering, vegetation cover and soil types, as well as increased atmospheric precipitation that might lead to suffosion in the ground. Analysis of historical mining documentations in Geographic Information System (GIS) help to determine precise location of old workings (Suh *et al.* 2016) whereas geophysical surveys can be used to confirm existence of underground voids (Pilecki 2008, Kotyrba and Kortas 2016).

This study is aimed at determining if mining-related ground movements, that can indicate sinkhole prone areas, persist within the area of influence of shallow underground mining of brown coal that ended over 5 decades ago. The spatiotemporal characteristics of ground movements were described based on repeated geodetic observations and analysed with reference to the known extent of mining activity and the results of gravimetric surveys in GIS.

2 AREA OF INTEREST

The research covers the "Pustkow" mining field of the "Przyjaźń Narodów - Szyb Babina" ("Babina") mine, located between the town of Łęknica and the villages of Nowe Czaple and Bronowice on the Polish-German border (Figure 1). Precisely the known extent of mining covered by geodetic monitoring. In terms of the geomorphology the area is situated in the SE part of post-glacial moraine structure spanning both sides of the Polish-German border and known as the Muskau Arch. The moraine had been formed during the South-Polish glaciation and was remodelled during successive glacial periods (Badura *et al.* 2003, Kupetz 1997). Glacial processes led to folding, upthrust and subsequent erosion of initially horizontal brown coal formations covered with Tertiary and Quaternary deposits (Urbański

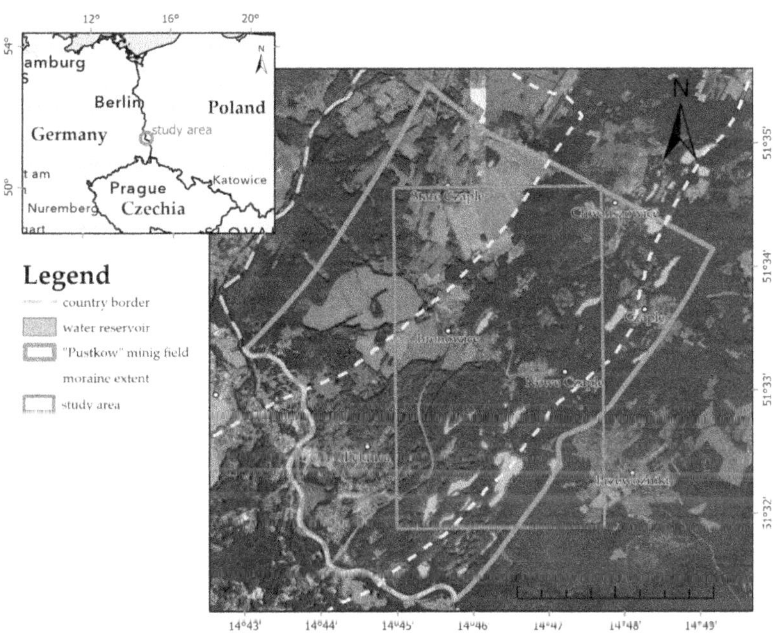

Figure 1. Geographic location of the Area of Interest - "Pustkow" mining field of the "Przyjaźń Narodów -Szyb Babina" ("Babina") mine.

2007). The presence of shallow brown coal deposits led to the start of coal mining in the western part of the Muskau Arch as early as 1843. In the now Polish part of this structure, mining began in 1854 (Koźma 2016). Initially, the steeply inclined and narrow coal seams were mined using underground methods. With time open-pit mining was introduced. Shallow underground mining utilised the room and pillar mining system. Due to the brown coal deposit inclination, mining levels were developed stepwise downward. The mining levels were connected by drifts used for transporting coal to lower galleries and to the main haulage level. Access to the underground mine was provided by inclines connecting it to the surface. The depth of the mining levels ranged from several to approx. 100 m below the ground (Gontaszewska 2015). Open-pit and to some extent underground mining continued until 1973 when the mines were closed for economic reasons (Greinert *et al.* 2015). The anthropogenic changes of the original glaciotectonic topography resulted in the development of a unique landscape with numerous post-mining lakes. The present-day geoheritage and geodiversity values of the area are promoted in the transboundary Muskau Arch UNESCO Global Geopark (Koźma and Migoń 2024).

2.1 *Status of ground deformation studies in the Muskau Arch*

Shallow underground mining led to characteristic surface deformations, such as elongated subsidence troughs and sinkholes. The troughs formed above underground galleries shortly after mining operations commenced. The troughs are characterized by widths from 50 to 150 m and lengths of up to 2 to 3 km. Their shape is related to the presence of brown coal outcrops (gisers), inclination of coal seams and the distribution of underground workings Within the troughs, numerous and clustered sinkholes are visible. These circular-shaped features typically have diameters and depths ranging from 2 to 5 m. Additionally, larger sinkholes of spherical shape and greater depth developed above former shafts (e.g., ventilation shafts) and access tunnels (Koźma 2016). Before the two National Science Centre projects (2019/33/B/ST10/02975 and 2021/43/B/ST10/02157), which include this study, no systematic geodetic monitoring of post-mining ground movements was conducted for the area. Previously published research concentrated on legacy of glacial and mining processes (Koźma and Migoń 2024), differential analysis of post-mining lake bed (Blachowski *et al.* 2025), terrestrial laser scanning inventory of newly formed sinkholes (Blachowski *et al.* 2024) and analysis of ground movements in the entire post-mining field (Walerysiak *et al.* 2025). This study presents ground movement monitoring results in an active sinkhole zone.

3 GEODATA AND METHODOLOGY

The integrated approach adopted for monitoring ground movements comprises the following complementary parts: development of a geodatabase of former mining activity and its effects

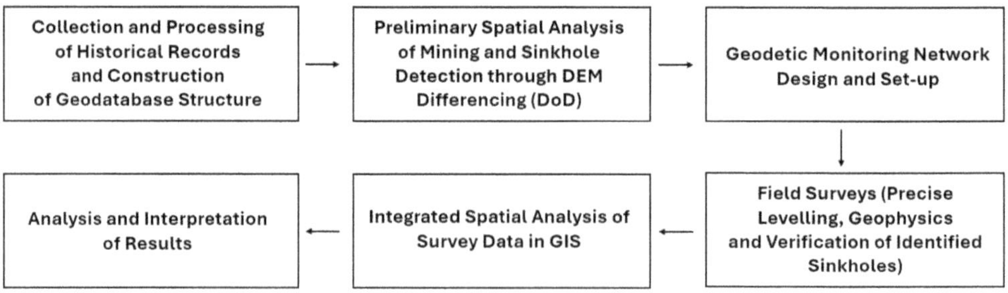

Figure 2. Schematic representation of the research methodology.

on the surface based on digital coding and analysis of available documentations, differential analysis of digital elevation models (DEM) from national open data sources and collected with Unmanned Aerial Vehicle (UAV) missions, cyclic geodetic observations, geophysical survey of subsidence zones identified based on the analysis of archival documentations, field reconnaissance, and analysis of the spatial congruence of mining maps and results of monitoring and differential calculations. The methodology has been presented in Figure 2.

In the first step, to facilitate monitoring and analysis of ground deformations, a comprehensive geodatabase has been developed with the objective to integrate information on historical mining operations, sinkhole locations, and geodetic levelling and geophysical survey results. The database was created through digital coding and structuring of the acquired data. The location of sinkholes was determined by differencing digital elevation models, a technique resulting in a DEM of Difference (DoD) (Wajs et al. 2024). These locations were verified in situ, and a documentation of each sinkhole was conducted and stored, encompassing its geometric characteristics, including depth, diameter, and volume. The geodatabase includes feature classes representing underground workings derived from archival mining maps. The results of periodic geodetic levelling measurements and geophysical survey have also been integrated in the geodatabase. An overview of geophysical surveys in the study area is described in (Blachowski *et al.* 2022). Integration of these data facilitated analysis of the spatial congruence of the locations of underground voids with sinkholes. The geodatabase and spatial analyses were performed in ArcGIS Pro software licensed to Wroclaw University of Science and Technology.

To determine if ground movements continue in the post-mining area a control geodetic levelling network was constructed. It comprises 99 controlled benchmarks, 26 in 39 main levelling lines spanning 36 km and 73 in 5 specific study networks located in 3 known zones of underground mining and reported sinkhole activity. Location of these study girds is shown in Figure 3. The main survey network points were stabilized as concrete posts and protected by PVC covers. The area was measured 6 times with the initial campaign in September 2020 adopted as baseline observations. The subsequent campaigns were executed

Figure 3. Study area with the location of the five detailed observation networks (1–5), including a new sinkhole in network 1: a) condition on September 20, 2023; b) condition on April 24, 2024 (after stabilisation works.

in April 2021, October 2021, May 2022, September 2023, and most recently in September 2024. The observations in the control-measurement network were referenced to national levelling network benchmarks located outside of the area of mining influence. A comprehensive account of the results of ground movements in the main levelling network for the study area is provided in Walerysiak *et al.* (2025). This study focuses primarily on the analysis and initial interpretation of ground movements in the detailed measurement networks. The adopted methodology encompassed 3 primary domains: (1) statistical analysis of vertical ground movements derived from levelling results, (2) an assessment of the spatial congruence between sinkhole location, known extent of underground voids and results of geophysical gravimetric mapping, (3) statistical analysis of the geometrical characteristics of sinkholes. The calculated statistics were presented in the form of column charts and tables, whereas the locations of sinkholes and underground workings on maps. Based on spatial analytical functions in GIS (i.e. measure distance, calculate distance between closest features in two data layers, extract pixel value by point) additional descriptive statistics relating location of sinkholes and underground voids, as well as sinkhole geometry were determined. This approach facilitated the quantification of the extent and nature of vertical ground movements in the study areas in the 5 year period of observations (2020-2024) with special attention on detailed measurement network 1 where recent sinkholes were identified (Figure 3).

4 RESULTS AND DISCUSSION

The results of benchmark monitoring in the 5 year period (September 2020–September 2024) for each of the detailed measurement networks have been presented in Figure 4 together with their configuration against the background of digitised extent of underground workings, location of vertical shafts (winze) and documented locations on sinkholes.

The greatest and the most consistent ground movements were observed in research network 1. An average benchmark subsidence of –8.9 mm was observed between September 2020 and September 2024, with the maximum displacement of –24.0 mm measured at controlled point 12 (Figure 4). Furthermore, 4 more controlled points located very close to probable underground voids exhibited movements in the range from -9.0 mm (2 and 3) to -19.0 mm (point 9). Figure 3a and 3b shows one of the recent sinkholes identified through differential analysis of subsequent DEMs collected with the UAV platform. Figure 3a shows the situation during a field investigation in September 2023, Figure 3b in April 2024 after the local authorities carried out stabilisation measures. This sinkhole appeared between 2021 and 2023 as it was not identified during the initial field works conducted in 2020 and is not present in the public DEM from 2020 (Blachowski *et al.* 2024). The depth of the underground workings above, which the sinkhole developed was estimated, during in-situ investigations, to be no more than 6 m. In networks 2 to 4, located in an horseshoe shaped underground mining field to the south of the detailed network 1 the registered vertical ground movements are smaller. In observation network 2 three out of 18 controlled points subsided by over 5 mm in the 5 year period, with the greatest movement of -9 mm observed in point S2. It is important to note that 2 sinkholes were discovered there in 2021. Differential analysis of public DEM from 2011 and 2020 (Blachowski *et al.* 2024) points to the development of these features after 2011. In network 3 just one controlled point S32 exhibits movement greater than 5.0 mm, with subsidence of -35.0 mm. It should be noted that it is the only point not directly located above probable underground voids and is located close to a public path. Research network 4 consists of 8 controlled points located above probable underground voids. The observed movements range from -1.0 mm (point S42) to -6.0 mm (point S41) and are generally statistically insignificant. In network 5, the observed height changes of 2 benchmarks, i.e. S45 equal to -45 mm and S55 equal to +35 mm, are most probably attributed to disturbance during earthworks related to the construction of a new tourist path. Consequently, these results were identified as outliers in the evaluation of ground movements within this network. The vertical movements observed on selected controlled points, especially in detailed measurement networks 1 and 2 affected by recent sinkhole activity, indicate ongoing ground deformation and probable instability in the these areas.

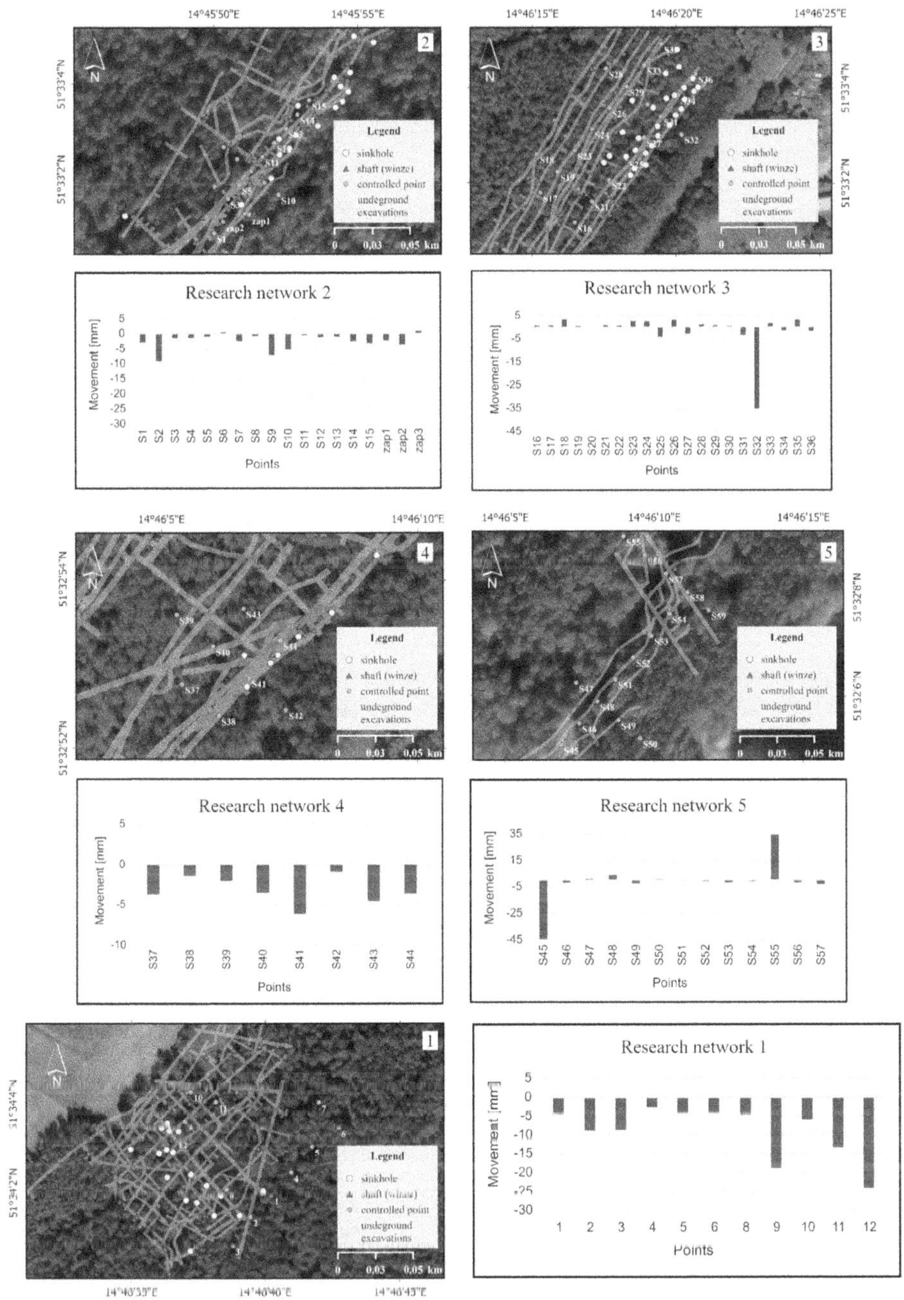

Figure 4. Location of controlled points and vertical ground movements observed in the detailed research networks.

The observed movements can be treated as indicators of sinkhole risk and need for continuous monitoring of the post-mining area as has been suggested among others by (Malinowska *et al.* 2019, Walczak *et al.* 2025). The underground mining zone where detailed research network 1 has been established was also covered by microgravimetric surveys to identify if gravity anomalies that may indicate subsurface voids are present (Blachowski *et al.* 2022). The map of the gravimetric anomalies is shown in Figure 5. The results confirm the existence of underground mining remnants determined through georeferencing and digitising historic mining plans. Furthermore, an increase of the observed gravity anomalies towards south-west can be seen. This is an area where another group of sinkholes has been identified. Unfortunately, mining plans of this section are missing. These data can be used to mark the zones posing the greatest risk of sinkhole development that should be the subject of periodic ground movement monitoring.

To complete the analysis of the sinkhole processes, detailed descriptive statistics of all the 230 sinkholes in the study area have been calculated. The results are presented in Table 1.

Table 1. Descriptive statistics of sinkhole geometry (depth, diameter, volume) in the study area.

Statistic	Minimum (all sinkholes)	Minimum (network 1)	Maximum (all sinkholes)	Maximum (network 1)	Average (all sinkholes)	Average (network 1)
Depth [m]	0.3	0.3	4.8	3.2	1.5	3.2
Diameter [m]	2.7	3.2	11.2	11.2	5.9	5.0
Volume [m^3]	5.5	8.9	370.4	370.4	51.7	50.1

The majority of sinkholes identified, through DEM differencing and confirmed as sinkholes during field investigations have approximately circular shape and occur in larger subsidence basins that developed above underground mining. The average depth of sinkholes in the study area is 1.5 m, whereas in the detailed measurement network this value is higher, 3.2 m. The greater depth of sinkholes in the network 1 can be attributed to brown coal seams located very close to the surface, with the shallowest level 5-6 m below the ground surface, the fact that this zone was the last one to be mined and the potential covering of the actual bottom of sinkholes by accumulated plant debris in older parts. The diameter of sinkholes ranges from 2.7 to 11.2 m in the entire study area. The volume has been calculated with the assumption that the sinkholes have a hemispherical shape. Table 2 contains calculated distances for the 18 sinkholes identified in the detailed measurement network 1 in relation to probable location of the closest underground voids, i.e. corridors and winze identified in historical mining plans. In the case of 11 sinkholes their distance from underground workings is less than 1 m, in 6 more cases up to 2.8 m and in one case 12.2 m.

Table 2. Descriptive statistics of the spatial relationship between sinkhole location and subsurface voids.

Statistic	Distance to winze (shaft) [m]	Distance to horizontal corridor [m]	Depth below the surface of the void closest to sinkhole
Min	1.0	0.0	15.5
Max	22.5	12.2	38.8
Average	9.2	1.4	27.7

As stated in the literature Airborne Laser Scanning derived DEM and Interferometric Synthetic Aperture Radar (InSAR) derived ground movements in sinkhole prone areas can be considered as potential indicators of increased sinkhole risk (Malinowska *et al.* 2019, Walczak *et al.* 2025). However, a thorough analysis of sinkhole susceptibility mapping requires additional information, especially characteristics of hydrogeological conditions, groundwater table fluctuations, geology, permeability, land cover types and precipitation records. This information is rarely available and this topic is outside of the scope of this paper, which concentrates on monitoring ground movements based on periodic geodetic levelling measurements. This method was selected to augment the DEM differencing results and as an alternative to InSAR technique, which was found out to be of limited applicability in the area, which is characterized by forest vegetation.

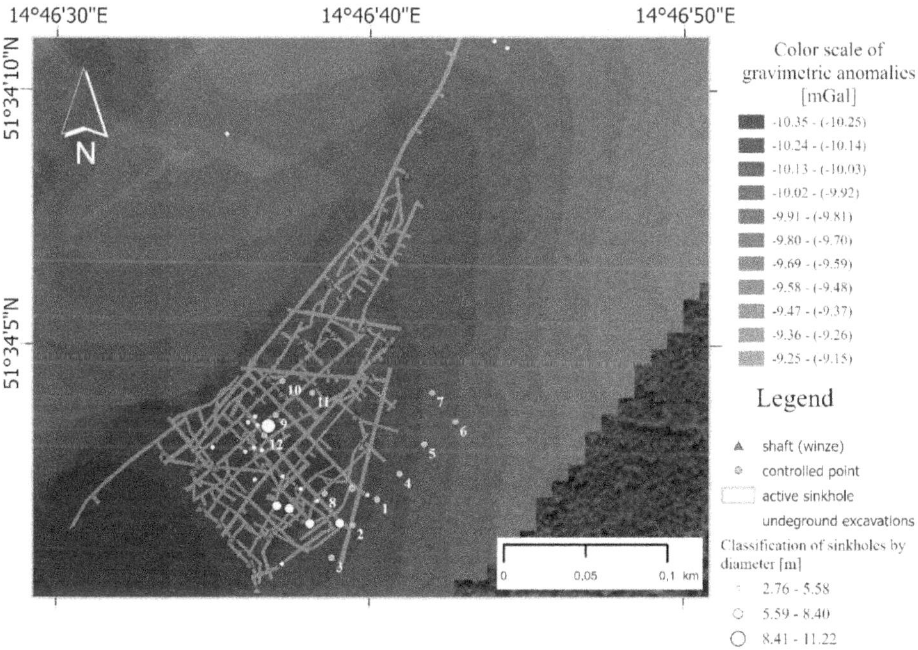

Figure 5. Location of controlled points and gravimetric anomalies observed in detailed research network number 1.

5 CONCLUSIONS

The presented results demonstrate that ground movements continue in the "Pustków" field of the "Babina" mine 5 decades after the end of mining. The adopted methodology integrating historical mining data, DEM differencing, geodetic levelling and geophysical survey results proved to be effective means for monitoring and analyzing ground stability in post-mining undergoing tourist development. Geodetic levelling results indicate an average subsidence of -8.9 mm, with a maximum value of -24 mm over a 5-year period, in the research site where new sinkholes were identified using DEM differencing. Gravimetric survey indicated gravity anomalies pointing to the existence of voids in the subsurface. GIS analysis confirmed spatial congruence of sinkholes with the probable location of underground workings determined from digitized historical mining maps. Statistical analysis of sinkhole characteristics show that such features in network 1 average 3.2 m deep, more than twice the

1.5 m average for the entire study area. These preliminary results provide a foundation for further research aimed at testing data-driven spatial statistical models that incorporate potential driving factors, including hydrogeological conditions and atmospheric precipitation, to assess sinkhole susceptibility and map risk zones of a study that precedes the building and testing of models based on data-driven spatial statistics and machine learning to map risk zones and sinkhole vulnerability.

ACKNOWLEDGMENTS

This work was supported by the Polish National Science Centre projects no 2019/33/B/ST10/02975 and no 2021/43/B/ST10/02157.

REFERENCES

Badura, J., Gawlikowska, E., Kasiński, J., Koźma, J., Kupetz, M., Piwocki, M. and Rascher, J., (2003). "Muskau Arch" Geopark – Suggested trans-boundary area of geodiversity conservation. *Przegląd Geologiczny* 51,pp.54–58.

Blachowski, J., Warchala, E., Koźma, J., Buczyńska, A., Bugajska, N., Becker, M., Janicki, D., Kujawa, P., Kwaśny, L., Wajs, J. and *others*, (2022). Geophysical Research of Secondary Deformations in the Post-Mining Area of the Glaciotectonic Muskau Arch Geopark—Preliminary results. *Applied Sciences*, 12, 1194. Available at: https://doi.org/10.3390/app12031194

Blachowski, J., Wajs, J., Walerysiak, N. and Becker, M., (2024). Monitoring of Post-Mining Subsidence using Airborne and Terrestrial Laser Scanning approach. *Archives of Mining Sciences*, 69(3), pp.431–446. Available at: https://doi.org/10.24425/ams.2024.151444

Caro Cuenca, M., Hooper, A.J. and Hanssen, R.F., (2013). Surface deformation induced by water influx in the abandoned coal mines in Limburg, The Netherlands observed by satellite radar interferometry. *Journal of Applied Geophysics*, 88, pp.1–11. Available at: https://doi.org/10.1016/j.jappgeo.2012.10.003.

Gontaszewska, A., (2015). Szkody górnicze. In: A. Greinert (ed.) *Wydobycie węgla brunatnego i rekultywacja terenów pokopalnianych w regionie lubuskim.* Zielona Góra: Uniwersytet Zielonogórski, pp.146–157. (in Polish)

Greinert, A. (ed.), (2015). *Lignite Mining and Reclamation of Post-Mining Areas in the Lubuskie Region - Wydobycie węgla brunatnego i rekultywacja terenów pokopalnianych w regionie lubuskim.* Zielona Góra: University of Zielona Góra Press. ISBN 978-83-937619-2-0. (in Polish)

Guzy, A. and Malinowska, A.A., (2020). State of the art and recent advancements in the modelling of land subsidence induced by groundwater withdrawal. *Water*, 12(7), p.2051. Available at: https://doi.org/10.3390/w12072051.

Kotyrba, A. and Kortas, Ł., (2016). Sinkhole hazard assessment in the area of abandoned mining shaft basing on microgravity survey and modelling — Case study from the Upper Silesia Coal Basin in Poland. *Journal of Applied Geophysics*, 130, pp.62–70. Available at: https://doi.org/10.1016/j.jappgeo.2016.04.007

Koźma, J., (2016). Anthropogenic landscape changes connected with the old brown coal mining based on the example of the Polish part of the Muskau Arch area. *Górnictwo Odkrywkowe*, 57(3), pp.5–13 (in Polish).

Koźma, J. and Migoń, P., (2024). Mużaków Rampart (Muskau Arch)—The legacy of glacial processes and mining in the UNESCO Global Geopark. In: P. Migoń and K. Jancewicz (eds.) *Landscapes and Landforms of Poland. World Geomorphological Landscapes*. Cham: Springer. Available at: https://doi.org/10.1007/978-3-031-45762-3_28.

Kratzsch, H., (1983). *Mining Subsidence Engineering*. 1st ed. Berlin, Heidelberg: Springer-Verlag. Available at: https://doi.org/10.1007/978-3-642-81923-0.

Kupetz, M., (1997). Geologischer Bau und Genese der Stauchendmoräne Muskauer Faltenbogen. *Brandenburgische Geowissenschaftliche Beiträge*, 4, pp.1–19 (in German).

Loewen, B., (2022). Coal, green growth and crises: exploring three European Union policy responses to regional energy transitions. *Energy Research & Social Science*, 93, Article 102849. Available at: https://doi.org/10.1016/j.erss.2022.102849.

Malinowska, A.A., Witkowski, W.T., Hejmanowski, R., Chang, L., van Leijen, F.J. and Hanssen, R.F., (2019). Sinkhole occurrence monitoring over shallow abandoned coal mines with satellite-based persistent

scatterer interferometry. *Engineering Geology*, 262, 105336. Available at: https://doi.org/10.1016/j.enggeo.2019.105336
Malinowska, A.A., Witkowski, W.T., Guzy, A. and Hejmanowski, R., (2020). Satellite-based monitoring and modeling of ground movements caused by water rebound. *Remote Sensing*, 12, p.1786. Available at: https://doi.org/10.3390/rs12111786.
Pilecki, Z., (2008). The role of geophysical methods in the estimation of sinkhole threat in the post-mining areas of shallow exploitation in the Upper Silesian Coal Basin, Poland. *Mineral Resources Management*, 24 (3), pp.27–40.
Strzałkowski, P., (2017). Proposal of predicting formation of sinkholes with an exemplary application. *Journal of Mining Science*, 53(1), pp.53–58. Available at: https://doi.org/10.1134/S1062739117011835
Strzałkowski, P., (2021). The influence of selected mining and natural factors on the sinkhole creation hazard based on the case study. *Environmental Earth Sciences*, 80, Article 117. Available at: https://doi.org/10.1007/s12665-021-09403-1.
Suh, J., Choi, Y. and Park, H.D., (2016). GIS-based evaluation of mining-induced subsidence susceptibility considering 3D multiple mine drifts and estimated mined panels. *Environmental Earth Sciences*, 75, Article 890. Available at: https://doi.org/10.1007/s12665-016-5695-1.
Urbański, K., (2007). The zones of glaciotectonical deformation in Muskau Arch region - Łuk Mużakowa jako złożona struktura glacitektoniczna. *Zeszyty Naukowe. Inżynieria Środowiska*, 134(14), pp.179–190. (in Polish with English abstract)
Walczak, S., Witkowski, W.T., Stoch, T. and Guzy, A., (2025). Detecting sinkholes and land surface movement in post-mining regions using multi-source remote sensing data. *Remote Sensing Applications: Society and Environment*, 38, p.101560. Available at: https://doi.org/10.1016/j.rsase.2025.101560.
Walerysiak, N. and Blachowski, J., (2025). Geodatabase of Sinkholes in the Post-Mining Area of the Brown Coal Mine "Babina" (W Poland). *EGU General Assembly 2025*, Vienna, Austria, 27 Apr–2 May 2025. EGU25-18583. Available at: https://doi.org/10.5194/egusphere-egu25-18583 [Accessed 15 Jun. 2025].
Walerysiak, N., Blachowski, J., Kujawa, P. and Wajs, J.W., (2025). Monitoring and analysis of Ground Movements in the Babina Post-Mining Area. *Civil and Environmental Engineering Reports*, 35(2), pp.258–279. Available at: https://doi.org/10.59440/ceer/203313 [Accessed 15 Jun. 2025].
Wódka, M., Kamieniarz, S., Wojciechowski, T., Przyłucka, M., Perski, Z., Sikora, R., Karwacki, K., Jureczka, J., Nadłonek, W., Krieger, W. and Zając, M., (2024). Post-mining deformations in the area affected by the former "Siersza" Hard Coal Mine in Trzebinia (southern Poland). *Geological Quarterly*, 68 (1). Available at: https://gq.pgi.gov.pl/article/view/33899.

Mine Surveying and Mining Area Protection – Sokoła-Szewioła and Jarczyk (Eds)
© 2025 The Author(s), ISBN: 978-1-041-10531-2

Comprehensive evaluation of mining-induced deformations and their consequences for a residential structure

I. Bryt-Nitarska
Strata Mechanics Research Institute of the Polish Academy of Sciences, Kraków, Poland

M. Pomykoł & K. Szafulera-Bałecka
Silesian University of Technology, Gliwice, Poland

ABSTRACT: The implementation of underground coal extraction is an anthropogenic activity that invariably generates consequences within a defined zone of influence. In the case of coal seam extraction, one of the effects is the transformation of the land surface and related changes in the building subsoil of existing structures. This article presents the process of identifying potential mining influences that may have affected a residential building, contributing to the observed structural damage and tilting of the building's frame. The analyses included both static influences from the extraction of a coal seam and dynamic effects associated with the high seismicity of the rock mass in the analysed area. Due to the presence of bed deposit tectonics and fault outcrops on the surface, a risk assessment of the occurrence of discontinuous deformations in the building subsoil was also performed. The technical condition of the building was diagnosed using the results of geodetic surveys of structural element inclinations. This comprehensive analysis made it possible to determine the relationship between the identified causes and the observed structural damage.

1 INTRODUCTION

Underground mining has, for years, exerted a significant impact on the natural environment and on the surface development of land. This issue has consistently remained within the focus of research centres in Poland and worldwide, as confirmed by extensive literature including textbooks, monographs, and scientific articles (Chudek 2010, Knothe 1984, Kowalski 2015, Kratzsch 1983, Kwiatek 1997, Peng 1992, Whittaker & Reddish 1989). In recent years, this subject has gained importance due to growing concern for the state of the environment and the safety of residents in mining areas (Strzałkowski & Szafulera 2020, Strzałkowski 2022, Strzałkowski & Ścigała 2017, Szojda & Strzałkowski 2025).

As numerous studies indicate, among the various forms of underground mining impact on the environment, the most serious threats are posed by surface deformations — both continuous and discontinuous — as well as rock mass tremors. In Polish coal basins, the prediction of surface deformations is commonly based on the theory of W. Budryk and S. Knothe (Knothe 1984), valued for its simplicity, well-established regional parameters, and the availability of computer tools supporting calculations and visualisation of results (Ścigała 2008).

Although discontinuous deformations occur less frequently than continuous deformations (Chudek *et al.* 1988, Kowalski 2015), their forecasting continues to pose a major challenge. Particular attention is now being paid to the growing number of linear type deformations (Malinowska & Hejmanowski 2016, Orwat & Gromysz 2021), associated, among other factors, with bed deposit tectonics (Ścigała & Szafulera 2020), the overlapping of extraction boundaries in several seams, and high values of horizontal strains (Kowalski 2015).

DOI: 10.1201/9781003655558-5

Simultaneously, high-energy rock mass tremors remain a significant threat to both the environment and infrastructure, with their number and intensity increasing alongside the growing depth of extraction. Their impact on buildings is assessed using specialist formulas (Mutke 1991 2018) and the GSIS-2017 scale (Mutke 2018).

This paper presents an example of a comprehensive analysis of the influence of mining activity and its effects on the land surface at the location of a residential building, taking into account continuous deformations, discontinuous deformations, and rock mass tremors.

2 CASE DESCRIPTION

Damage was observed in a residential building located within a mining area in the Upper Silesian Coal Basin (see Figure 1a). The building was constructed in 2007 (Figure 1b), and since 2016, progressive structural damage has been recorded, which is linked to the effects of underground coal mining. From the time of its construction, the building has remained within the zone of mining-induced ground movement.

Although the structure was not located directly above an active mining panel, it was subjected to significant influences from mining operations in the vicinity. The nearest longwall panel was mined in 2014–2015, approximately 140 meters from the property. Additionally, the building lies within a zone of dynamic impact from mining-induced seismic tremors and relatively close to a tectonic fault zone, increasing the potential for discontinuous ground deformations and amplifying seismic effects.

It is important to note that, under Polish regulations, designating a mining area does not automatically prohibit surface development, including residential construction. However, buildings erected within such areas are required to be structurally adapted to withstand potential mining-related impacts. In the case described, appropriate structural reinforcements were implemented, which helped to limit the extent of damage resulting from continuous ground deformation.

3 ANALYSIS OF GEOLOGICAL AND MINING CONDITIONS

3.1 *Geological structure of the rock mass*

The geological structure of the rock mass in the area of the analysed property was determined based on maps of the thickness of Quaternary and Tertiary formations and the profile of a borehole located approximately 750 metres to the northwest of the building. The rock mass consists of two principal parts: Qverburden — Quaternary formations (approx. 70 m): soil, clay, sand; Tertiary formations (approx. 85 m): clay shales, sand, gypsiferous shales, marly and sandy clays, limestone, and mudstones. Carboniferous formations (coal-bearing layers) — Orzesze Beds (down

a) 

b) 

Figure 1. a) Location of the Upper Silesian Coal Basin (USCB) on the map of Poland; b) General view of the building.

to approx. 690 m): mainly clay shales, less frequently sandy and sandstone shales, with numerous coal seams of the 300 group; Ruda Beds (from the roof of seam 401 down to approx. 850 m): predominantly clay shales, with a smaller share of sandstones and coal seams of the 400 group.

3.2 *Tectonics*

At a distance of approximately 210–560 m southeast of the analyzed building, there is an outcrop of a fault zone consisting of three diagonal faults, displacing the Carboniferous strata towards the southeast by approximately 40 m, 50 m, and 70 m, respectively (Figure 2). According to the Carboniferous roof map, the building is located within the hanging wall of this dislocation.

3.3 *Performed underground coal extraction*

In the analyzed area, underground coal extraction began in 1979, prior to the construction of the building. The mining works were carried out using the longwall system with roof caving in the Orzesze and Ruda beds, at depths ranging from approximately 480 m to 900 m. The shallowest working, located directly beneath the building, reached a depth of about 550 m, which excludes the risk of discontinuous surface deformations (sinkholes) (Chudek *et al.* 1988). Within the property and its immediate surroundings, goafs are present in the rock mass, whose potential reactivation cannot be ruled out. Since the building was erected, underground coal extraction has been conducted in two coal seams, affecting a total of five plots. The mining works were carried out using the longwall system with roof caving at depths ranging from approximately 788 m to 852 m, with an extraction height from about 1.8 m to 2.9 m. Figure 2 shows a sketch illustrating the edges of the accomplished extraction workings in relation to the location of the building. Basic information regarding the geological and mining conditions of the extraction conducted after the construction of the building is presented in Table 1.

Table 1. Basic data on the geological and mining conditions of the accomplished extraction in the area of the building.

Seam	Longwall	Beginning of mining	Completion of mining	Thickness [m]	Depta [m]	Distance from the object [km]	Direction of mining
405/2	59	15-05-08	15-12-09	2.9	822	0.27	NW
405/1	50a	10-04-11	30-04-11	2.1	852	0.45	NW
405/1	50b	01-05-11	30-07-12	1.8	810	0.38	NE
405/1	51	30-06-14	15-05-15	1.8	822	0.14	N
405/2	63	22-03-16	31-12-16	2.6	788	0.54	N

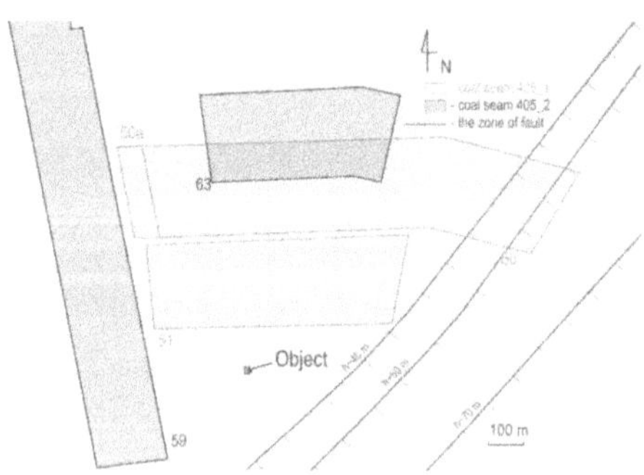

Figure 2. Sketch of the accomplished coal extraction conducted since the construction of the building.

4 IMPACT OF THE ACCOMPLISHED COAL EXTRACTION ON THE SURFACE

4.1 *Geodetic surveys*

In the analyzed area, surface deformations related to the operation of the mining plant were monitored using geodetic surveys at dispersed measurement points. The nearest benchmark, point A, is located approximately 130 m southeast of the property. Figure 3 presents the subsidence trend for this point over time. The measurements confirmed that the analyzed area was within the range of mining influences from coal seams 405/1 and 405/2 (with a maximum subsidence of w = -0.48 m), with the intensity of deformation depending on the distance between the mining front and the measurement point. The greatest subsidence increments were recorded in 2014–2015 during the extraction of seam 405/1. Stabilization of the rock mass movements occurred around 2018.

For the diagnostic assessment of the technical condition of the building, the results of geodetic measurements of the tilting of structural elements were used. The methods for determining these tilts can be divided into two main groups: direct measurements (e.g., using inclinometers, tiltmeters, or laser devices) and indirect measurements (e.g., leveling rods, small angle measurements, and technical leveling) (Rola 1997). In order to determine the tilt values of the building's edges, two indirect methods were applied: the technical leveling method and the small angle measurement method.

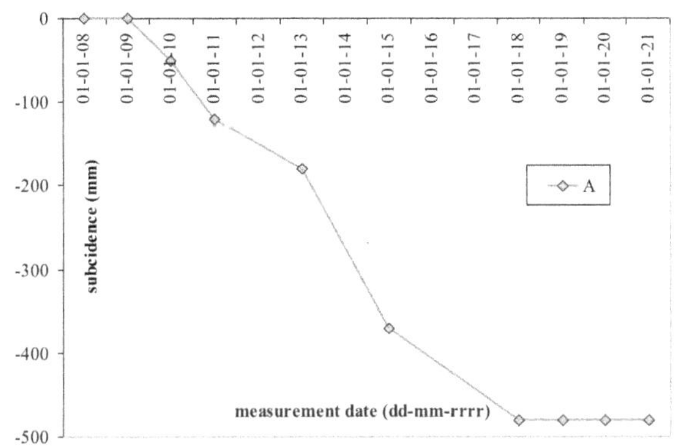

Figure 3. Subsidence of measurement point A over time.

The use of precise leveling from a central point enabled the measurement of height differences between points on the plinth located at the ground floor level of the building. As a result of a rigorous adjustment of these observations (GeoNET), the elevation values were determined in a local reference system. Based on this data, a contour map of the plinth (Figure 4a) was prepared, which allowed for the determination of the average inclination of the plinth plane. It was found to be inclined towards the north, with an inclination value of approximately 3.5 ‰, corresponding to a maximum height difference of 83 mm over a length of about 23 m.

To determine the inclination components of the building's vertical edges in two mutually perpendicular directions, the so-called small-angle measurement method was used. The obtained values and directions of the components made it possible to calculate the resultant vector of displacements for each edge (Figure 4b). These revealed slight discrepancies in direction, with inclination values ranging from 3 to 6 ‰. However, both the directions (tending northward) and the magnitudes of the vertical edge inclinations were consistent with the tendencies observed in the measurements of the plinth inclination.

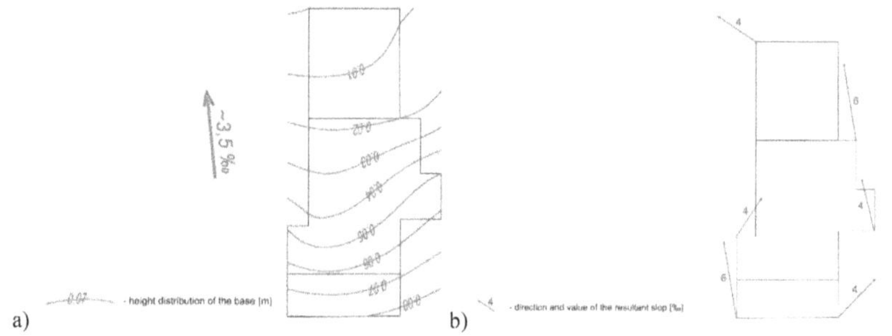

Figure 4. a) Distribution of the plinth elevations [m]; b) Resultant inclinations of the building's edges [‰].

Significant discrepancies are often observed not only between the inclination values of individual vertical edges of a building (especially in masonry structures located in areas repeatedly affected by mining influences) but also between the directions of their inclination vectors. In such cases, it is difficult to associate the inclination of the entire building structure with the resultant of the inclination vectors of its individual vertical edges. Therefore, when assessing the degree and direction of inclination, it is essential to determine the tilt of other structural edges of the building, not only its vertical edges.

4.2 *Calculation results*

The calculations were performed using a program (Ścigała 2012) based on the theory developed by W. Budryk and S. Knothe (Knothe 1984), conducting a computer simulation of longwall panel operations. Immediate manifestation of mining influences was assumed ($c \rightarrow \infty$). The calculations were carried out for both the final deformation state and time-extreme values. The calculation point was located at the geometric center of the residential building.

The following values of theoretical parameters were adopted for the computations:

- roof control coefficient $a = 0.8$,
- tangent of the angle of the main influence range $tg\beta = 2.0$,
- extraction boundary $d = 0$ m,
- proportionality coefficient of horizontal displacement to tilt B. Following E. Popiołek and J. Ostrowski (Popiołek 2009), the value $B = 0.32r$ (where r is the radius of the main influence range) was assumed, along with a deviation coefficient $n = 0.7$.

The calculation results are summarized in Table 2.

Table 2. Summary of maximum values of selected surface deformation indices for the geometric center of the building.

Calculation variants	Type of calculations	Subsidence, w [mm]	Maximum tilt, T_{max} [‰]	Mining area category	Horizontal strain, ε_{max} [‰]	Mining area category
1a (>2008)	Final	-376	2.7	II	1.6	II
1b (>2008)	TE*	-376 25-12-16	2.7 25-12-16	II	1.6 25-12-16	II

*Temporally Extreme, the dates corresponding to specific values of the deformation indicator provided in Table 2 should be directly associated with the simulated position of the active edge of the mining front on the given day.

5 FAULT INFLUENCE ON DEFORMATION DISTURBANCES

As stated by M. Chudek (Chudek *et al.* 1988), the area endangered by discontinuous deformations in the vicinity of a fault outcrop can be determined using the following relationship (1):

$$S = 1.2 \cdot h_n \cdot ctg\alpha \qquad (1)$$

where: h_n – thickness of the overburden [m]; α – natural angle of repose of overburden layers [degrees].

According to geological and mining documentation:

- The building is located at a distance of approximately 210–560 m from the fault zone, which comprises three faults with throws ranging from about 40 to 70 m.
- At a distance of 80–160 m east of the building, a series of linear discontinuous deformations is present, mainly resulting from the activation of the fault zone caused by mining-induced disturbances of rock mass equilibrium. The influence of tensile deformations cannot be excluded either (Table 2).
- Calculations indicate that the area endangered by discontinuous deformations in the fault outcrop zone extends to approximately 220 m, covering the analysed property.

6 IMPACT OF ROCK MASS TREMORS ON THE SURFACE

The assessment of the influence of rock mass tremors on the surface was based on the analysis of events with energy $\geq$ 1 E+05 J, recorded within a 3 km radius from the building between 2015 and the first quarter of 2025. Seismic activity measurements were carried out at a monitoring station located approximately 2300 m northwest of the property.

Using formulas by Mutke (1991, 2018), values of PGA (Peak Ground Acceleration) and PGV (Peak Ground Velocity) were calculated for the location of the property based on 76 tremors. An amplification factor of k = 1.4 was adopted for the calculations. Figure 5 presents the measured PGA and PGV values at the monitoring station as well as the computed accelerations and vibration velocities induced by tremors at the property location.

The analysis of the measurements and calculations indicates the following findings:

- The largest tremor, with an energy of 1 E+08 J, occurred in May 2022.
- The highest measured surface acceleration and velocity values at the monitoring station were: PGA = 613.5 mm/s^2 and PGV = 16.3 mm/s for tremor no. 43 in December 2022. The duration of the intensive vibration phase was 2.2 s (PGA) and 3.3 s (PGV). These vibrations correspond to intensity degree II according to the GSIS-2017 scale (Mutke 2018).
- The maximum theoretical PGA value at the building location, calculated using Mutke's formulas, was 168.9 mm/s^2 (for tremor no. 20 in February 2018) and PGV was 5.2 mm/s. These vibrations correspond to intensity degree I. Similar values were noted for tremor no. 39 in October 2020. All other tremors were classified as intensity degree 0.
- The calculated PGA and PGV values (Figure 5) are simulations and may differ from actual values, particularly for tremors no. 39, 43, and 50. Notably, for tremor no. 39, the epicentral distance between the station and the tremor was similar to the distance between the epicenter and the building. At the station, PGA = 498.1 mm/s^2 and PGV = 13.8 mm/s were measured, while at the building, values of 162 mm/s^2 and 5.0 mm/s were computed, respectively.
- Research by Pilecka, Stec *et al.* (2017) indicates the potential amplification of tremor effects due to the presence of tectonic faults — a significant factor here, as the building is located near a fault, on the epicentral side.

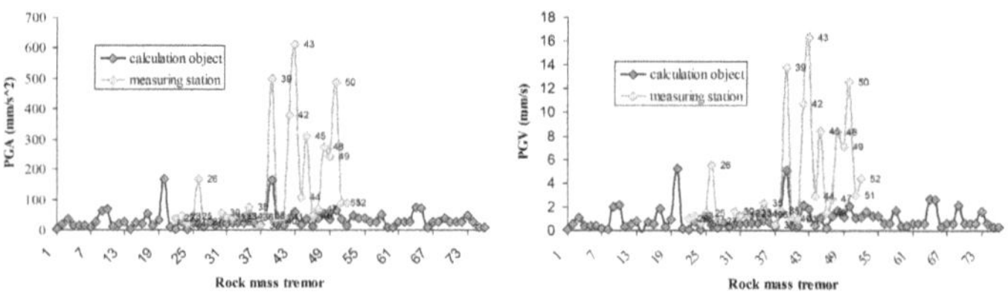

Figure 5. Results of PGA and PGV measurements from the monitoring station and calculated values of surface acceleration and vibration velocity induced by tremors at the building location.

7 ASSESSMENT OF THE EFFECTS OF MINING INFLUENCES

7.1 *Building structure characteristics*

The building was erected in 2007 as a detached, two-segment structure with varying heights (Figure 1b). The southern residential segment, measuring 17.20 m in length and 14.30 m in width, is two storeys high. The garage segment is single-storey, measuring 7.60 m in length and 6.80 m in width. The building has no basement, and the load-bearing walls of both segments are founded on reinforced concrete strip footings and foundation walls, connected with diagonal reinforced concrete ties. The above-ground part was constructed in masonry wall technology using ceramic blocks. The ceiling above the ground floor in the residential part is a monolithic reinforced concrete slab with a perimeter beam on the external walls. Above the first floor, there is a partially flat roof and a partially gable roof supported on knee walls. A flat roof was constructed above the garage. The external walls are insulated with a layer of polystyrene and finished with structural plaster. The property is enclosed by sections of freestanding walls and posts.

7.2 *Building damage condition*

The identified damage to the building was classified based on the location within structural elements, an assessment of the crack intensity, and their morphology. On this basis, the following categories of damage were distinguished in the building:

- Damage in load-bearing walls – numerous significant horizontal shear cracks in the masonry (Figure 6a), cracks radiating from the corners of window openings (Figure 6b), diagonal fissures in walls stabilizing the structural system (Figure 6c), and cracking along the fire wall of the flat roof above the garage.
- Cracks in partition walls and ceramic claddings, running vertically or horizontally on walls or floors, in some areas accompanied by local surface detachment of cladding elements.
- Architectural cracks in wall plasters, vertical or diagonal on wall surfaces, typically initiating from the corners of window and door openings, horizontal cracks in the masonry beneath window sills, and fissures at the junctions of ceilings with walls and at the joints of plasterboard linings or elements infilling the wooden roof structure.

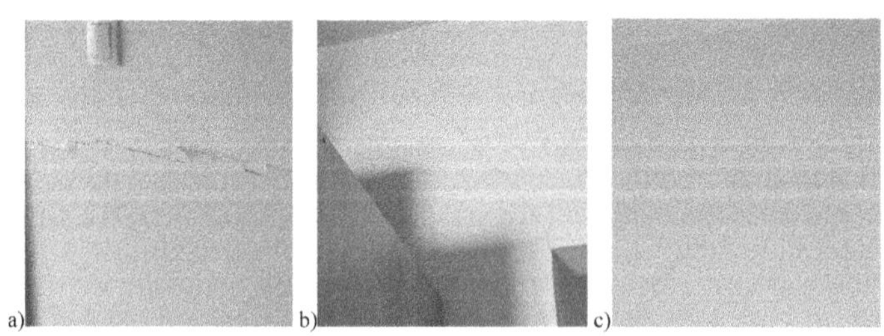

Figure 6. a) Significant horizontal shear crack in the load-bearing wall masonry; b) Example of a crack propagating from the corner of a door opening; c) Diagonal crack in a structural bracing wall.

Throughout the entire building, problems with the operation of window and door joinery were observed. In the surroundings of the building, minor deformations of the paving block surfaces and a loss of verticality in the mounting posts of the entrance gate and wicket gate were recorded. Cracks in the plaster and detachment of the terrace occurred along the junction between the main building structure and the garage, as well as along the surface-mounted terrace.

7.3 *Assessment of the causes of damage*

In the context of mining operations, the effects on a building represent a set of indicators of varying intensities, and the degree of their harmfulness depends on structural factors. To determine the actual causes of the damage identified in the building, it was necessary to carry out a diagnostic procedure taking into account the material-structural conditions and the adopted design solutions (Kwiatek 1997). The analysis also considered the impacts of completed mining operations, determined based on calculation analyses verified by the results of surface subsidence measurements and mining-induced seismic activity intensity. As a result of continuous ground deformations, the subsoil beneath the building experienced horizontal strain of a spreading nature, with a maximum value of 1.6 ‰. This could have caused the observable minor separation at the junctions between the garage and the main building. However, the width of the cracks did not exceed 20 mm — a value that can be calculated as the possible maximum resulting from the specified horizontal ground strain. No damage was found in the load-bearing walls of the building or the garage segment indicating that tensile stresses in the foundation footings had been exceeded. Therefore, no signs of potential activation of discontinuous deformations beneath the structure were observed. The recorded detachment of the surface-mounted terrace, deformations in the fencing, and irregularities in the paving surfaces could have resulted from tensile stresses in the subsoil caused by the formation of the subsidence trough. The average inclination of the building's structure, determined by geodetic measurements, was 3.5 ‰ and sloped towards the trough. This value slightly exceeded the analytically determined ground tilt (T_{max}=2.7 ‰), but for low-rise, single-family buildings, inclination at this level does not significantly increase static force values. The damage observed in the building originated from horizontal forces generated by the kinematic excitation to which the structure was subjected during the propagation of seismic waves from a mining tremor. The horizontal inertial forces, acting with varying directions, led to visible shear failures in the load-bearing walls and diagonal cracks on the surfaces of internal, secondary walls. The extent of the damage indicates insufficient spatial stiffness of the building's structure (Cholewicki 2004, Maciąg 1999). The variation in the flat roof structure over the upper floor made it particularly vulnerable to seismic-induced bending-torsion loads, which manifested in the shearing of window and door pier corners and internal wall junctions. Additionally, the lack of a continuous peripheral tie beam at the knee wall level accounts for the insufficient lateral stiffness of the structure at the roof support level. Under mining-induced tremor loads, even a slightly tilted structure experiences amplified bending-torsion stresses. In this case, it caused the masonry deformation criterion (Federowicz 1997) to be exceeded, resulting in distortions in window and door openings.

8 CONCLUSIONS

Since the construction of the building, mining operations have affected the surface in the area of the property several times. The greatest impact was caused by the extraction of seam 405/1 with panel 51, conducted from the third quarter of 2014 to May 2015. Theoretical deformations at the location of the property could have reached the second category of mining impact area, both in terms of tilt (for the final state and temporary-extreme values) and horizontal strain.

No shallow mining was carried out beneath the property, meaning there is no risk of discontinuous deformations in the form of sinkholes. However, mined-out workings are present directly beneath and near the property, creating a minor risk of old workings' reactivation. The level of risk is low because mining beneath the building occurred in only two seams. The property is located within a zone endangered by discontinuous deformations due to the outcrop of a fault zone — as confirmed by earlier occurrences of linear surface deformations within a distance of 80–160 m.

Based on calculations supported by available measurements, mining tremors in the vicinity of the property between 2015–2025 mainly reached intensities of 0 and I degrees according to the GSIS-2017 scale. An exception was the October 2020 tremor, which could have generated surface vibrations of II degree intensity.

The values of tilts in structural elements obtained through technical leveling (plinth) and small-angle measurements (vertical edges) oscillate around an average of approx. 4 ‰. Additionally, the similar direction of these tilts (north) indicates that they can be associated with the overall tilt direction and value of the entire building's structure.

Based on the diagnostic procedure, which included a comprehensive analysis of geological and mining conditions and possible values of impacts resulting from completed mining operations, it was determined that the decisive factor in the state of damage to the building's structure was the dynamic impact from mining-induced tremors.

Taking into account the most unfavorable calculated values of surface vibration accelerations for the building (PGA value of 168.9 mm/s^2) and the recorded damage effects, it should be concluded that significant amplification of tremor effects occurred at the building's location due to the presence of tectonic faults — particularly since the building was situated on the epicentral side.

In the analyzed area, mining-induced seismic shocks caused damage to many buildings; however, the severity and nature of this damage varied. These differences resulted primar-ily from distinct material and structural solutions, as well as from varying spatial stiffness of individual buildings. This confirms that the effects of dynamic impacts are largely dependent on the layout of structural stiffening elements (such as load-bearing walls and floors), which determines a building's susceptibility to deformation caused by vibrations.

REFERENCES

Cholewicki A. (2004). Design Guidelines for Buildings (or Their Reinforcements) Subject to Mining-Induced Seismic Impacts (Instruction ITB No. 325/1995 and No. 391/2003. *Second Scientific and Technical Conference: Issues of Designing and Protecting Construction Facilities in Mining Areas.* Rudy Raciborskie (in Polish).

Chudek M. (2010). *Rock Mass Mechanics with Fundamentals of Environmental Protection Management in Mining and Post-Mining Areas.* Publishing House of the Silesian Univ of Technology (in Polish).

Chudek M., Janusz, W. and Zych, J. (1988). A study on the state of knowledge concerning the formation and prediction of discontinuous deformations caused by underground mineral exploitation. *Scientific Papers of the Silesian University of Technology*, Mining Series.141, Gliwice (in Polish).

Fedorowicz J. and Kawulok M. (1997). Deformation Criteria for the Resistance of Buildings in Mining Areas. *GIG Scientific Papers.* Series: Conferences, No. 20, Protection of Surface and Building Structures against Mining Damage, Katowice (in Polish).

GeoNET AlgoRes-Soft. (2006). Technical Documentation of the Geodetic Calculation System. Rzeszów (in Polish).

Knothe S. (1984). Forecasting the Effects of Mining Operations. "Śląsk" Publishing House, Katowice, 1984 (in Polish).

Kowalski A. (2015). Surface Deformations in the Upper Silesian Coal Basin. *GIG Publishing House.* Katowice (in Polish).

Kratzsch H. (1983). *Mining Subsidence Engineering.* Springer-Verlag, Berlin, Heidelberg, New York.

Kwiatek J. (1997). Protection of construction objects in mining areas. *Publishing House of Central Mining Institute*, (in Polish).

Maciąg E: (1999). Diagnosing Buildings Affected by Mining-Induced Tremors. *Scientific and Technical Conference, Problems of Designing and Protecting Buildings in Mining Areas*, Rudy Raciborskie, (in Polish).

Malinowska A. and Hejmanowski R. (2016). The impact of deep underground coal mining on earth fissure occurrence. *Acta Geodyn. Geomater.* Vol. 13 (184) 2016. S. 321–330.

Mutke G. (1991). A Method for Predicting Ground Vibration Parameters Generated by Mining Tremors in the Upper Silesian Coal Basin Area. *Unpublished doctoral dissertation*, Central Mining Institute, Katowice (in Polish).

Mutke G. red. (2018). *Principles for Applying the GSIS-2017 Mining Seismic Intensity Scale for the Prediction and Assessment of the Impact of Mining-Induced Tremors on Buildings and for the Classification of Their Dynamic Resistance.* Katowice, Central Mining Institute (GIG), Series: Instructions No. 23 (in Polish).

Orwat J. and Gromysz K. (2021). *International Conference on Applied Sciences (ICAS 2020)* Journal of Physics: Conference Series 1781; 012013 https://doi.org/10.1088/1742-6596/1781/1/012013 1 Occurrence consequences of mining terrain surface discontinuous linear deformations in a residential building. 37.

Peng SS. (2008). *Coal Mine Ground Control, 3rd ed.* West Virginia University: Morgantown, WV, USA, 750.

Pilecka E., Stec K. and Szermer-Zaucha R. (2017). Analysis of the Impact of the High-Energy Tremor of January 29, 2015 at the "Rydułtowy-Anna" Coal Mine and Local Tectonics on Mining Damage to Buildings. *Mining Review (Przegląd Górniczy).* No. 10/2015, pp. 51–58 (in Polish).

Popiołek E. (2009). *Protection of Mining Areas.* Kraków, AGH University of Science and Technology Press (in Polish).

Rola F. (1997). *Engineering and Industrial Geodesy*. Publications AGH. Kraków (in Polish).

Strzałkowski P. (2022). Predicting mining areas deformations under the condition of high strength and depth of cover. *Energies.* https://www.mdpi.com/1996-1073/15/13/4627.

Strzałkowski P. and Szafulera K. (2020). Occurrence of linear discontinuous deformations in Upper Silesia (Poland) in conditions of intensive mining extraction - case study. *Energies.* Vol. 13, Iss.8.s.1-16. DOI: 10.3390/en13081897. Impact Factor: 2.702 (140 pkt.).

Strzałkowski P. and Ścigała R. (2017). The causes of mining induced ground steps occurrence - case study from Upper Silesia in Poland. *Acta Geodyn. Geomater.* 14(3), 305–312. https://doi.org/10.13168/AGG.2017.0013.

Szojda L. and Strzałkowski P. (2025). Numerical analysis of building damage caused by the exposure of a terrain threshold as a result of underground mining-induced subsidence. *Scientific Reports.* 15, Art. 1. https://doi.org/10.1038/s41598-025-96648-6.

Ścigała R. (2013). The identification of parameters of theories used for prognoses of post mining deformations by means of present software. *Archives of Mining Sciences*. Strata Mechanics Research Institute of the Polish Academy of Sciences vol. 58, nr 4, s. 1347–1357, DOI:10.2478/amsc-2013-0093.

Ścigała R. and Szafulera K. (2020). Linear discontinuous deformations created on the surface as an effect of underground mining and local geological conditions - case study. *Bull. Eng. Geol. Environ.* Vol. 79, Iss. 4. s. 2059-2068. DOI: 10.1007/s10064-019-01681-1. Impact Factor: 3.041 (100 pkt.)

Whittaker B.N. and Reddish D.J. (1989). *Subsidence – Occurrence, Prediction and Control.* Amsterdam-Oxford-New York-Tokyo: Elsavier.

Mine Surveying and Mining Area Protection – Sokoła-Szewioła and Jarczyk (Eds)
© 2025 The Author(s), ISBN: 978-1-041-10531-2

Natural and anthropogenic rock mass deformations as threats to sustainable development

L. Florkowska & I. Bryt-Nitarska
Strata Mechanics Research Institute of the Polish Academy of Sciences, Kraków, Poland

ABSTRACT: Surface deformations of the rock mass, driven by natural geodynamic activity and anthropogenic processes such as underground mining, constitute a major constraint on sustainable land use and the long-term stability of urban infrastructure. This study examines two case study areas in Poland: urban zones impacted by deep coal mining in the Upper Silesian Coal Basin and landslide-prone regions in the Carpathian foothills. The analysis addresses structural integrity, land degradation, socio-economic consequences, and planning constraints. Findings are interpreted within the framework of the UN Sustainable Development Goals (SDG 9 and SDG 11), underscoring the need to integrate geotechnical expertise into spatial planning, infrastructure design, and risk governance. Beyond technical assessments, the study highlights the normative imperative of sustainable development as an intergenerational obligation. It calls for anticipatory, interdisciplinary, and systems-based approaches to surface deformation risk management, essential for enhancing territorial resilience and ensuring equitable spatial transformation in deformation-prone regions.

1 INTRODUCTION

Permanent and sustainable development is a rational direction of planning and shaping human activity. This applies equally to the industrial, economic and social spheres (Gawor 2010). However, industry-related activity is especially important when it comes to a direct impact on people's environment. This is why appropriate planning and conduct of building and industrial projects is a necessary condition for sustainable development. This also applies to problems of protecting the earth's surface as an important resource in the form of farmland and land for development

1.1 *Hazards coming from land surface degradation*

One of the basic human needs is a safe place of living and a stable, friendly life environment. That is why any factors which threaten human settlements both impede the satisfying of present basic needs and make sustainable development impossible. One such factor is the deformation of surface rock mass strata. Regardless of the hazards (natural or anthropogenic) these deformations destroy the original structure of the land surface and have a very negative impact on buildings and technical infrastructure of residential areas

1.2 *Technical aspects of substrate degradation*

Any activities carried out in real property, including buildings, situated in areas where natural geodynamic processes can be activated or deformations can arise as a result of human activity always carries a high risk. This applies to structural safety of buildings and

DOI: 10.1201/9781003655558-6

infrastructure and to ensuring the required serviceability conditions. Factors which increase the risk of extensive deformations of land surface usually are: an unfavourable hydrological situation, heterogeneity of the substrate or bad development planning. If the geological situation is not assessed appropriately and if necessary security measures (geotechnical and structural) are not introduced at the design making stage, a building may suffer degradation and damage. In extreme situations a state of failure or even a building disaster may occur.

1.3 *Economic and societal aspects of substrate degradation*

The destruction/destroying of buildings and technical infrastructure on areas affected by rock mass deformations has an obvious negative impact on their economic and societal situation. This is a factor which disturbs the efforts to achieve sustainable development. People who live in areas affected by land surface deformations bear the cost of damaged buildings and of damage to farmland and recreational areas. At the same time the degradation of soil makes it difficult, and sometimes permanently impossible, for this land to be developed with buildings or with technical or transport infrastructure.

1.4 *Sustainable use of resources – protecting the earth's surface*

Balancing the benefits which arise from using the resources and the losses caused in the process is a necessary element of responsible economic policy. This is especially true when we realise that the damage caused is mostly irreversible. The land which can be used for development or farming are valuable both in the context of the present and in terms of planned development of residential or urbanized areas and food production. The earth's surface is, therefore, an important resource, which requires rational management and protection

2 STATE OF THE ART IN SURFACE DEFORMATION AND SUSTAINABILITY CHALLENGES

Surface deformations resulting from both natural and anthropogenic processes—particularly those associated with mining activities—pose substantial challenges to the realization of sustainable development goals. While mineral extraction delivers undeniable economic value, it frequently results in irreversible landscape alterations, soil degradation, and the destabilization of built environments (Florkowska *et al.* 2016, Zabuski *et al.* 1999). These effects are particularly acute in regions with a legacy of intensive underground mining, where they undermine the structural resilience and socio-economic stability of urban and peri-urban areas.

Research by Florkowska and Bryt-Nitarska (2015, 2017) underscores the structural and social impacts of mining-induced ground deformation, noting its detrimental influence on property integrity, community well-being, and local planning capacity. Mining damage is also linked to a decline in property value (Bryt-Nitarska 2016) and limits the redevelopment potential of degraded land (Florkowska 2016). In relation to Sustainable Development Goals—especially SDG 9 (Industry, Innovation and Infrastructure) and SDG 11 (Sustainable Cities and Communities)—these studies emphasize the importance of integrating environmental, spatial, and social dimensions into sustainable territorial governance (Florkowska *et al.* 2018).

At the international level, increasing attention is being directed toward the hazards posed by natural mass movements. Krishnapriya *et al.* (2024) developed a rapid run-out assessment methodology for debris flows in Wayanad, India, highlighting the synergistic role of monsoonal rainfall and unregulated hillside development in escalating landslide risk. Similarly, Ding *et al.* (2025) conducted a landslide susceptibility assessment in China's Tongguan District, identifying slope angle, proximity to faults, and anthropogenic drainage modification as key contributing factors. These findings reinforce the need for terrain-sensitive land-use planning and risk mapping in geodynamically active regions.

Anthropogenic subsidence presents another significant threat to urban sustainability. In Maceió, Brazil, salt mining operations resulted in gradual but unmitigated ground subsidence over more than a decade, culminating in a major sinkhole and the evacuation of over 60,000 residents (Vassileva *et al.* 2021). This event exemplifies the systemic risks associated with insufficient monitoring and delayed institutional response. In a similar vein, Ohenhen *et al.* (2025) documented the progressive sinking of critical infrastructure along the U.S. East Coast due to groundwater withdrawal, warning of the growing vulnerability of transportation and industrial systems in low-lying coastal zones.

The global transition to renewable energy has introduced a new paradox: while decarbonization is necessary for climate resilience, it has intensified demand for transition minerals such as lithium, cobalt, and rare earth elements. Sonter *et al.* (2020) demonstrated that this demand exerts increased pressure on ecologically sensitive and biodiverse regions, threatening ecosystem integrity. Owen *et al.* (2023) further noted that mineral extraction projects, particularly in the Global South, often proceed without adequate regulatory oversight, resulting in land-use conflicts, displacement, and the marginalization of land-connected communities.

Technological advances in geomechanics and monitoring systems have enabled earlier detection and mitigation of ground deformation risks. Zhang *et al.* (2025) proposed a novel computational approach for estimating residual deformation in mined-out areas, enhancing the predictive modeling of subsidence hazards. Similarly, Lu *et al.* (2025) developed an angular displacement-based method for assessing the mechanical behavior of rock strata, which holds potential for improving risk diagnostics in subsidence-prone environments.

Beyond hazard identification, post-mining land restoration has emerged as a crucial component of sustain-able spatial management. Zhong *et al.* (2025) evaluated the effectiveness of ecological rehabilitation in former resource-extraction cities, identifying key drivers of successful transformation. Complementarily, Ruiz *et al.* (2023) introduced the concept of "technosoils"—engineered soils derived from mining waste—as an innovative solution for environmental recovery and carbon sequestration. These practices support the implementation of SDG 9 and SDG 11 by promoting circular economy principles and improving adaptive capacity in post-industrial landscapes.

From a systemic and philosophical standpoint, Gawor (2010) and Mierzejewska (2017) stress the importance of integrative, interdisciplinary approaches to urban development, in which natural constraints are treated as essential components of long-term planning. Konarski (2014) highlights the intersection of resource policy and environmental ethics, calling for governance frameworks that reconcile energy needs with ecological and societal responsibilities.

3 METHODOLOGY

This study investigated the effects of rock mass deformations—both natural (landslides) and anthropogenic (mining-induced subsidence)—on spatial development and infrastructure stability in the context of sustainable development. A descriptive-analytical approach was applied, combining case study analysis, technical documentation, and field data from urban areas affected by surface deformations.

Two types of deformation were examined: subsidence from underground coal mining in the Upper Silesian Coal Basin, and landslides in southern Poland. For the mining cases, archival data from the Strata Mechanics Research Institute (Polish Academy of Sciences) were used, including subsidence maps, damage inventories, and displacement vectors analyzed relative to building orientation. In the landslide case, damage indicators (e.g., fissures, structural failure) were recorded and supported by geological and hydrological assessments to evaluate hazard extent and development constraints.

Deformation impacts were assessed across four dimensions: structural integrity, redevelopment potential, socio-economic consequences, and relevance to SDG 9 and SDG 11. Much of the material came from expert reports prepared for legal proceedings, supplemented where possible with published studies.

To contextualize the findings, comparative examples from Brazil, India, China, and the USA were reviewed, enabling identification of both shared risk patterns and location-specific differences.

4 MAIN CAUSES OF LAND SURFACE DEFORMATION

There Deformation of the Earth's surface results from natural, anthropogenic, and combined processes, manifested as subsidence, uplift, or slope instability.

4.1 *Natural causes*

Land surface deformation results from diverse natural processes, including gravitational slope movements, tectonic activity, volcanism, karst dissolution, and permafrost thaw. Landslides in the Polish Carpathians exemplify the role of soil saturation and seismic triggering (Zabuski *et al.* 2009), while GNSS and InSAR observations document ongoing crustal deformation in the Zagros Mountains, Iran (Hessami *et al.* 2024). Volcanic unrest at the Campi Flegrei caldera, Italy, illustrates ground uplift driven by magma and hydrothermal pressurization (Giudicepietro *et al.* 2025), whereas sinkhole-related subsidence in Florida highlights the impact of karst processes (Muili *et al.* 2025). In high-latitude environments, degradation of ice-rich permafrost on the Qinghai–Tibet Plateau demonstrates how climate warming induces subsidence and slope instability through thermokarst development (Li *et al.* 2024).

4.2 *Anthropogenic causes*

Anthropogenic activities significantly contribute to land surface deformation through multiple mechanisms. Mining creates underground voids that trigger subsidence and surface collapses, as observed in Upper Silesia, Poland (Bryt-Nitarska 2016; Florkowska 2016). Groundwater over-exploitation compacts aquifers and induces basin-wide subsidence, exemplified by the Himalayan foreland basin (Sahadevan and Pandey 2023). Structural loading from high-rise construction causes settlement in compressible soils, with Shanghai providing a clear case where building loads and groundwater history jointly intensified subsidence (Yuan *et al.* 2020). Subsurface engineering such as tunneling alters stress regimes and induces deformation, as documented in Hangzhou, China (Li *et al.* 2023). In addition, land use and land cover changes, hydroengineering projects such as dam construction, and fluid injection activities—including wastewater disposal—further destabilize slopes or induce seismicity, with cases from Hubei Province (Chen *et al.* 2019), the Three Gorges Reservoir (Cojean and Caï 2019), and Oklahoma (Skoumal *et al.* 2024) demonstrating the global scope of human-driven deformation.

4.3 *Conjunctive causes*

Natural and anthropogenic factors frequently interact, amplifying land surface deformation. Landslides are often reactivated by human activities, with slope cutting and rainfall jointly driving failures such as the Dashuitian landslide in Wushan County, China (Zhang *et al.* 2021). Permafrost degradation is similarly accelerated by infrastructure, where roads and buildings intensify thaw settlement and thermokarst hazards on the Qinghai–Tibet Plateau (Hjort *et al.* 2022). Subsurface excavation can destabilize pre-existing landslides in tectonically active regions, as demonstrated by failures near the Val di Sambro tunnels in the Northern Apennines, Italy (Bandini *et al.* 2015). Urban expansion on inherently weak slopes further increases instability, exemplified by Cluj-Napoca, Romania, where development coincides with predisposing geology (Muntean *et al.* 2024). Finally, groundwater over-extraction intensifies natural compaction in subsiding basins, with global case studies highlighting the scale of aquifer-related subsidence (Tzampoglou *et al.* 2023).

5 DESCRIPTION OF ROCK MASS DEFORMATION

Surface rock mass strata deformations pose the most serious threat in densely built-up areas with well-developed infrastructure systems. Deformation processes which take place in the subsoil have an impact on buildings, utilities, roads and civil engineering structures. The destruction of the original subsoil structure also makes the areas less suitable as sites for new development projects and makes them more expensive to maintain due to the fact that additional geotechnical security measures are necessary. In the case of deformations caused by mining, their time of occurrence, character and extent are generally predictable, so the existing buildings can be secured and new buildings can be properly designed (preventive measures in construction). It is also possible to adjust mining operations in such a way that negative impact on the surface is minimised (preventive measures in mining).

Natural geodynamic threats are more serious. These are phenomena which are difficult to predict and are often of violent nature, which makes it impossible to take any immediate rescue measures. This is the reason why on landslide areas preventing deformation processes is the key element in risk mitigation. Geodynamic processes can be activated both by natural factors (e.g. heavy rain and water saturation of the soil) and by anthropogenic factors. In built-up areas mass movement activation is usually a consequence of both these factors acting together. Wrong management (most frequently excessive loading of slopes with buildings and disturbing the natural run-off of water) with conjunction with adverse weather conditions cause emergency situations and landslide disasters.

5.1 *Natural factors - landslides*

Mass movements of surface strata are translational or rotational, they can also be a combination of the two. Complex landslide forms assume irregular shapes due to complicated climatic processes which occur in complex geological structures (Zabuski et. al. 1999). The rate of soil mass displacements can vary greatly: from several meters per minute (extremely quick movements) to several millimeters per year. For example, the landslides in the Carpathian flysch occur due to a lump of soil or a block of rock sliding on a stable surface contiguously to the geologically formed slide surface. The existence, the path and the shape of the slide surface determine the depth and shape of the landslide. Often in the profile of slopes above stable rocky ground there are several potential slide surfaces running at different depths. A slide can be activated on any of them. In this situation, to ensure the stability of a building the foundation must be anchored in stable ground, below slide strata or/and the slope must be stabilised geotechnically.

5.2 *Surface deformations due to mining*

The form and progress in time of mining-induced surface deformations depend on mining methods used and properties of the rock mass. The shape of surface deformation is connected with the shape of mining panels and with repeated exploitation carried out on several levels, sometimes on more than ten. Traditionally, under-ground coal deposits are extracted using longwall mining. Underground mining invariably causes the formation of a regular subsidence trough on the surface. It arises as a result of rocks moving in the direction of the void left after coal is extracted. He extent of the trough depends on the size of the mining panel, the thickness of mined-out coal and local geological conditions. Surface deformations are a rheological process, due to the exploitation's progress in time and because of rheological properties of the rock mass, which cause a certain delay in the effects becoming visible on the surface. Coal mining at medium and great depths is accompanied by a gradual shift of displacements towards the surface. In the mining trough formation process the surface undergoes a number of transient states of deformation. The process finishes once rock mass movements have ceased completely after more than ten months, and sometimes even after several years in the case of coal mining.

6 IMPACTS OF NATURAL OK AND ANTHROPOGENIC ROCK MASS MOVEMENTS

Rock mass deformations, regardless of their cause, exert a significant impact on built structures. In mining regions, continuous subsidence within depression troughs is frequently accompanied by discontinuous deformations, flooding, or rises in groundwater levels, reflecting disruption of the natural hydrological balance. Mining activities also cause a long-term decline in deep groundwater levels, leading to the disappearance of springs and streams, while altered hydrogeological conditions may trigger surface deformations and flooding even at considerable distances from the exploitation site.

Both natural landslides and mining-induced subsidence impose additional stresses on buildings, damaging their structural elements and threatening overall stability. Slow-moving landslides (Figure 1) or mining-related subsoil creep (Figure 2) may severely compromise load-bearing capacity, rendering buildings unsafe for further use (Florkowska and Bryt-Nitarska 2017). Such processes accelerate infrastructure deterioration, reduce property values, and heighten the perception of risk among residents in affected areas (Bryt-Nitarska 2016).

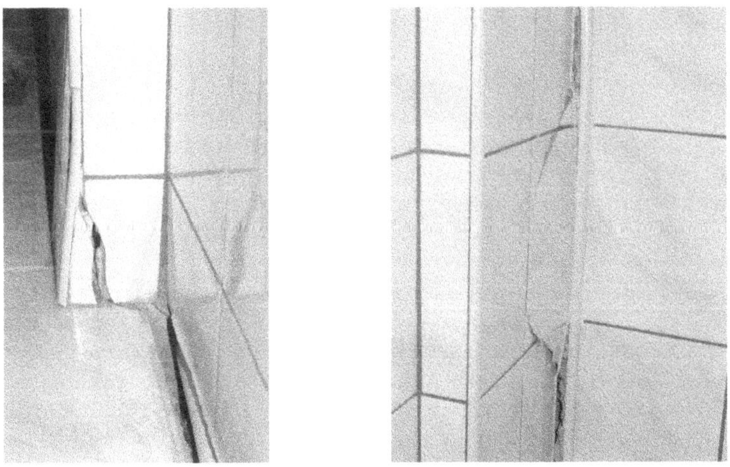

Figure 1. Damage to building walls resulting from a slow landslide process (Lesser Poland Region). Fot. I. Bryt-Nitarska.

Figure 2. Cracks in load-bearing walls due to mining-induced subsoil deformations in Upper Silesian Coal Basin (Poland). Fot. I. Bryt-Nitarska (Florkowska *et al.* 2016).

7 THE IMPACT OF DEFORMATION THREATS ON SUSTAINABLE DEVELOPMENT

In September 2015 in New York during the global Sustainable Development Summit 2030 fundamental assumptions of a new strategy of changing the perception of global development were elaborated. The final summit document called "Transforming Our World: the 2030 Agenda for Sustainable Development" was signed by 150 UN member states. The 2030 Agenda contains 17 main Sustainable Development Goals (SDGs).

Two of the seventeen main SDGs included in the Agenda can be directly linked to the impact of natural and anthropogenic rock mass deformation on environment changes and on the deterioration of people's living conditions. They are: Goal 9: Build resilient infrastructure, promote inclusive and sustainable industrialization and foster innovation and Goal 11: Make cities and human settlements inclusive, safe, resilient and sustainable.

This means that until 2030 in the area of social infrastructure sustainable urbanization fostering social inclusion will have to be intensified and people will have to be able to participate in an integrated and sustainable planning and development of human settlements in all the countries. As far as infrastructure at risk of excessive state of deformation and damage is concerned, measures to protect buildings on the surface must be introduced in a systemic manner. They must take into account existing and planned buildings. The present approach, in which the states of deformation and damage in excess of allowable levels are accepted must be rede-fined (Florkowska and Bryt-Nitarska 2015, 2018).

In mining areas achieving the goals stated as building cities and infrastructure which are resilient to the effects of disasters and building sustainable cities and settlements is more difficult than anywhere else. At the same time achieving these SDGs in these very areas is extremely important for further development of the region and for improving living conditions for those people who encounter problems connected with mining damage which exist only in mining areas.

Other consequences of damage resulting from earth's surface degradation are economic and social problems of the people who live in devastated areas. Surface degradation, deterioration, damage to and accelerated technical wear of buildings in mining areas are factors which reduce the value of the property and on a regional scale make the areas less attractive to investors. This has a negative influence on people's financial situation and in extreme cases carries a risk of poverty and marginalization. The result is that regional inequality and a feeling of social injustice appear and become more marked in a situation when the regions which ensure the country's energy security are themselves adversely affected by mining. Because of these negative phenomena the principles of sustainable development must be incorporated into the economic and social life in risk areas.

8 DISCUSSION

The findings confirm that both natural and anthropogenic ground deformations pose substantial risks to urban infrastructure. Case studies from Upper Silesia and Małopolska (Bryt-Nitarska 2016; Florkowska *et al.* 2016, 2018) demonstrate that deformation—regardless of origin—can lead to significant and lasting structural and socio-economic impacts. These outcomes are consistent with global observations, such as groundwater-related subsidence in the U.S. (Ohenhen *et al.* 2025) and salt mining-induced displacement in Brazil (Vassileva *et al.* 2021).

Importantly, even technically managed deformations can cause severe damage when spatial planning fails to account for geotechnical behavior. Landslides in southern Poland and India (Krishnapriya *et al.* 2024, Florkowska *et al.* 2020) further show that unmonitored terrain can become hazardous under changing climatic and hydrological conditions.

These patterns highlight the need to integrate geotechnical risks into broader land-use policy. As noted by Konarski (2014) and Owen *et al.* (2023), resource extraction intersects with governance and environmental vulnerability, requiring systemic and inclusive planning. To achieve SDG 9 and SDG 11, urban resilience strategies must address ground stability as a core component of infrastructure development.

9 SUMMARY AND CONCLUSIONS

The Earth's surface constitutes a finite and irreplaceable natural resource whose rational management and protection are imperative for intergenerational equity. Undegraded land suitable for construction, infrastructure deployment, and sustainable investment is of particular strategic value in densely urbanized regions. Therefore, the prevention of land degradation in the form of geomechanical deformations and structural dam-age must be regarded not only as a technical necessity but also as a foundational condition for sustainable territorial development.

The present study demonstrates that the most significant threats to surface integrity arise from two interrelated factors: underground mineral extraction and natural mass movement processes, particularly landslides. Both are often exacerbated by inappropriate land-use planning and insufficient recognition of geotechnical constraints. Damage resulting from ground deformation and subsoil degradation encompasses a wide range of technical, environmental, and socio-economic consequences. As shown in case studies from Upper Silesia and southern Poland, such impacts may include irreversible land transformation, structural failure of buildings, increased maintenance costs, and a decline in both property value and investment attractiveness.

These risks underscore the need to integrate surface protection into all levels of spatial and infrastructure planning—particularly in mining, post-mining, and landslide-prone areas. Such integration should be codified within urban policy frameworks and engineering standards, especially in regions with a legacy of subsurface disturbance or terrain instability. In addition, land surface management must go beyond short-term economic interests and reflect a broader commitment to environmental responsibility and social resilience.

In the context of the 2030 Agenda for Sustainable Development, especially SDG 9 and SDG 11, geohazard management emerges as a critical enabler of long-term urban sustainability. Moreover, drawing on the philosophical foundations of sustainable development as articulated by Gawor (2010), the land should be perceived not merely as an economic asset but as a common good—integral to ecological systems, cultural continuity, and ethical stewardship. This perspective reinforces the importance of aligning engineering, spatial, and environmental disciplines within an integrative and precautionary planning paradigm.

Based on the conducted analyses, the following key conclusions can be drawn:

- The Earth's surface must be treated as a strategic and vulnerable development asset, essential to socioeconomic stability and environmental sustainability.
- Geomechanical risk factors, such as mining-induced subsidence and landslides, must be systematically assessed and incorporated into spatial and construction planning processes.
- Surface protection and geohazard mitigation should be formalized in legal and technical regulations, particularly in areas historically affected by extractive industries or active mass movement processes.
- Effective management of deformation-prone regions requires long-term monitoring systems, interdisciplinary collaboration, and institutional mechanisms grounded in scientific evidence.
- Resilient and inclusive development in at-risk territories demands the adoption of adaptive design solutions, circular land-use models, and nature-based approaches to land rehabilitation.

Future research should prioritize the development of predictive models for deformation dynamics, the advancement of early warning technologies, and the assessment of ecosystem-based restoration strategies. Ultimately, a holistic, ethically grounded and science-informed approach to land surface governance is indispensable for reconciling development imperatives with the principles of eco-development, ensuring both structural safety and environmental justice for present and future generations.

This research was supported by the statutory research funding of the Strata Mechanics Research Institute, Polish Academy of Sciences.

REFERENCES

Bandini, A. *et al.* (2015). Tunnelling-induced landslides: The Val di Sambro tunnel case study, *Engineering Geology* 196: 71–87 https://doi.org/10.1016/j.enggeo.2015.07.001.

Bryt-Nitarska, I. (2016). A building's situation in a mining area as a factor of its value. *Bull. Miner. Energy Econ. Res. Inst. PAS* 94: 119–125.

Chen, L. *et al.* (2019). The influence of land use and land cover change on landslide susceptibility: a case study in Zhushan Town, Xuan'en County (Hubei, China), *Nat. Hazards Earth Syst. Sci.*19: 2207–2228. https://doi.org/10.5194/nhess-19-2207-2019.

Cojean, R. and Caï, Y.J. (2011) Analysis and modeling of slope stability in the Three-Gorges Dam reservoir (China) — The case of Huangtupo landslide. *J. Mt. Sci.* 8: 166–175 https://doi.org/10.1007/s11629-011-2100-0.

Ding, D., Wu, Y., Wu, T. and Gong, C. (2025). Landslide susceptibility assessment in Tongguan District, Anhui, China using information value and certainty factor models. *Scientific Reports* 15: 12255. https://doi.org/10.1038/s41598-025-93704-z.

Florkowska, L. and Bryt-Nitarska, I. (2015). Social aspects of mining damage. *Mining Review* 71(11).

Florkowska, L. and Bryt-Nitarska, I. (2017). An example of the impact of geodynamic processes on the state of a building's structure. Part 1. The analysis of the state of damage. *Trans. of SMRI PAS* 19(1): 19–24.

Florkowska, L. and Bryt-Nitarska, I. (2018). The requirements for implementing Sustainable Development Goals (SDGs) and for planning and implementing Integrated Territorial Investments (ITI) in mining areas. *E3S Web of Conferences* 36: 01004. https://doi.org/10.1051/e3sconf/20183601004.

Florkowska, L. (2016). Issues of mining damage in the context of Sustainable Development Goals. *Bull. Miner. Energy Econ. Res. Inst. PAS* 94: 105–117.

Florkowska, L. *et al.* (2018). Reactivation of the old landslide caused by the land development—the case study. In Litvinenko V. (ed.), *Geomechanics and Geodynamics of Rock Masses1: 499-504; Proc. of Int. Eur. Rock Mech. Symp., EUROCK 2018; Saint Petersburg; Russian; 22-26 May. CRC Press/Balkema*

Florkowska, L., Bryt-Nitarska, I. and Maj, A. (2016). *Mining damage in buildings. An outline of the problem.* Kraków: Arch. of Min. Sc. Monographs, Issue 17. SMRI PAS.

Florkowska, L.*et al.* (2020). Monitoring and Assessing the Dynamics of Building Deformation Changes in Landslide Areas. *Buildings* 10(1): 3. https://doi.org/10.3390/buildings10010003

Gawor, L. (2010). Philosophy of Sustainable Development: Preliminaries. *Probl. of Sust. Dev.* 5(2): 69–76.

Giudicepietro, F *et al.* (2025) Burst-like swarms in the Campi Flegrei caldera accelerating unrest from 2021 to 2024. *Nat Commun* 16, 1548. https://doi.org/10.1038/s41467-025-56723-y

Hessami,K. *et al.* (2024) Active deformation within the Zagros Mountains deduced from GPS measurements. *Journal of the Geological Society* 163: 143–148 DOI 10.1144/0016-764905-031

Hjort, J. *et al.* (2022). Impacts of permafrost degradation on infrastructure. *Nat Rev Earth Environ* **3**, 24–38 https://doi.org/10.1038/s43017-021-00247-8

Konarski, W. (2014). Mineralne zasoby energetyczne i działania polityczne: wprowadzenie do wzajemnych zależności i ich wybrana egzemplifikacja. *Problems of Sustainable Development* 9(1): 63–70.

Krishnapriya, V.K. *et al.* (2024). A rapid run-out assessment methodology for the 2024 Wayanad debris flow. *npj Natural Hazards* 1: 41. https://doi.org/10.1038/s44304-024-00044-5.

Li, J. *et al.* (2024). Monitoring surface deformation of the Qinghai–Tibet permafrost region using an improved MT-InSAR method based on spatial–temporal constraints. *International Journal of Digital Earth 17*(1). https://doi.org/10.1080/17538947.2024.2406383

Li, Y. (2023). Additional Stress of Soil and Surface Settlement during Tunnel Shield Construction. *Buildings* 13(6):1437. https://doi.org/10.3390/buildings13061437.

Lu, C. *et al.* (2025). Deformation properties of rock strata based on angular displacement and its hydrogeological significance. *Scientific Reports* 15: 6698. https://doi.org/10.1038/s41598-025-91295-3.

Mierzejewska L. (2017). Sustainable Development of a City: Systemic Approach. *Pr. of Sus. Dev.* 12(1): 71

Muili, O. and Babaie H. (2025). Sinkhole susceptibility analysis using machine learning for west central Florida, *Applied Computing and Geosciences* 27: 100262 https://doi.org/10.1016/j.acags.2025.100262.

Muntean O./L. *et al.* (2024). Environmental impact assessment of uncontrolled urban sprawl in a peri-urban area: A case study of Cluj-Napoca municipality, Romania. *Present Env. and Sust. Dev.* 18 2. https://doi.org/10.47743/pesd2024182023

Ohenhen, T.E. *et al.* (2025). Critical infrastructure is sinking along the US East Coast. *Nature Cities* 1(4): 210–220. https://doi.org/10.1038/s44284-025-00240-y.

Owen, J.R. *et al.* (2023). Energy transition minerals and their intersection with land-connected peoples. *Nature Sustainability* 6: 203–211. https://doi.org/10.1038/s41893-022-00994-6.

Ruiz, F. *et al.* (2023). Constructing soils for climate-smart mining. *Communications Earth and Environment* 4: 2019. https://doi.org/10.1038/s43247-023-00862-x.

Sahadevan, D.K. and Pandey A.K. (2023). Groundwater over-exploitation driven ground subsidence in the himalayan piedmont zone: Implication for aquifer health due to urbanization. *Journal of Hydrology* 617 C: 129085. https://doi.org/10.1016/j.jhydrol.2023.129085.

Skoumal, R.J. *et al.* (2024) Reduced Injection Rates and Shallower Depths Mitigated Induced Seismicity in Oklahoma. *The Seismic Record* 4: 279–287. https://doi.org/10.1785/0320240030

Sonter, L.J. *et al.* (2020). Renewable energy production will exacerbate mining threats to biodiversity. *Nature Communications* 11: 4174. https://doi.org/10.1038/s41467-020-17928-5.

Tzampoglou, P. *et al.* (2023). Selected Worldwide Cases of Land Subsidence Due to Groundwater Withdrawal. *Water* 15 6: 1094. https://doi.org/10.3390/w15061094

Vassileva, M. *et al.* (2021). A decade-long silent ground subsidence hazard culminating in a metropolitan disaster in Maceió, Brazil. *Scientific Reports* 11: 7704. https://doi.org/10.1038/s41598-021-87033-0.

Yuan, L. *et al.* (2020) Land subsidence induced by the engineering-environmental effect in Shanghai, China. *Arab J Geosci* 13: 251. https://doi.org/10.1007/s12517-020-5224-0.

Zabuski L., *et. al.* (2009). Landslide processes in a flysch massif - case study of the Kawiory landslide, Beskid Niski Mts. (Carpathians, Poland) *Geological Quarterly* 53 (3): 317–332

Zabuski, L., Thiel, K. and Bober, L. (1999). *Landslides in the flysch of the Polish Carpathians: geology, modelling, stability calculations*. Gdańsk: Wyd. IBW PAN.

Zhang, C. *et al.* (2021) Reactivation mechanism of a large-scale ancient landslide. *Landslides* 18: 397–407. https://doi.org/10.1007/s10346-020-01538-9.

Zhang, H. *et al.* (2025). A new numerical method for calculating residual deformation in mined-out areas considering water–rock interaction and its application. *Scientific Reports* 15: 11207.

Zhong, A. *et al.* (2025). Evaluation of the effect of ecological restoration in mineral resource cities and analysis of driving factors. *Scientific Reports* 15: 14543. https://doi.org/10.1038/s41598-025-99043-3.

 ISBN: 978-1-041-10531-2

Analysis of building reinforcement against discontinuous deformation in mining terrain

K. Gromysz
Silesian University of Technology, Gliwice, Poland

ABSTRACT: A numerical analysis of a two-storey building with a small footprint indicated that discontinuous step-type mining-induced deformation beneath the structure results in detachment from the ground over more than half of its plan area. Consequently, this detachment alters the distribution of internal forces within the structural elements. To ensure compliance with Ultimate Limit States, the required reinforcement for the analyzed structure comprises a reinforced concrete foundation slab with a thickness of 0.35 m and reinforced concrete cores integrated within the walls. Numerical analysis was conducted to determine the deformation behaviour of the building under discontinuous deformation, identify the most unfavorable location of this deformation, and determine the necessary reinforcement to guarantee the safety of the structure.

1 INTRODUCTION

The design of building reinforcement against mining ground deformation requires, on the one hand, the adoption of a ground deformation model and, on the other hand, appropriate structural solutions for the building. In the case of well-recognized building design in areas of continuous deformation, the model of mining ground deformation is a subsidence trough (Kowalski 2015). The values of the indicators describing this subsidence trough are adopted on the basis of forecasts made by the mining company (Popiołek 2009). When designing building reinforcement, it is assumed that the form of deformation of the structure corresponds to the form of deformation described by the subsidence trough (Ledwoń 1983). As a result, the horizontal load on the strip foundation (Kawulok 2015) resulting from both the deformation of the terrain and the adopted foundation system (Szojda 2019) is determined, and then the internal forces and the designed reinforcement are calculated. With regard to ring beams, forces are determined based on two conditions: uneven vertical reaction transmitted from the ground (wall bending) (Barczyk and Świądrowski 1969) and the load-bearing capacity of the wall for horizontal delamination (Wachniewski 1954). Similarly, in the case of linear structures, it is assumed that they transfer deformations corresponding to the form of a subsidence trough with mining terrain deformation indicators values (Kliszczewicz 2023).

However, the situation of structures affected by discontinuous deformations is insufficiently understood. This is due in particular to the fact that the form of deformation of a structure differs from that of a mining area. This means that the adopted reinforcement against deformations applied in the building have a significant impact on the deformations experienced by the building. At the same time, they also have a local impact on the deformation of the mining area itself. For this reason, when designing structures subjected to discontinuous deformation, it is necessary to adopt a soil-structure model and a deformation model. At the same time, based on the standard information on impacts provided by the

DOI: 10.1201/9781003655558-7

mining company, it is not possible to adopt a model of the building's load under discontinuous deformation. In particular, this information does not specify the form of deformation, the value of the fault or its exact location. At the same time, the building reinforcement determined on the basis of calculations are the basis for preparing a differential cost estimate between the situation of a building in which reinforcement have been designed to absorb mining ground deformation and a building located outside the mining ground deformation. This cost estimate is the basis for the investor to apply for reimbursement of the costs of the reinforcement. The problem addressed is therefore of both cognitive and utilitarian importance. The computational analysis presented in this article was carried out in relation to a specific case of a segment with a small footprint subjected to the impact of discontinuous step-type mining ground deformation.

2 DESCRIPTION OF THE SEGMENT

The analysed residential segment has no basement, is two-storey with a ventilated flat roof with a rectangular floor plan of 12.64 m × 6.84 m (Figure 1) and a height of 8.70 m from ground level to the roof ridge (Figure 2). The building is founded on a reinforced concrete foundation slab with a rectangular floor plan measuring 12.64 m × 6.89 m. The ceiling above the ground floor is reinforced concrete, monolithic, 0.14 m thick, supported by walls via ring beams with a cross-section (*b*/*h*) of 0.24 m/0.25 m.

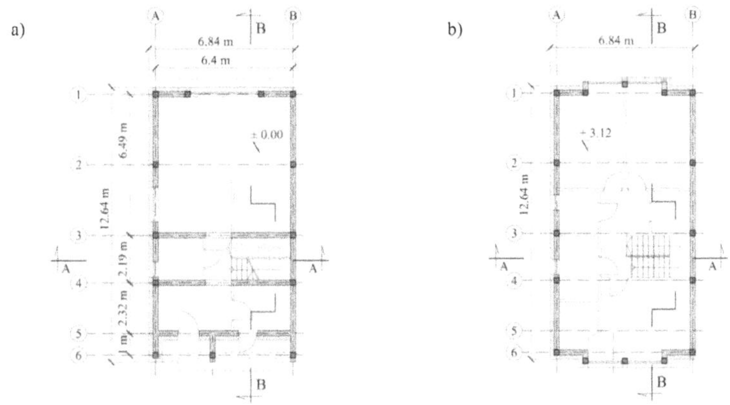

Figure 1. Plan of the building: a) ground floor, b) first floor.

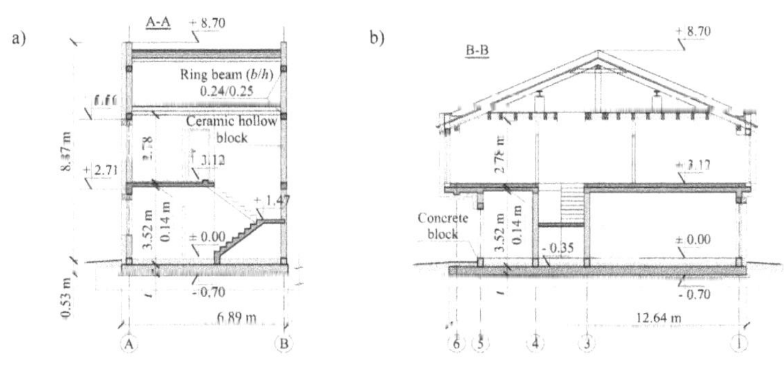

Figure 2. Building cross-sections: a) transverse A-A, b) longitudinal B-B.

Ring beams were also installed at the level of the ceiling above the first floor, except that the structure of this ceiling consists of wooden beams with a cross-section (b/h) of 0.12/0.26 m spaced at 0.625 m intervals. The wooden roof truss is gabled with a slope of 23 degrees. The walls are made of 0.24 m thick ceramic hollow bricks, with two layers of concrete blocks laid directly on the foundation slab, which is important for the calculation analysis. The lintels in the external load-bearing walls are ring beams with increased height in the area of the openings. In the segment models analysed below, the thickness (t) of the reinforced concrete foundation slab was varied, assuming it to be either 0.18 m or 0.35 m. In addition, reinforced concrete cores with a cross-section of 0.24 m × 0.24 m were introduced or not introduced into the walls of the models.

The segment is part of a semi-detached building consisting of four segments: two residential and two garages. The entire building has a T-shaped floor plan, which can be inscribed in a rectangle measuring 18.59 m × 13.83 m.

3 MODELS OF THE SEGMENT LOADED WITH DISCONTINUOUS DEFORMATION OF THE MINING TERRAIN OF THE STEP TYPE

The segment model was developed in RFEM 6 using member elements, surface elements and, in the first stage of analysis, solid elements. The foundation slab was represented by a surface element with a thickness t, which was assumed to be 0.18 m or 0.35 m, depending on the reinforcement analysed. The ceiling above the ground floor and the staircases were also represented by surface elements with a thickness of 0.14 m. The ring beams running horizontally above the ground floor and first floor were represented by beam elements with a cross-section (b/h) of 0.24 m / 0.25 m. Vertical cores in walls representing structural reinforcement, where present, were assumed to be bar elements with a cross-section of 0.24 m / 0.24 m. The above reinforced concrete elements were assigned the parameters of C20/25 concrete, for which the modulus of elasticity E is 30 GPa, the shear modulus G is 12.5 GPa and Poisson's ratio $\nu = 0.2$. For foundation walls with a height of 0.35 m and made of concrete blocks, $E = 27$ GPa, $G = 11.25$ GPa and $\nu = 0.2$ were assumed.

With regard to load-bearing above-ground walls, it was assumed that they are made of ceramic hollow blocks with thin mortar joints, groups '1' and '4' with $f_b \geq 2.4$ MPa. Based on the Eurocode - Part 1-1 standard, the characteristic compressive strength of the masonry was determined.

$$f_k = K f_b^{0,85} \tag{1}$$

where $K = 0.45$ and f_b – normalized compressive strength of masonry elements – assumed according to catalogue data (construction groups 3 and 2) equal to 15 MPa. On this basis, $f_k = 4.50$ MPa was determined. The design compressive strength of the masonry f_d, assuming $\gamma_d = 1.7$, is $f_d = 2.65$ MPa. The shear modulus of elasticity of the masonry, E, was determined from the formula

$$E = 1000 f_k. \tag{2}$$

Therefore, the modulus of elasticity E of the wall is 4.50 GPa. Poisson's ratio was assumed to be 0.25.

The roof truss and ceiling elements above the first floor were assumed to be bars made of softwood, for which $E = 6.4$ GPa and $G = 0.48$ GPa.

A uniformly distributed load of 4.0 kN/m^2 was applied to the foundation slab, the ceiling above the ground floor and the stairs, which corresponds to the total permanent load and the live load. This approach to the model load results from its non-linearity (detachment of supports and inelastic behavior of the material in calculations not discussed in detail here, considering the boundary surfaces of the wall material).

The purpose of the calculations is to analyse the impact of the adopted reinforcement on the stress of the model after the occurrence of deformation of the mining terrain of the step type and the impact of this deformation on the amount of calculated reinforcement of the foundation slab. The calculations are of the comparative character. Therefore, four structural solutions were analysed, designated as A, B, C, D (Table 1). Model A is characterised by the fact that no reinforcements have been introduced. In this model, the thickness of the foundation slab is 0.18 m and no reinforced concrete cores have been introduced in the walls of the ground floor and first floor. The weight of this model, considering the permanent and live loads, is 2947 kN.

In model B, 10 additional reinforced concrete cores with a cross-section of 0.24 m / 0.24 m were introduced in the walls of the ground floor and first floor. These cores are located at the intersections of axes A and B with axes numbered 1, 2, 3, 4, 6. The weight of this model is 3038 kN.

In model C, the thickness of the foundation slab was increased to 0.35 m, but no cores were introduced into the walls (model weight 3317 kN). In model D, the foundation slab was thickened to 0.35 m and 10 reinforced concrete cores were introduced (model weight 3409 kN). A summary of the analysed structural solutions is presented in Table 1.

Table 1. Computational models of the segment analysed.

Model identification	A	B	C	D
Thickness of the foundation slab	0,18 m	0,18 m	0,35 m	0,35 m
Reinforced concrete cores in walls	not present	0,24/0,24 m	not present	0,24/0,24 m
Model weight	2947 kN	3038 kN	3317 kN	3409 kN

Due to the need to map the impact on the segment of discontinuous deformation of the mining area, as well as the local impact of the building on this deformation, the foundation conditions were mapped in the model in two phases. In the first phase, information from four geotechnical boreholes was entered into the calculation programme – plastic clay lies under a 0.5 m thick layer of sand. For sand, E = 45 MPa, G = 17.6 MPa, ν = 0.28, and for slightly plastic clay, E = 12 MPa, G = 4.3 MPa, ν = 0.4. On this basis, a soil body with a projection of 40 m / 70 m and a thickness of 18 m was defined (Figure 3a). A model of a residential segment with a weight of 3409 kN was placed on this soil body and an average settlement of 6.05 mm was determined. With a slab projection area of 87.10 m^2, the global stiffness of the ground is

$$k_{\text{tot}} = \frac{3409\text{kN}}{6,05\text{mm}} = 563\frac{\text{MN}}{\text{m}} . \tag{3}$$

The second phase of foundation modelling assumed that the residential segment rests on 112 evenly distributed elastic elements. The stiffness of each such element is therefore

$$k = \frac{563 \quad \frac{\text{MN}}{\text{m}}}{112} \approx 5000\frac{\text{kN}}{\text{m}} . \tag{4}$$

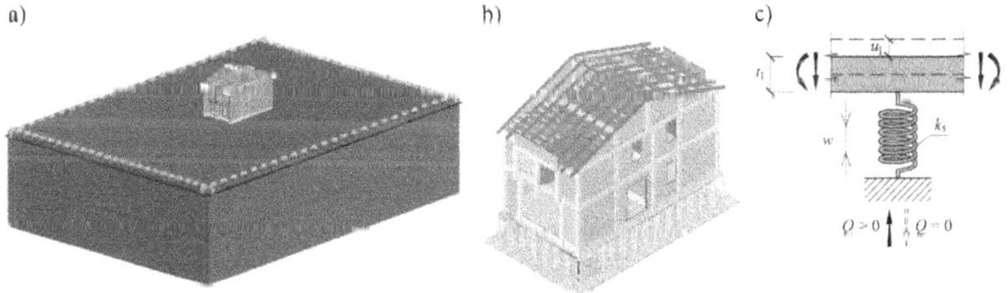

Figure 3. Building and foundation model: a) building founded on a soil massive, b) building founded on single-sided supports, c) definition of a single-sided support.

The supports for the elastic elements are single-sided supports, characterised by the fact that when detachment occurs, they are excluded from load transfer, and when compressed, they meet the condition of fixation (Figure 3c). This model represents the non-linear behaviour of the ground. Placement model of the building on 112 evenly distributed supports allows vertical displacements to be applied to these supports and thus modelling of discontinuous deformation of the mining terrain. Modelling the occurrence of discontinuous deformation consists in applying a shortening of value w to the relevant elastic elements, where w is the assumed height of the mining terrain step. The calculations differentiated between the position of the step and its height. The calculations were performed for an elastic model of the wall material, whose parameters are described above, and a non-linear model, whose boundary surface is based on the Drucker-Prager (D-P) criterion. In the (D-P) models, different values of masonry compressive strength f_c (from 1 to 4 MPa) and tensile strength f_t (from 0.1 to 0.4 MPa) were assumed. Due to editorial limitations on the length of the article, the following is limited to reporting only the results of calculations for the elastic model of wall material, with a degree of w = 0.1 m and one stiffness value k of single-sided supports described above for four models A, B, C, D (Table 1).

4 SELECTED CALCULATION RESULTS

As presented above, the calculations involved four segment models, differing in the presence and type of reinforcements (models marked as A, B, C, D, characterised in Table 1). These models were loaded with mining terrain deformations in the form of a step with a height of w. The results obtained are presented for two different situations. The first is a model not loaded with the effects of discontinuous deformation, and the second is a model in which the discontinuous deformation coincides with the wall in axis 3 at a single fixed value of w = 0.1 m (Figure 4). The results characterizing the stress in the model are presented in the form of a distribution of bending moments in the x and y directions (mx and my) in the α-α cross-section drawn at a distance of 2.94 m from the longer edge of the slab. In addition, the figures and tables show the computationally required lower and upper reinforcement of the foundation slab in two directions (x, y).

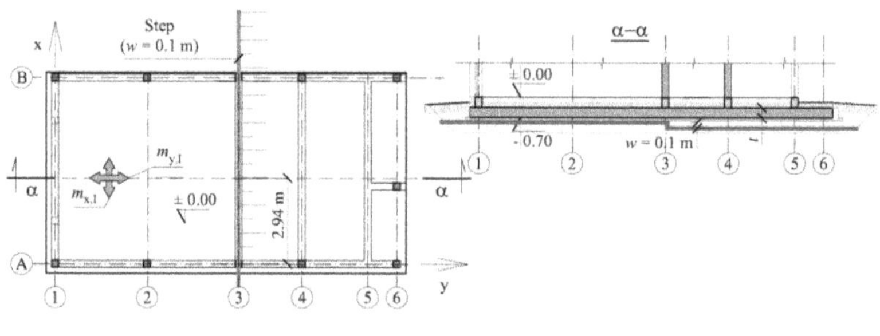

Figure 4. Foundation slab projection with the location of discontinuous deformation (step w = 0.1 m), the location of the α-α cross-section and the m_x, m_y moments in this cross-section.

4.1 *Model without discontinuous terrain deformation*

The distribution of bending moments in the foundation slab, to which the dead load and live load are applied (model not loaded with discontinuous ground deformation), is characterised by the fact that extreme negative moments occur in the area bounded by axes 1-3 and A-B ($m_{x,(1\text{-}3)}$, $m_{y,(1\text{-}3)}$ – Figure 5). This is due to the fact that this is the area of the slab with the largest projection (6.49 m / 6.40 m) and its load is the reaction of the ground. The extreme values of moments in the α-α cross-section depend on the analysed building structure.

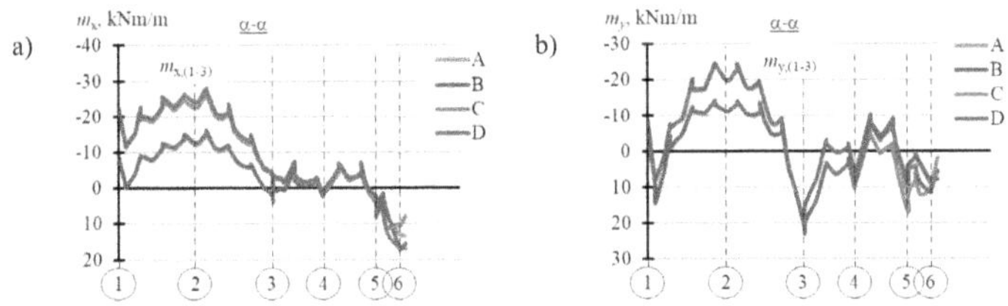

Figure 5. Bending moments in the foundation slab in models A, B, C and D in cross-section α-α - model not subject to step-type deformation: a) moments in the "X" direction, b) moments in the 'Y' direction.

The highest moment values were obtained for model D ($m_{x,(2-3)}$ = – 28.09 kNm/m and $m_{y,(2-3)}$ = – 24.74 kNm/m) and the lowest for model A ($m_{x,(2-3)}$ = – 15.20 kNm/m and $m_{y,(2-3)}$ = – 13.9 kNm/m). This means that higher bending moment values occur in the thicker slab (t = 0.35 m), which is characterised by greater stiffness and weight. At the same time, the influence of the cores on the value of the moments is insignificant. The occurrence of local extremes in the graphs in Figure 5 results from the adopted method of supporting the model plate with single-sided nodal supports.

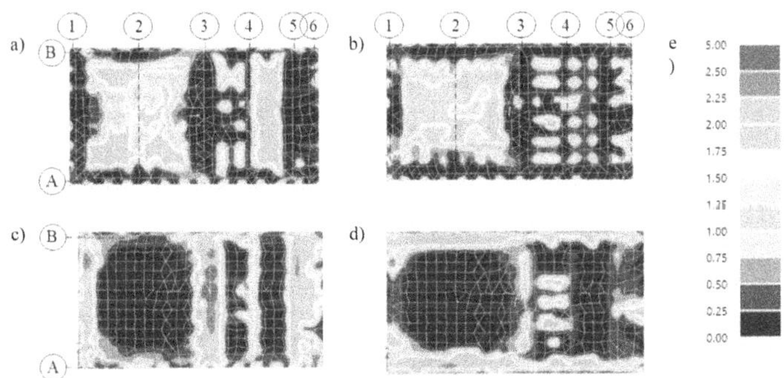

Figure 6. Calculated reinforcement for a 0.18 m thick foundation slab determined for model A in the absence of step-type deformation loads: a) upper reinforcement in the 'Y' direction, b) upper reinforcement in the 'X' direction, c) lower reinforcement in the 'Y' direction, d) lower reinforcement in the 'X' direction, e) scale representing the amount of calculated reinforcement in cm^2/m.

For model A (slab thickness 0.18 m) without continuous deformation of the mining terrain, the calculated reinforcement for bending was also determined. It amounts to 5 cm^2/m (Figure 6), which means that the percentage of reinforcement for a slab thickness of 0.18 m is up to 0.36%. However, on 30-40 per cent of the slab's projected area, design reinforcement is

Table 2. Designated reinforcement area in a 0.18 m thick foundation slab for model 'A' with no building load due to a fault.

Calculated reinforcement [cm^2/m]	Percentage of the slab area where the calculated reinforcement requirement occurs			
	Top reinforcement		Bottom reinforcement	
	Direction y	Direction x	Direction y	Direction x
from 0 to 0.25	31.68	31.89	36.81	40.33
from 0.25 to 0.50	12.60	15.62	11.42	18.33

(*continued*)

Table 2. Continued

Calculated reinforcement [cm²/m]	Percentage of the slab area where the calculated reinforcement requirement occurs			
	Top reinforcement		Bottom reinforcement	
	Direction y	Direction x	Direction y	Direction x
from 0.50 to 0.75	8.54	10.29	8.75	9.26
from 0.75 to 1.00	6.62	8.28	5.61	6.63
from 1.00 to 1.25	8.01	6.96	4.82	5.04
from 1.25 to 1.50	8.56	6.27	5.19	4.14
from 1.50 to 1.75	8.51	5.59	6.66	4.46
from 1.75 to 2.00	15.27	12.17	14.51	6.54
from 2.00 to 2.25	0.17	2.84	3.16	4.79
from 2.25 to 2.50	0.02	0.05	1.49	0.23
from 2.50 to 5.00	0.03	0.05	1.59	0.11
Mean cm²/m	0.84	0.78	0.86	0.66

not required (Table 2). The average calculated reinforcement depending on the direction (x, y) and position (top, bottom) ranges from 0.66 cm²/m to 0.84 cm²/m. These are small amounts of reinforcement.

4.2 *Model loaded with discontinuous deformation of the mining terrain*

The distribution of bending moments in the foundation slab of a building model subjected to discontinuous deformation depends on the deformation pattern, the value of the fault (w), the reinforcement structure, the stiffness of the ground, the material model of the walls, and other parameters. Figure 7 shows the distributions of bending moments m_x and m_y in the α-α cross-section obtained for four models A, B, C, D, $w = 0.1$ m and other material parameters identical to those of the model not loaded with discontinuous deformation. Extreme negative moments occur in the field bounded by axes 1-3 and A-B ($m_{x,(1\text{-}3)}$, $m_{y,(1\text{-}3)}$). This is due to the fact that this is the area of the slab on which the weight of the building rested after the fault occurred and the foundation detached from the ground over a significant area (the space between axes 3-5). The extreme values of moments in the α-α cross-section depend on the analysed building structure. The highest (extreme) values of moments were obtained for model D (m_x = – 93.16 kNm/m and m_y = – 118.97 kNm/m), and the smallest for model A (m_x = – 73.29 kNm/m and m_y = 96.55 kNm/m), as shown in Figure 7. In this figure, the fault line (w) is marked on an independent scale on the right-hand side.

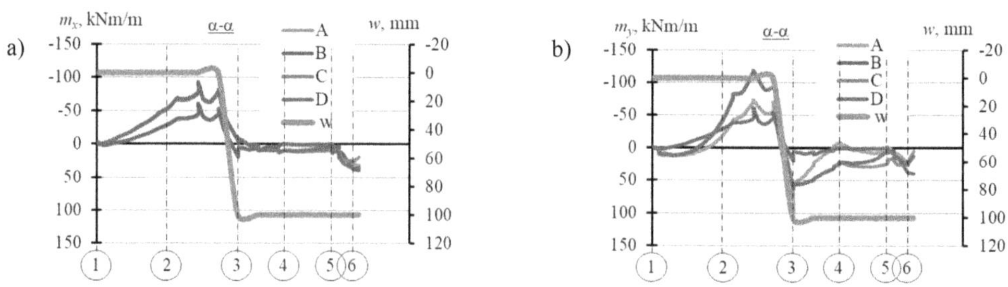

Figure 7. Bending moments in the foundation slab in models A, B, C and D in cross-section α-α – model loaded with step-type deformation: a) moments in the 'X' direction, b) moments in the 'Y' direction.

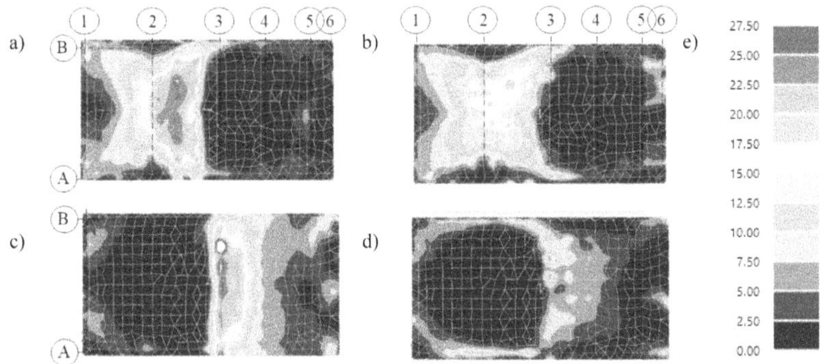

Figure 8. Calculated reinforcement for a 0.35 m thick foundation slab determined for model D with the building loaded with step-type deformation: a) upper reinforcement in the 'Y' direction, b) upper reinforcement in the 'X' direction, c) lower reinforcement in the 'Y' direction, d) lower reinforcement in the 'X' direction, e) scale representing the amount of reinforcement in cm^2/m.

This means that lower bending moment values occur in a thinner slab (t = 0.18 m). At the same time, it is not possible to design reinforcement for this thin slab for bending moments of 100 kNm/m. Therefore, a foundation slab thickened to 0.35 m is a necessary element for reinforcing the analysed structure against mining ground deformation. At the same time, further calculations involving wall stress analyses (Drucker-Prager model) show that it is necessary to introduce additional cores in order to limit the deformation of load-bearing walls (model D). The cores ensure that the walls function as supports, detached from the foundation slab.

The calculated reinforcement of the foundation slab in model D ranges from zero to 27 cm^2/m, with the maximum value referring only to 0.42% of the projected area of the 0.35 m thick slab (Table 3). The average calculated reinforcement, depending on the direction (x, y) and position (top, bottom), ranges from 3.37 cm^2/m to 6.09 cm^2/m. The average percentage of reinforcement for a 0.35 m thick slab is therefore up to 0.19%.

The presented results indicate that the occurrence of discontinuous step-type deformation increased the values of extreme bending moments in the foundation slab nearly five times compared to the situation of the model not loaded with this deformation.

Table 3. Designated reinforcement area in a 0.35 m thick foundation slab for model D with discontinuous deformation of the mining terrain of step type

	Percentage of the slab area where the calculated reinforcement requirement occurs			
	Top reinforcement		Bottom reinforcement	
Calculated reinforcement [cm^2/m]	Direction y	Direction x	Direction y	Direction x
from 0 to 2.5	41.57	35.95	41.8	52.57
from 2.5 to 5.0	15.92	16.78	20.73	21.12
from 5.0 to 7.5	1.54	11.74	15.36	17.85
from 7.5 to 10	8.23	10.88	5.27	7.12
from 10 to 12.5	5.71	9.6	3.95	0.82
from 12.5 to 15	3.67	8.65	2.68	0.26
from 15 to 17.5	3.67	5.78	3.18	0.13
from 17.5 to 20	3.82	0.6	2.59	0.08
from 20 to 22.5	3.8	0.03	3.17	0.03
from 22.5 to 25	2.65	0	0.83	0.02
from 25 to 27.5	0.42	0	0.42	0.01
Mean cm^2/m	5.96	6.09	5.54	3.37

5 CONCLUSIONS

Numerical simulations of the non-basement, two-storey segment model of a building with a modest footprint of 12.61 m $\times$ 6.84 m have demonstrated that the step-type deformation of the mining terrain results in the detachment of the structure from the ground at approximately half of its footprint. This alteration in the static scheme of the segment significantly increases the stress values transmitted from the ground to the foundation slab. Consequently, the internal force distribution within the structural elements is altered, leading to a nearly fivefold increase in the extreme values of bending moments within the slab. The most significant internal forces within the structure arise when ground deformation propagates through the center of the object, oriented parallel to the shorter dimension. To ensure that the analyzed structure complies with the Alitame Limit States conditions, it is imperative to implement a reinforced concrete foundation slab with a thickness of 0.35 m, in conjunction with reinforced concrete cores embedded within the walls. These cores serve to mitigate wall deformation and maintain functionality following the detachment of a portion of the building from the ground, as substantiated by an analysis of the wall utilizing a non-linear model that incorporates the effects of a boundary surface within the material.

REFERENCES

Barczyk, J. and Świądrowski, W. (1969). *Projektowanie zabezpieczeń istniejących budynków murowanych przeciw wpływom eksploatacji górniczej*. Zabrze: Wojewódzkie Biuro Projektów w Zabrzu.

Kawulok, M. (2015). *Szkody górnicze w budownictwie*. 2. wyd. Warszawa: Instytut Techniki Budowlanej.

Kliszczewicz, B. (2023). Infrastruktura techniczna na terenach górniczych - problemy projektowe, wykonawcze i eksploatacyjne. in: *Wyzwania budownictwa na terenach górniczych, pogórniczych i zdegradowanych*. Gliwice: Politechnika Śląska 145–66.

Kowalski, A. (2015). *Deformacje powierzchni w Górnośląskim Zagłębiu Węglowym*. Katowice: Główny Instytut Górnictwa.

Ledwoń, J. (1983). *Budownictwo na terenach górniczych*. Warszawa: Arkady.

Popiołek, E. (2009). *Ochrona terenów górniczych*. Kraków: Wydawnictwa AGH.

Szojda, L. (2019). *Aspekty konstrukcyjne zabezpieczania budynków na terenach górniczych*. Gliwice: Politechnika Śląska.

Wachniewski, W. (1954). Zabezpieczenie istniejących budynków murowanych na terenach odbudowy górniczej. *Inżynieria i Budownictwo* (11):328–30.

Measurement of water surface slope using tachymetric observations

M.F. Gruszczyński, J. Błotnicki, W. Sowa, A. Pałęcki & K. Stasch
Wrocław University of Environmental and Life Sciences, Poland

M. Popczyk
Silesian University of Technology, Gliwice, Poland

ABSTRACT: Accurate determination of the water surface slope (WSS) is essential for hydraulic calculations, yet it remains challenging in lowland rivers with gentle gradients, particularly over short reaches. Existing methods such as leveling, satellite data, TLS, or photogrammetry show significant limitations in small streams. This study presents a novel tachymetric approach that employs a robotic total station to continuously track a specially designed floating target drifting with the current. Field tests conducted on the Ślęza River demonstrated that the method enables continuous WSS profiling and achieves high precision, with measurement accuracy up to 2.5 mm. Compared to the reference method, the approach offered superior accuracy and eliminated the need for an operator holding a staff on the water surface. The results confirm the potential of this technique for precise WSS determination in lowland streams over short distances. The developed technique is potentially very useful in mining damage areas where, over short distances, the subsidence of the ground surface also causes a change in the slope of the streams.

1 INTRODUCTION

The water surface slope (WSS) is a key parameter in hydraulic calculations (Ayalew *et al.* 2022, Muste *et al.* 2019), including applications of the Chézy-Manning equation (de Chézy 1775, Manning 1891), and in the calibration of CFD models (Benefice *et al.* 2022). However, in lowland streams with very gentle gradients, often less than one per mille, accurate measurement of WSS is challenging (Mirosław-Świątek *et al.* 2016) and existing methods such as geometric leveling, satellite techniques, TLS, or photogrammetry have notable limitations (Bandini *et al.* 2020, Lyu and Tian 2024, Sulistioadi *et al.* 2015). In the areas of mining damage, a change in the (WSS) inclination is observed as a result of surface subsidence (Ciszewski *et al.*2022, Wade *et al.* 2008). Therefore, this study presents a tachymetric method designed to combine high accuracy with the capability to measure over short reaches while minimizing the influence of the human factor.

The technique presented in this study was developed to improve the accuracy of determining differences in the water surface slope, enable continuous measurements, and eliminate the human factor associated with an operator holding a staff on the water surface. The observations are carried out using a robotic total station that tracks a floating target drifting with the river current. The applied tachymetry, performed with electronic total stations, represents another classical surveying technique.

2 MATERIALS

2.1 *Water surface slope measurement techniques*

Geometric leveling is the most commonly used reference method for measuring the water surface slope (WSS). It involves directly determining the geometric height difference between

DOI: 10.1201/9781003655558-8

points on the water surface using a level and leveling staffs. Accurate determination of WSS is essential in hydrological and engineering studies. *In situ* geometric methods employing surveying instruments form the fundamental and often reference group of measurement techniques, characterized by high accuracy.

Measurements of the water surface slope can be carried out using optical, digital, or laser levels (Fitzpatrick *et al.* 1998). A level, together with a tripod and a leveling staff, constitutes the basic surveying setup used to determine height differences. A level transfers the horizontal plane from its position to the point where the leveling staff is placed (Pietraszek *et al.* 2021). In fluid mechanics, the slope of the energy line is one of the most important hydraulic parameters describing free-surface flow. Under the assumption of steady-state conditions, the energy line is parallel to the water surface, and, under specific circumstances – including laboratory conditions – also to the channel bed (Roohi *et al.* 2020). Accurate determination of the energy line slope by measuring the slope of the flowing water surface is particularly important for hydraulic calculations using the Chézy-Manning equation. The water surface slope is also a key parameter in CFD modeling and in the calibration of such models (Wang *et al.* 2022).

In water surface slope measurements, a key challenge is the accurate determination of the difference in water surface elevation between measurement cross-sections. In lowland streams and power channels, the water surface slope is small, often less than one per mille. Accurately determining this elevation difference is difficult and subject to considerable error (Marchand *et al.* 1984). Positioning a leveling staff or rod at the water - shore interface during leveling is problematic and introduces errors of approximately 1 cm, which is often assumed to represent the accuracy of water surface elevation observations (Połoska-Wróbel *et al.* 2016). Furthermore, the magnitude of this error depends on the experience of the survey team and the precision of the measuring equipment (Castaing *et al.* 1990).

To avoid difficult and time-consuming field surveys, many studies have employed satellite data (Góraj *et al.* 2019, Scherer *et al.* 2023, Wang *et al.* 2025). However, this approach is applicable only to large rivers, where the width of the water surface allows for the identification of several measurement points within a cross-section (Li *et al.* 2024).

Other modern approaches include terrestrial laser scanning (TLS/LiDAR) and photogrammetry for example using UAV platforms, which have been applied in riverine studies (Błotnicki, Gruszczyński, *et al.* 2023, Błotnicki, Jarzembowski, *et al.* 2023). TLS involves scanning the water surface with a laser beam to generate dense three-dimensional point clouds, which offer very high resolution but limited accuracy due to partial beam penetration and reflections from air bubbles, often leading to underestimated elevations (Brach and Chormański 2014, Pleterski *et al.* 2023). Photogrammetry, particularly when combined with Structure from Motion (SfM) algorithms, enables the reconstruction of 3D surfaces from 2D images and has shown agreement with laser scanning within ±20 mm under laboratory conditions (Lüthi *et al.* 2018, Pleterski *et al.* 2023).

Tachymetry, implemented using total stations, represent another classical surveying technique (Hirt 2010). They allow for both reflectorless and reflector-based measurements, enabling the determination of the spatial coordinates (X, Y, Z) of multiple points on the water surface from a single instrument position. The slope is subsequently calculated from the height differences and distances between the measured points. Similar to geometric leveling, this technique offers high accuracy but is a point-based method that is time-consuming and requires direct access to the watercourse

2.2 *Research site*

The Ślęza River in Wrocław was selected as the site for implementing the developed measurement technique. Field surveys were conducted at km 12+040 of the river, in Wrocław, on the southern side of Grabiszyński Park, downstream of the weir crossed by Eugeniusza Romera Street (Figure 1a). Figure 1b shows a detailed layout of the hydrographic network within the catchment.

At the nearest Ślęza gauging station, located 5.5 km upstream of the studied reach, the mean discharge (SSQ) for the period 2006-2021 is 2.30 $m^3 \cdot s^{-1}$. Measurements were carried out on 5 August 2025 at a gauge reading of 221 cm, corresponding to a discharge of approximately 2.4 $m^3 \cdot s^{-1}$. While the gauge did not record any changes in water level or discharge during the measurement period, a rise in the water surface was observed in the study area. An increase in water level of approximately 8 cm was recorded near the weir. This reach had previously been the subject of academic research, which makes it a well-documented study site. The decision was also influenced by the river's location within the city limits of Wrocław, the shallow depth of the channel facilitating safe measurements, and the compact embankment development, with the main channel symmetrically aligned to the longitudinal axis of the riverbed.

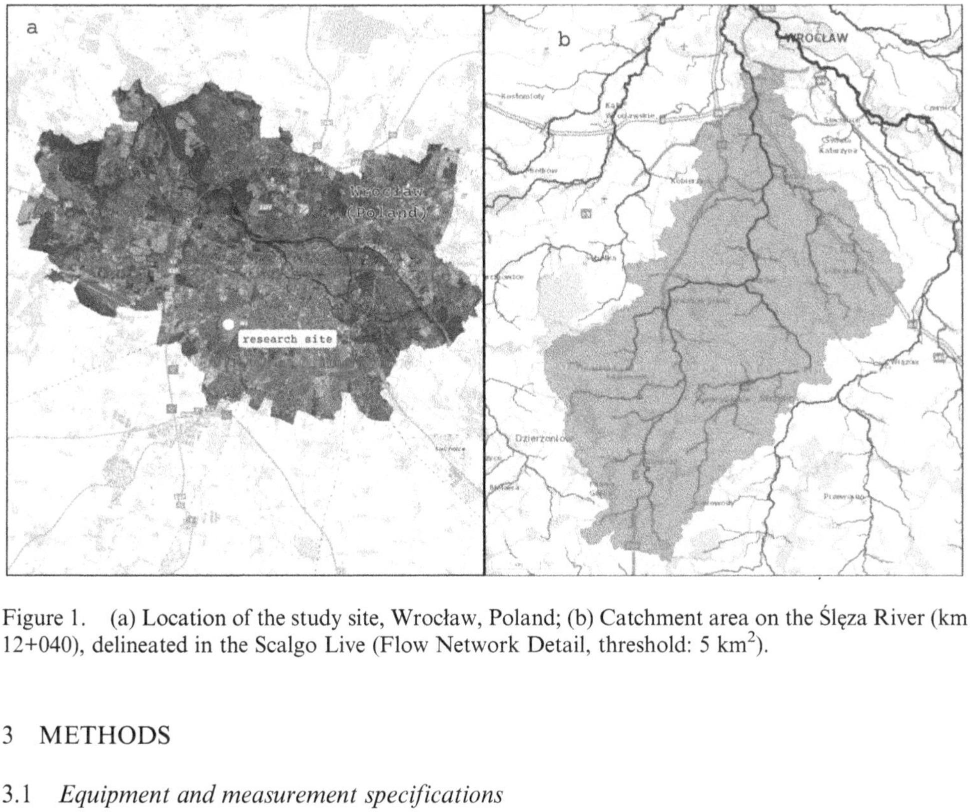

Figure 1. (a) Location of the study site, Wrocław, Poland; (b) Catchment area on the Ślęza River (km 12+040), delineated in the Scalgo Live (Flow Network Detail, threshold: 5 km^2).

3 METHODS

3.1 *Equipment and measurement specifications*

This study employed professional surveying instruments and accessories typically used in high-precision engineering measurements. The selected equipment ensured the high measurement accuracy required to meet the objectives of the study. The following subsections describe the instruments along with their key parameters relevant to this research.

3.1.1 *Trimble S8 0.5" precision total station*

This study employed the Trimble S8 robotic total station, an advanced surveying instrument designed for high-precision engineering measurements. Its technical specifications and specialized functions provide the highest levels of accuracy and performance, which are essential for tasks requiring maximum data reliability, such as geodetic deformation monitoring or precision construction surveying. The key features of the instrument relevant to the present measurements include:

- High-accuracy Trimble DR HP rangefinder (EDM) - enables precise distance measurements to surveying prisms.
- MagDrive servomotors - allow the instrument to track moving targets at high speeds (up to 115°/s). This technology was essential for the continuous observation of the survey prism drifting with the river current.
- FineLock technology - enables precise detection and targeting of the correct prism with sub-millimetre accuracy (less than 1 mm) at distances of up to 300 m. This feature ensured that the instrument automatically locked onto the prism mounted on the floating target, while ignoring other potential reflections in the surroundings.

3.1.2 *Leica sprinter 100M level with barcoded staffs*

The Leica Sprinter 100M is a digital level whose key feature is the automation of the measurement process. The use of barcode-reading technology on the leveling staff helps eliminate gross and observer-induced errors, directly enhancing the reliability and repeatability of results. The integrated software enables basic calculations, such as determining height differences or performing line leveling, allowing for preliminary checks and data processing directly at the survey site. The Leica Sprinter 100M level is characterized by 1,5 mm/km accuracy.

3.2 *Reference technique - Water surface levelling*

Geometric leveling, performed with a digital level and barcoded staffs, was used as the reference technique for the tachymetric measurements. The measurements were tied to the survey control network, whose benchmark elevations had been previously determined using classical geometric leveling in two directions, ensuring the highest accuracy and consistency of the reference system. The instantaneous water surface elevation was measured using the forward leveling method, which involved direct readings from the barcoded staff, with its base kept in constant contact with the water surface.

A key methodological challenge and the main source of uncertainty was the staff placement error, resulting from unstable contact with the dynamic water surface. To minimize this effect, measurements were taken in series comprising several readings. From each series, only three measurements that met rigorous real-time evaluation criteria were retained for further analysis. These criteria included a subjective assessment by the surveyor of the stability of the staff's contact with the water, verification of its vertical alignment, and final approval of reading differences by the observer. Additionally, to ensure the most stable conditions possible, measurements were conducted at locations characterized by minimal surface wave action and calm flow.

3.3 *Design of a floating target for a robotic total station*

To develop the measurement technique, a mobile floating target was constructed for a robotic total station. The passive prism, used to determine position in three axes (X, Y, Z), was mounted on a Riverboat SP trimaran, originally employed as a platform for an ADCP. The trimaran measured 114 × 81 cm, and its draft with the centerboards folded was approximately 3 cm. A tribrach and a 10 cm extension were mounted between the float and the prism. Elevating the prism by this amount significantly facilitated targeting in areas with dense vegetation. No stability issues resulting from the raised center of gravity were observed with this configuration.

3.4 *Measurement procedure using a floating target*

The measurement procedure with a floating target for a robotic total station begins by leveling the prism mounted on the float using a tribrach. This step should be carried out outside the main current, on a relatively calm water surface (Figure 2a). After leveling, the float is rotated 180° to verify the accuracy of the leveling. The float is then placed in the river current, allowing the target to drift freely with the flow. During the drift, the target's trajectory cannot be continuously monitored, which may complicate and prolong the measurement process. During the measurement campaign, the target is continuously monitored by the total station, which automatically tracks the prism (Figure 2b). It is crucial to position the total station in relation to the anticipated drift trajectory so that the target remains within the instrument's field of view at all times. If the measurement is interrupted, automatic tracking must be manually reinitiated once the target becomes visible again. When conducting surveys on natural streams, obstacles such as patches of vegetation should be anticipated. In some cases, the target became immobilized on a reed bed, causing visible deviation from the vertical, which required filtering such data from the measurement database.

Dynamic water surface measurements were carried out using tachymetry. The methodology involved the use of a robotic total station, which continuously tracked the target in automatic tracking mode. A precise control network established on the riverbank served as the reference framework for the measurements, with the total station's position tied to this network to ensure a uniform coordinate system. To maintain full data consistency, the elevations of the control points were additionally determined using geometric leveling, allowing the tachymetric results to be accurately linked with the reference measurements.

Figure 2. a) Leveling the prism on the water surface; b) Automatic tracking of the prism by the total station.

The main phase of the measurement involved recording long series of observations to the floating target. Each series consisted of at least 500 continuous measurements taken in a single telescope position. Upon completing each series, a control measurement was carried out on the control points to verify the stability of the station. In the tachymetric survey, the XY coordinates of the prism on the water surface were determined using the polar method, while the prism's elevation relative to the local vertical datum was determined using trigonometric leveling.

The measurements were taken continuously to the prism mounted on the float. This enabled the determination of water surface elevation differences between individual measurement points, which in turn allowed for calculating the water surface slope in the stream with high accuracy.

4 RESULTS

The results from two series of measurements taken with a Trimble S8 total station, featuring an angular accuracy of 0.5″, are presented below. The map illustrates the locations of the measurement points, the control network, and the cross-sections where the reference measurements were conducted (Figure 3). The analysis used data from two measurement series covering cross-sections II and IV.

In the first stage, measurements were taken to the surveying prism mounted on the float, drifting with the river current in two measurement series. To analyze the suitability of the method, measurement results from points located in cross-sections II and IV were used. This choice was dictated by the limited visibility of the water surface along the entire reach of the studied stream.

Table 1 summarizes the results of water surface elevation measurements (relative elevations - without accounting for the offset of the prism height above the water surface). For each cross-section, six measurements taken at the closest distance to the cross-section line were selected. The mean value (adopted as the water surface elevation in the cross-section for further calculations) and the standard deviation were then calculated.

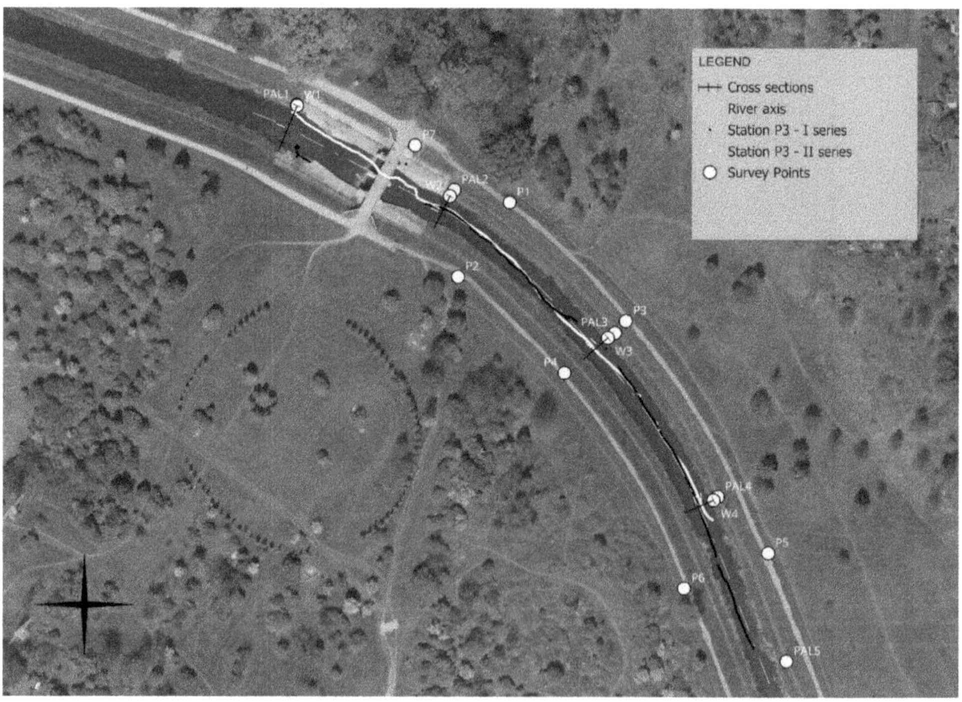

Figure 3. Location of the control network and cross-sections and track of measurement points.

Table 1. Results of measurements to the prism on the float at the cross-sections.

Cross-section	Height [m]						Avg. Height [m]	Std Dev [m]
SERIES I								
II	119.2056	119.2057	119.2053	119.2054	119.2054	119.2054	119.2055	0.0002
IV	119.2148	119.2148	119.2151	119.2143	119.2144	119.2153	119.2148	0.0004
SERIES II								
II	119.2192	119.2196	119.2205	119.2206	119.2201	119.2200	119.2200	0.0005
IV	119.2316	119.2303	119.2324	119.2316	119.2317	119.2311	119.2315	0.0007

The standard deviation from a series of six measurements was used as the control indicator. In this study, these values were an order of magnitude lower than the assumed accuracy of the method.

The reference method consisted of geometric leveling (for the leveling traverse) and forward leveling (for points on the water surface). For cross-sections II and IV, the average values from three readings taken with a staff placed on the water surface were used. The measurement results are presented in Table 2.

Table 2. Results of water surface elevation measurements using forward leveling.

Cross section	Height [m]			Avarage Height [m]	Std Dev [m]
II	118.8990	118.8990	118.8986	118.8989	0.0002
IV	118.9143	118.9142	118.9138	118.9141	0.0003

Based on the obtained data, the differences in water surface elevation between cross-sections II and IV and the corresponding slopes were calculated. The distance between the cross-sections was determined along the longitudinal axis of the river channel; for cross-sections II and IV, it was 138.30 m. The elevation differences and slope values are summarized in the table below (Table 3).

Table 3. Calculated water surface elevation differences and slopes – various methods.

Method	Series	Cross-section	Avarage Height [m]	dH [m]	i [%]
Trigonometric Leveling	I	II	119.2055	0.0093	0.007
	I	IV	119.2148		
	II	II	119.2200	0.0114	0.008
	II	IV	119.2315		
Geometric & forward leveling	-	II	118.8989	0.0152	0.011
	-	IV	118.9141		

This study applied the trigonometric leveling method using a high-precision Leica S8 robotic total station with an angular accuracy of 0.5″. The following measurement accuracies were adopted to estimate the accuracy of the elevations determined with this method:

- Vertical angle measurement accuracy: 0.5″ (i.e., 0.00015 gon)
- Spatial distance measurement accuracy: 2 mm + 2 ppm, which, for the assumed target distances (maximum 100 m), is no more than 2.2 mm
- Aiming accuracy to a 360° prism: 2 mm

Because the measurements were taken from a single station to the same target, the accuracies of instrument height and prism height measurements were omitted, as they have no influence on the resulting elevation differences.

The theoretical accuracy of the height difference measurement is expressed by formula (1):

$$m_{\Delta h} = \pm\sqrt{(\cot z \cdot m_d)^2 + \frac{d^2}{\sin^4 \cdot z} m_z^2} \tag{1}$$

Where: $m_{\Delta h}$ = standard deviation of height difference; m_d = standard deviation of horizontal distance measurement; m_z = standard deviation of zenith angle measurement; d = horizontal distance between the instrument and the prism; z = zenith angle.

The maximum angle and distance values (102 gon and 75 m) were used in the analysis. Based on the assumed measurement accuracies and maximum measured values, the theoretical accuracy of this method is 0.2 mm. However, taking into account the aiming accuracy to the prism, it can be assumed that the elevation accuracy for the adopted measurement is no greater than 2.5 mm.

The reference method for assessing the suitability of the tachymetric approach was geometric leveling. With the measuring equipment used in this study, its accuracy is 1.5 mm/km, which, for the applied leveling route, yields about 1 mm.

For comparison, the foreleveling method was also applied. Its accuracy is lower due to the use of a single staff and potential staff zero error, which was eliminated here by using the same staff for both cross-sections. Another issue is the challenging placement of the staff on the water surface, which was mitigated by repeated readings and joint evaluation by the surveyor and staff holder. The accuracy of this method can thus be assumed at no worse than 3 mm.

To provide a concise assessment, Table 4 summarizes the accuracy parameters of the three methods used in the study: trigonometric leveling, geometric leveling, and foreleveling, including equipment characteristics, measurement setup, and estimated accuracy with notes on error sources.

Table 4. Assumed accuracy of height determination for the applied methods.

Method	Equipment / Measurement Setup	Assumed accuracy of height determination	Notes
Trigonometric leveling	Robotic total station Trimble S8, 0.5″ angular accuracy, target $\leq$ 100 m	$\leq$ 2.5 mm	Includes aiming accuracy (2 mm) at 360° prism; instrument/prism height error omitted
Geometric leveling	Precision leveling instrument, 1.5 mm/km	$\approx$ 1 mm	Based on leveling route used for reference points
Foreleveling	Single leveling staff placed on water surface	$\leq$ 3 mm	Staff zero error eliminated by using the same staff;

The presented method for determining differences in water surface elevation enables the detection of stream level changes over short sections with an accuracy of up to 2.5 mm. The results from two measurement series confirm this precision - the difference between the calculated height differences in cross-sections II and IV was 2.1 mm. Future studies should consider verifying these results with another, more precise reference method, such as hydrostatic leveling.

The reference method used in this study, due to errors associated with positioning the staff at the water-shore interface, exhibited lower accuracy than the tested technique. Direct comparison between the trigonometric leveling results and those obtained from method is not fully justified.

5 CONCLUSIONS

The tachymetric method applied to determine water surface elevation differences over short stream sections provides high measurement accuracy. Nevertheless, the technique requires further verification using more precise methods, as well as refinement of measurement procedures. Potential future actions include:

- conducting measurements from two independent tachymetric stations to enable cross-verification of results,
- performing precise leveling and improving the accuracy of measurements to a staff positioned on the water surface (e.g., by increasing the number of observations, measuring in multiple cross-sections, and on both banks),
- carrying out independent trigonometric leveling measurements to a geodetic prism placed at the water surface level (without the use of a float),
- employing hydrostatic leveling as a reference method.

The developed technique eliminates the influence of human error associated with positioning a staff at the water-bank interface. A key advantage is the ability to obtain a continuous WSS profile along the stream, enabling the correction of local errors and substantially improving the accuracy of determining the water surface slope. The developed technique will be used in areas affected by subsidence of the ground surface due to underground mining.

REFERENCES

Ayalew, A.T., Koshuma, A.E., Ayele, M.A., and Dirate, D.D. (2022) 'An Integrated Approach of Profile Studies for Surface Water Velocity on a Steep Slope in a Rectangular Channel', *Advances in Science and Technology Research Journal*, 16(2), 74 – 80, available: https://doi.org/10.12913/22998624/145965.

Bandini, F., Sunding, T.P., Linde, J., Smith, O., Jensen, I.K., Köppl, C.J., Butts, M., and Bauer-Gottwein, P. (2020) 'Unmanned Aerial System (UAS) observations of water surface elevation in a small stream: Comparison of radar altimetry, LIDAR and photogrammetry techniques', *Remote Sensing of Environment*, 237, available: https://doi.org/10.1016/j.rse.2019.111487.

Benefice, G., Duron, L., Villanueva, A., and Yang, R. (2022) 'Calibration of 1D and 2D Fluvial Models with a Metamodel Based Optimization', Springer Water, 773 – 794, available: https://doi.org/10.1007/978-981-19-1600-7_48.

Błotnicki, J., Gruszczyński, M., Jarzembowski, P., and Popczyk, M. (2023) 'Application of close-range remote sensing for automatic identification of ice jams in rivers in the area of the inlet to the fishway', *Zeszyty Naukowe / Akademia Morska w Szczecinie*, 75(147), 97–106, available: https://doi.org/10.17402/578.

Błotnicki, J., Jarzembowski, P., Gruszczyński, M.F., and Popczyk, M.K. (2023) 'The Use of UAV for Measuring the Morphology of Ice Cover on the Surface of a River: A Case Study of the Low Head Dam and Fishway Inlet Area in the Odra River', *Water*, 15(22), 3972, available: https://doi.org/10.3390/w15223972.

Brach, M. and Chormański, J. (2014) 'Terrestrial Laser Scanning (TLS) as a detection method of the natural river valley microtopography – case study of the Upper Biebrza', Annals of Warsaw University of Life Sciences, *Land Reclamation*, 46(4), 276–278, available: https://doi.org/10.1515/sggw-2015-0001.

Castaing, B., Gagne, Y., and Hopfinger, E.J. (1990) 'Velocity probability density functions of high Reynolds number turbulence', *Physica D: Nonlinear Phenomena*, 46(2), 177 – 200, available: https://doi.org/10.1016/0167-2789(90)90035-N.

Ciszewski, D. and Sobucki, M. (2022) 'River response to mining-induced subsidence', *Catena*, 214, 106303, available: https://doi.org/10.1016/j.catena.2022.106303

de Chézy, A. (1775) 'Thesis on the velocity of the flow in a given ditch'.

Fitzpatrick, F.A., Waite, I.R., D'Arconte, P.J., Meador, M.R., Maupin, M.A., and Gurtz, M.E. (1998) Revised Methods for Characterizing Stream Habitat in the National Water Quality Assessment Program - U.S. *Geological Survey Water-Resources Investigations Report* 98-4052, U.S. Geological Survey, Raleigh, North Carolina.

Góraj, M., Wróblewski, C., Ciężkowski, W., Jóźwiak, J., and Chormański, J. (2019) 'Free water table area monitoring on wetlands using satellite and UAV orthophotomaps – Kampinos National Park case study', *Meteorology Hydrology and Water Management*, 7(1), available: https://doi.org/10.26491/mhwm/95086.

Hirt, C. (2010) 'Prediction of vertical deflections from high-degree spherical harmonic synthesis and residual terrain model data', *Journal of Geodesy*, 84(3), 179 – 190, available: https://doi.org/10.1007/s00190-009-0354-x.

Li, D., Qin, C., Xue, Y., Xu, M., and Liu, Y. (2024) 'Extraction of small river cross-section in mountainous based on ICESat-2 data, *Qinghua Daxue Xuebao/Journal of Tsinghua University*, 64(12), 2122 – 2131, available: https://doi.org/10.16511/j.cnki.qhdxxb.2024.21.021.

Lüthi, B., Philippe, T., and Peña-Haro, S. (2018) 'Method and system for determining the velocity and level of a moving fluid surface (European Patent No. EP3018483B1)', available: https://patents.google.com/patent/EP3018483B1.

Lyu, H. and Tian, F. (2024) 'Satellite-based water surface slope over a small mountain river in northern China', *Journal of Hydrology*, 639, available: https://doi.org/10.1016/j.jhydrol.2024.131576.

Manning, R. (1891) 'On the flow of water in open channels and pipes', *Transactions of the Institution of Civil Engineers* (Ireland).

Marchand, J.P., Jarret, R.D., and Jones, L.L. (1984) Velocity Profile, Water-Surface Slope, and Bed-Material Size for Selected Streams in Colorado, U.S. *Geological Survey*, Lakewood, Colorado, available: https://doi.org/10.3133/ofr84733.

Mirosław-Świątek, D., Michałowski, R., Szporak-Wasilewska, S., Ignar, S., and Grygoruk, M. (2016) 'Unraveling uncertainties of water table slope assessment with DGPS in lowland floodplain wetlands', *Environ. Monit. Assess.*, 188(11), available: https://doi.org/10.1007/s10661-016-5642-3.

Muste, M., Bacotiu, C., and Thomas, D. (2019) Evaluation of the slope-area method for continuous streamflow monitoring, in *Proceedings of the IAHR World Congress*, 121 – 130, available: https://doi.org/10.3850/38WC092019-1860.

Pietraszek, Z., Żelazko, J., Okruszko, T., Mirosław-Świątek, D., and Osuch, P. (2021) 'Przyrząd do precyzyjnego pomiaru spadku zwierciadła wody (Patent No. PL238530B1)'.

Pleterski, Ž., Hočevar, M., Bizjan, B., Kolbl Repinc, S., and Rak, G. (2023) 'Measurements of Complex Free Water Surface Topography Using a Photogrammetric Method', *Remote Sensing*, 15(19), 4774, available: https://doi.org/10.3390/rs15194774.

Połoska-Wróbel, A., Michalec, B., Wałęga, A., Cupak, A., and Michalec, A. (2016) 'Wyznaczenie krzywej natężenia przepływu w przekroju cofki zbiorników wodnych w Zesławicach', *Acta Sci. Pol.. Formatio Circumiectus*, 15(1), available: DOI: http://dx.doi.org/10.15576/ASP.FC/2016.15.1.113

Roohi, M., Soleymani, K., Salimi, M., and Heidari, M. (2020) 'Numerical evaluation of the general flow hydraulics and estimation of the river plain by solving the Saint–Venant equation', *Modeling Earth Systems and Environment*, 6(2), 645 – 658, available: https://doi.org/10.1007/s40808-020-00718-9.

Scherer, D., Schwatke, C., Dettmering, D., and Seitz, F. (2023) 'ICESat-2 river surface slope (IRIS): A global reach-scale water surface slope dataset', *Scientific Data*, 10(1), 359, available: https://doi.org/10.1038/s41597-023-02215-x.

Sulistioadi, Y.B., Tseng, K.-H., Shum, C.K., Hidayat, H., Sumaryono, M., Suhardiman, A., Setiawan, F., and Sunarso, S. (2015) 'Satellite radar altimetry for monitoring small rivers and lakes in Indonesia', *Hydrology and Earth System Sciences*, 19(1), available: https://doi.org/10.5194/hess-19-341-2015.

Wade, S.A. (2008) 'Stream Flow Characterization over Longwall Coal Mines in Pennsylvania, Ohio, and West Virginia', Master's thesis, West Virginia University, Eberly College of Arts and Sciences, Department of Geology and Geography, Morgantown, WV, available: https://doi.org/10.33915/etd.2644

Wang, X., Wang, F., Yang, J., and Yu, P. (2022) 'Modelling the hydraulic characteristics of diaphragm fishways and the effects on fish habitats', *Frontiers in Environmental Science*, 10, available: https://doi.org/10.3389/fenvs.2022.977295.

Wang, Y., Morton, Y.J., Minear, J.T., Putnam, A., Conrad, A., Axelrad, P., Nerem, R.S., Warnock, A., Ruf, C., Moreira, D.M., and Talpe, M. (2025) 'Measuring river slope using spaceborne GNSS reflectometry: Methodology and first performance assessment', *Remote Sensing of Environment*, 318, 114597, available: https://doi.org/10.1016/j.rse.2025.114597.

Mine Surveying and Mining Area Protection – Sokoła-Szewioła and Jarczyk (Eds)
© 2025 The Author(s), ISBN: 978-1-041-10531-2

Assessment of mining impacts at the location of a building situated in the area of a former mine – A case study

P. Jarczyk & K. Szafulera-Bałecka
Silesian University of Technology, Gliwice, Poland

ABSTRACT: The article presents an assessment of post-mining hazards on a property located above shallow workings of a former hard coal mine, whose operations ceased in 1999 and has been undergoing flooding since 2000. The aim of the study was to determine the risk of discontinuous deformations (sinkholes) due to the geological structure of the overburden, changes in hydrogeological conditions, and the history of mining activities. The analysis considered geological and mining documentation, geodetic survey data from 2021–2023, and calculations based on the Chudek–Olaszowski and Kotyrba methods. It was found that the effects of continuous deformation subsided around 1994, and the minor surface subsidence observed in recent years ($\pm$15 mm) is related to rising groundwater levels and suffosion processes in the overburden. Classification results indicate a low risk for former longwall workings but a high risk for corridor (gallery) workings, categorizing the area as B_{23}. Continuous monitoring, geophysical investigations, and exploratory boreholes are recommended, emphasizing the need for responsible land use in post-mining areas, even in the absence of current deformation symptoms.

1 INTRODUCTION

Mining activity, both historical and contemporary, exerts a significant impact on the environment, causing numerous changes in the rock mass structure and hydrogeological conditions. Particularly dangerous are the effects of former mining operations conducted at shallow depths, resulting in numerous shallow exploitation and corridor workings, often with incomplete technical documentation (Chudek *et al.* 1988, Kowalski 2015). In post-mining areas, discontinuous deformation phenomena such as sinkholes remain a serious engineering and environmental problem, especially in built-up areas (Fajklewicz *et al.* 2004, Popiołek *et al.* 2005; Tihansky 2015). An important factor intensifying the risk of sinkhole formation is the change in groundwater conditions within the rock mass, which can lead to the reactivation of abandoned workings. Infiltration of surface and groundwater through fractures caused by mining may result in the leaching of backfill materials and the initiation of suffosion processes, leading to the formation of voids that, under favorable conditions, may collapse (Baryakh *et al.* 2011, Strzałkowski 2019, Strzałkowski *et al.* 2021). As a consequence, irregular surface subsidences may appear, and in extreme cases—sinkholes that pose a real threat to the safety of buildings and human life (Strzałkowski 2021).

However, the sinkhole phenomenon is not a problem confined solely to post-mining areas in Poland. It occurs worldwide, often taking the form of sudden and unpredictable surface deformations. Among the most well-known cases are sinkholes in the U.S. state of Florida, where specific geological conditions, karst formations, and a high water table promote the formation of sinkholes with diameters of several dozen meters and depths exceeding 30 meters (Waltham *et al.* 2018). One of the most notable incidents occurred in Guatemala City in 2010, when heavy rainfall and weakened ground led to the formation of a sinkhole approximately

DOI: 10.1201/9781003655558-9

20 meters in diameter and 30 meters deep, engulfing buildings and parts of streets (Paris 2020). In China and Russia as well, in areas of both abandoned and active coal mines and karst regions, sinkholes pose significant threats to buildings and infrastructure (Lei *et al.* 2022). In Europe, the most vulnerable areas include karst regions of Spain and Italy, where intensive land use and changes in hydrogeological conditions contribute to the reactivation of natural voids (Gutiérrez *et al.* 2008). In many cases, sinkholes cause substantial material losses and present a direct danger to human life, highlighting the necessity of continuous monitoring and research into their formation mechanisms and mitigation strategies.

In recent years, geoinformatics tools and spatial analysis methods have gained increasing importance in addressing the impacts of sinkholes. A notable example is the study by Blachowski and Walerysiak (2025), which presents a comprehensive sinkhole database and hazard maps for the post-mining area of the former Babina lignite mine in western Poland. Such datasets facilitate the identification of areas with elevated risk of discontinuous deformations and provide valuable input for spatial planning and risk assessment. This approach may also be effectively adapted to other regions affected by the consequences of underground mining.

In light of the above, this article attempts to identify potential mining impacts in the area of a specific building located in a post-mining zone. The study includes an analysis of geological and mining conditions, tectonic disturbances, and the history of mining operations in the immediate vicinity of the structure. It also considers the possibility of changes in groundwater conditions that could lead to the reactivation of former workings and the formation of voids at various levels of the rock mass. Based on available data and the results of calculations performed, the degree of risk of discontinuous deformations was assessed, and the site was classified into the appropriate post-mining category (Strzałkowski *et al.* 2019). The findings serve as a basis for formulating further research and design recommendations aimed at protecting the analyzed structure from future ground deformation effects.

2 THE CASE UNDER ANALYSIS

The analyzed case concerns damage observed in a detached single-family residential building (Figure 1) and its surroundings, located in a post-mining area previously subjected to underground extraction activities. The building was commissioned in 1990, following the commencement of construction around 1989. During the construction phase, the structure was reinforced to meet the requirements of Category IV static resistance.

In 2022—several decades after the cessation of mining operations and the decommissioning of the mine—sudden and concerning damage began to manifest within the building. These damages primarily took the form of cracks in internal walls and ceilings, as well as deterioration of the external façade. Additionally, localized ground subsidence was observed in the vicinity of the building. The onset of these damages coincided with recorded fluctuations in the groundwater table within Quaternary strata.

Figure 1. View of selected internal damages within the residential building (author's own photograph, 30-09-25).

This phenomenon, in the context of historical shallow-depth mining operations and the incomplete technical documentation of excavations, suggests the possible reactivation of former voids or weakened zones within the rock mass. Such reactivation may have led to discontinuous subsoil deformations and secondary ground movements.

3 ANALYSIS OF GEOLOGICAL AND MINING CONDITIONS IN THE VICINITY OF THE STRUCTURE

3.1 *Lithology and stratigraphy*

The geological structure of the rock mass was identified based on the thickness map of Quaternary formations (WUG 2024) and the geotechnical borehole log No. 1 – Figure 2. The rock mass in the area consists of overburden layers and Carboniferous formations. Stratigraphically, the overburden is composed of Quaternary deposits. The thickness of the Quaternary layer directly beneath the structure is approximately 25 meters. It comprises a 20-centimeter layer of topsoil and medium sand with isolated gravel admixtures, slightly dusty and interbedded with silt. The presence of clay layers cannot be ruled out, as indicated by data from a geotechnical borehole drilled in the central part of the property.

The Carboniferous formations, from a stratigraphic perspective, include the Łaziska Beds (Group 200 coal seams) and the Orzesze Beds (Group 300 coal seams). The Łaziska Beds extend to a depth of approximately 270 meters and are primarily composed of sandstones, with minor occurrences of sandy shales and hard coal seams: 207, 208, 209, 210, and 214. These seams are spaced at intervals of approximately 20 meters and range in thickness from about 1.4 to 3.3 meters. The Orzesze Beds begin at the roof of seam 301, at a depth of approximately 275 meters, and are predominantly developed as mudstones and claystones.

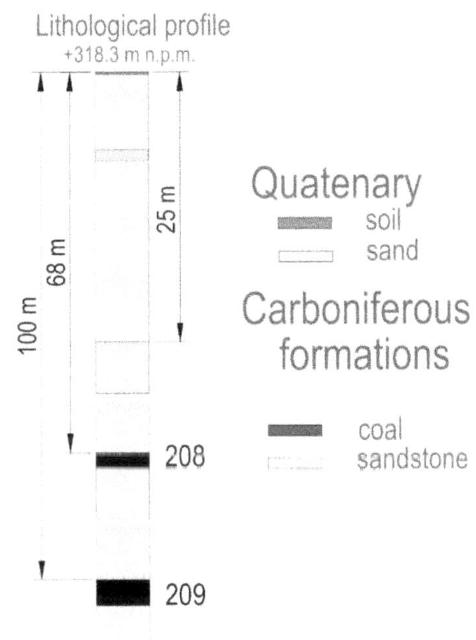

Figure 2. Lithological profile illustrating the geological structure in the study area (WUG 2024).

3.2 *Tectonics*

According to the Carboniferous roof map (WUG 2024), no tectonic fault outcrops occur directly beneath the property. The site is located between two main faults of the former mine – Figure 2.The first, or more precisely, the fault zone known as Fault S I, has its outcrop on the Carboniferous roof approximately 260 meters south of the building's geometric center. This fault, along with its two branches, runs approximately in an east–west (latitudinal) direction and causes a displacement of the rock layers of about 60 meters, while the branches produce displacements of approximately 10 meters. The second main fault, Fault S II, has its outcrop on the Carboniferous roof about 110 meters north of the building's geometric center. This fault also follows a near-latitudinal direction and causes the rock layers to be displaced northward by approximately 80 meters. Apart from these faults, no other tectonic structures have been identified in the area.

3.3 *Conducted mining operations*

The first mining operation was conducted in 1963–1964 in seam 208, at an average depth of approximately 68 m and a thickness of 1.4–2.6 m. A shortwall system with hydraulic backfilling was applied. In 1971–1972, extraction continued in seam 210 at a depth of about 125 m, with a thickness of 2.5–2.6 m, using the longwall method with backfilling. In 1981–1982, seam 209 was mined directly beneath the location of the structure, at a depth of approximately 100 m and a thickness of 2.3–2.5 m, also using the longwall system with backfilling. After the construction of the structure, mining was carried out again in 1991–1994 in seam 214, at a depth of 194–196 m and a thickness of 3.1–3.4 m, with voids also eliminated by means of hydraulic backfilling.

Figure 3 presents a schematic illustration of the extent of mining operations in relation to the location of the structure, conducted at a distance of 1.4 r, where *r* denotes the radius of the main influence zone [m]. Basic information concerning geological–mining conditions and extraction parameters is summarized in Table 1.

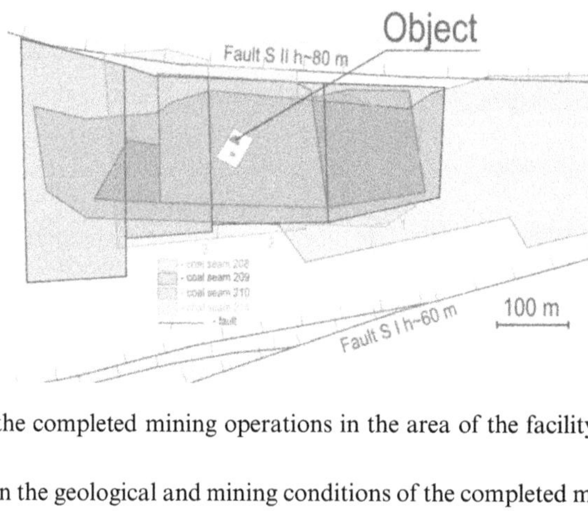

Figure 3. Sketch of the completed mining operations in the area of the facility.

Table 1. Basic data on the geological and mining conditions of the completed mining operations in the vicinity of the property.

Seam	Longwall	Start of extr.	End of extr.	Seam thickness [m]	Depth [m]	Dist. from the object [km]	Direct. relative to object	Mining system
208	-	1963	1964	1.4	68	0	Safe	ph
209	-	1981	1982	2.4	100	0	Safe	ph
210	-	1971	1972	2.5	125	0	Safe	ph

(continued)

Table 1. Continued

Seam	Longwall	Start of extr.	End of extr.	Seam thickness [m]	Depth [m]	Dist. from the object [km]	Direct. relative to object	Mining system
214	1	01-01-1991	30-09-1991	3.1	196	0.1	E	ph
214	2	01-07-1991	31-03-1992	3.4	196	0	Safe	ph
214	3	01-01-1992	31-03-1994	3.1	194	0.03	W	ph

Explanations: ph – hydraulic backfill.

3.4 *Hydrogeological conditions*

Since December 2000, following the decommissioning of the mine, a flooding process has been underway, resulting in a systematic rise in the water table within the goaf areas. According to hydrogeological documentation, since the second half of 2022, there has been a change in the groundwater level of the Quaternary aquifer in the area surrounding the closest piezometer (Piezometer A) relative to the building. Until June 2022, this piezometer remained dry. Monthly monitoring of the impact of the mine flooding recorded significant fluctuations in the groundwater table, with monthly rises reaching several tens of centimeters. As of the July 2024 measurement, the water table at the piezometer location was recorded at +308.48 meters above sea level. A graph illustrating the changes in groundwater level over time at the piezometer location is presented in Figure 4.

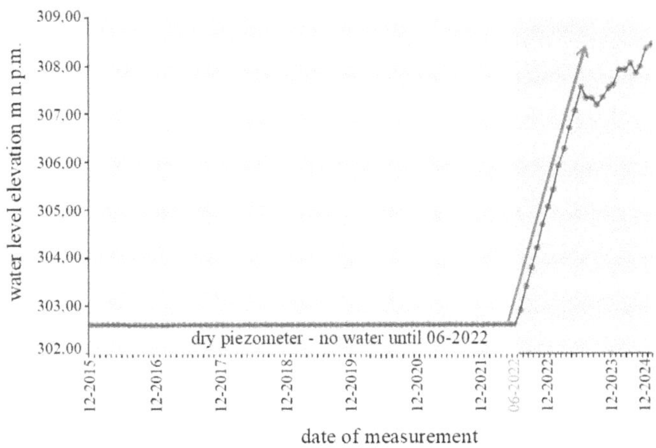

Figure 4. Graph of groundwater level elevation changes over time at the piezometer location.

Geotechnical investigations conducted in May 2023 indicated a change in soil moisture to a wet condition at a depth of approximately 8.5 meters below ground level, corresponding to an elevation of +309.8 meters above sea level. This observation suggests the presence of a free groundwater table at a depth ranging from approximately 9.5 meters (+308.8 m a.s.l.) to 10.5 meters (+307.8 m a.s.l.). These conclusions are consistent with the measurements obtained from the piezometer.

3.5 *Discontinuous deformations*

According to geological and mining documentation, the shallowest mining operations in the vicinity of the property were conducted in seam 208, at a depth of approximately 68 m beneath the building. At a depth of about 100 m, workings in seam 209 are also present; however, a preliminary assessment indicates a very low probability of sinkhole formation in this area due to the depth of the void.

In engineering practice, the boundary for shallow mining workings relevant to the risk of discontinuous deformations is assumed at a depth of H = 120 m. This value includes a safety margin, since, based on previous research and operational experience in the Upper Silesian Coal Basin, the threshold depth at which sinkhole hazards associated with shallow workings are typically considered is H = 100 m.

Figure 5 presents a fragment of the map of seam 208 with the location of the workings relative to the property. The analysis indicates that:

- within the property boundaries in seam 208, two types of workings occur: goaf areas backfilled with hydraulic stowing (marked in purple) and roadway workings (a dip gallery, a diagonal gallery, and their intersection),
- while the method of backfilling the goaf areas is documented, no information is available regarding the backfilling of roadway workings, which suggests that voids may still be present at the full height of these workings.

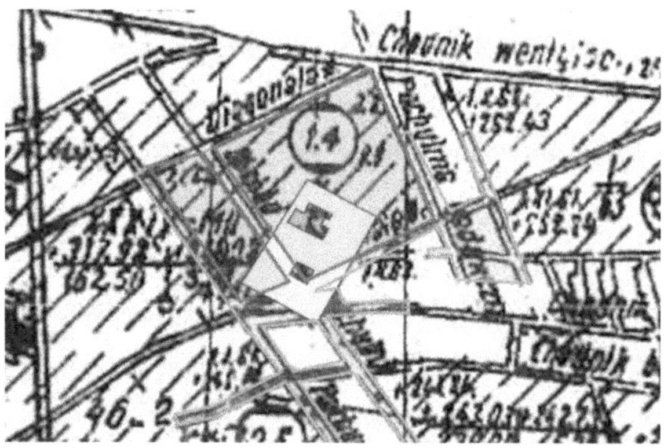

Figure 5. Section of the seam 208 map – blue indicates corridor workings; purple indicates post-mining goafs.

4 IMPACT OF COMPLETED MINING OPERATIONS ON THE SURFACE

4.1 *Geodetic measurements*

Geodetic measurements were carried out in the analyzed area to monitor changes in elevation coordinates within the post-mining terrain of the decommissioned mine. Observations were conducted over two time intervals: a quarterly period (June 2024 – September 2024) and an overall period (since 2008).

Measurements were taken, among other locations, at the site of the property, using two wall benchmarks: the first, designated as point A, was stabilized on a residential building, while the second, point B, was installed on a garage wall. The results of subsidence measurements for the observation points are shown in Figure 6.

Analysis of the measurement results allows for the formulation of the following conclusions:

- At the surface in the location of the property, during the period from September 2021 to September 2023, both positive and negative values of terrain subsidence were recorded. Such deformations are not characteristic of continuous subsidence. Their causes may be attributed to rising groundwater levels and variations in soil moisture, potentially linked to the flooding process of the decommissioned mine workings.

- During the analyzed period, the greatest increases in subsidence were observed between Q3 and Q4 of 2021 — 11 mm for point B and 10 mm for point A — as well as between Q1 and Q2 of 2022, when the values were -8 mm for point A and -7 mm for point B.
- From September 2023 to September 2024, the recorded subsidence values for the measurement points were either zero or within the margin of measurement accuracy, and therefore should be considered negligible.

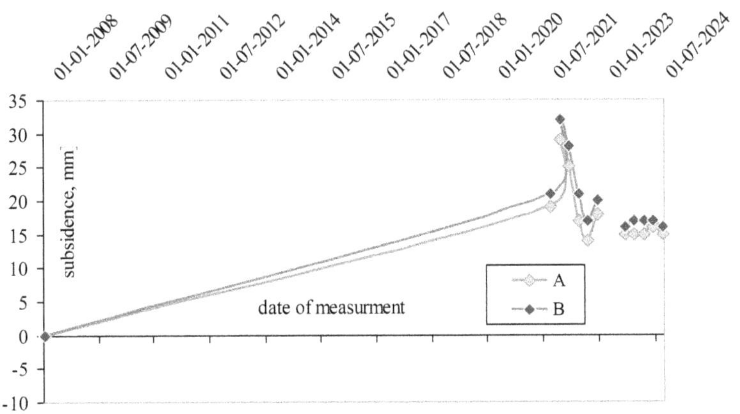

Figure 6. Time-dependent subsidence of monitoring points.

4.2 *Result of calculations*

Continuous Deformations

The calculations were carried out using the software described by Ścigała (2012), based on the theory of W. Budryk and S. Knothe (Knothe 1984). The simulation of the longwall panel progression was performed using a computational model, assuming instantaneous influence manifestation ($c \to \infty$) for the final deformation state. Only mining activity conducted after the construction of the structure was considered.

The following parameter values from the theory were adopted for the calculations: roof control coefficient: a = 0.8, tangent of the main influence angle: tg b = 2.0, mining boundary offset: d = 0 m and proportionality coefficient between horizontal displacements and ground slope: B. Following E. Popiołek and J. Ostrowski (Popiołek 2009), B = 0.32 r, where r is the radius of the main influence range. Additionally, a deviation coefficient of n = 0.7 was adopted. The results of the calculations are as follows:

- Subsidence: w = –663.3 mm
- Maximum tilt: Tmax = 0.2 mm/m
- Maximum horizontal strain: εmax = –0.6 mm/m (compression)

It was also determined that the last mining operation affecting the surface in the vicinity of the property was Longwall 3 in seam 214. The duration of the final phase of rock mass movements was estimated based on the findings presented in Strzałkowski (2021), and it was assumed to last approximately 3 months. Since the panel face was located within the vicinity in March 1994, mining-induced influences ceased around July 1994. No further mining operations causing continuous deformations have been conducted in the area since that time.

Sinkhole hazard assessment

- M. Chudek – J. Olaszowski Method (Chudek *et al.* 1988) According to the adopted method, the formation of a sinkhole requires that the collapse zone above a void reaches the layers of the rock mass composed of unconsolidated rocks. The maximum height of this collapse zone is defined by the following relationship:

$$w_z = w\frac{4(k_r + 1) - \pi(k_r - 1)}{2\pi(k_r - 1)} \tag{1}$$

where: k_r – rock loosening coefficient (for the calculations, k = 1.1 was assumed for sandstone, based on the results of laboratory tests conducted by Chudek and Staroń (1969), w – height of the original void (workings), in meters.

In the analyzed case, assuming a void height of w = 1.4 m for extraction workings and w = 3.0 m for corridor workings, the height of the collapse zone, calculated using formula (1), amounted to w_z = *12 m* for extraction workings and w_z = 38 m for corridor workings.

Taking into account the geological structure of the rock mass, in which the overburden thickness is h = *25 m*, it should be noted that in both cases, the collapse zone did not reach the overburden layers. As a result, the necessary condition for the formation of a sinkhole was not met.

Another component of sinkhole hazard assessment in the analyzed method involves determining the *Z-index* (Chudek *et al.* 1988) and the probability of its occurrence P (Chudek 2010). These values are calculated based on the following formula:

$$Z = \frac{H - h}{w} \tag{2}$$

$$P(z) = 1,34 - 0,036 \cdot Z + 0,00019 \cdot Z^2 \tag{3}$$

where: H – depth to the roof of the working, in meters; h – thickness of the overburden, in meters; w – height of the working, in meters.

According to formula (2), the value of the *Z-index* was 30.1 for extraction workings and 14.3 for corridor workings. This corresponds to a sinkhole formation probability of P = 0.42 for extraction workings (3) and P = 0.86 for corridor workings.

Based on the obtained results, the area where the property is located was classified as Category *C* for extraction workings and Category *B* for corridor workings — with respect to the degree of sinkhole hazard. As a result, the sinkhole risk level is assessed as low for extraction workings and high for corridor workings (Chudek *et al.* 1988).

- A. Kotyrba Method (Kotyrba 2005) The assessment of the sinkhole hazard level can also be carried out by determining the so-called *void migration potential index* (Kotyrba 2005), defined as the ratio of the depth to the roof of the void (H) to its height (w). According to the sinkhole hazard scale adopted for the Upper Silesian Coal Basin (Kotyrba 2005), the index is interpreted as follows:

- below 10 – indicates a high hazard level,
- from 10 to 20 – indicates a medium hazard level,
- above 20 – indicates a low hazard level.

For the analyzed geological and mining conditions in the area of the property, the sinkhole hazard level was assessed as low. The calculated values of the index were 22.6 for extraction workings and 48.5 for corridor workings.

5 POST-MINING AREA CLASSIFICATION

Based on the identified hazard of sinkhole occurrence associated with former mining operations in seam 208, and the absence of documented cases of discontinuous deformations in the past within the vicinity of the property, the area at the location of the assessed property should be classified as category B_{21} (Strzałkowski *et al.* 2019), defined as a post-mining transformed area with a low risk of discontinuous deformations.

However, considering the geological structure of the overburden, which is dominated by sands—soils susceptible to suffosion processes that, under variable hydrogeological conditions, may migrate toward mining voids—it is justified to increase the hazard level and classify the area as category B_{23}, i.e. a post-mining transformed area with a high risk of discontinuous deformations.

The validity of this classification is further supported by the results of geodetic measurements carried out at the property site, which demonstrated that between September 2021 and September 2023 the area had not reached a state of stabilization.

In accordance with current guidelines and recommendations for the spatial development of post-mining areas classified as category B23, in order to reduce the risk of further surface deformations and damage, it is advisable to conduct geophysical surveys to assess the condition of the rock mass. Any anomalies identified should subsequently be verified through exploratory boreholes.

6 CONCLUSIONS

The article presents an analysis of hazards occurring in post-mining areas, using the example of a property located in the vicinity of a decommissioned coal mine. Particular attention was paid to the influence of existing mine workings on surface stability, as well as to the geomechanical processes occurring within the rock mass due to the long-term flooding of underground voids. Calculations and an assessment of the current condition of the area were conducted based on geodetic survey data, along with an analysis of the geological structure of the overburden. The combination of these factors with changing hydrogeological conditions increases the risk of surface deformations. Based on the collected data, analyses, and calculations, the following conclusions were drawn:

(1) The analyzed property is located above shallow mine workings. This, combined with the ongoing process of mine flooding and rising water levels within the rock mass, creates a real risk of reactivation of post-mining voids and the formation of sinkholes.
(2) The overburden is primarily composed of loose sands. These, together with variable hydrogeological conditions, promote suffosion processes and the migration of soil toward old mine workings, which may lead to localized surface deformations.
(3) Based on the conducted calculations and survey results, the area should be classified as Category B_{23}, which indicates a high post-mining hazard requiring special caution in future land use planning.
(4) To ensure safety and minimize the risk of discontinuous deformations, it is essential to carry out detailed geophysical investigations and exploratory drilling in order to assess the condition of the rock mass and identify any voids or anomalies.
(5) Similar post-mining areas, even in the absence of visible surface deformation, should be regarded as potentially hazardous. A lack of systematic monitoring and subsurface assessment may lead to sudden, unpredictable deformations that pose a threat to the safety of people, buildings, and infrastructure.

In the context of residual post-mining hazards and dynamically changing hydrogeological conditions, neglecting monitoring and assessment of the rock mass may result in serious risks to public safety and private property. Therefore, before any planned development in such areas, detailed geotechnical and engineering investigations are necessary, along with the implementation of land use restrictions in accordance with the identified post-mining hazard category. It is also recommended to maintain ongoing surface monitoring and conduct regular geodetic and hydrogeological surveys, which can serve as an effective early warning system for potentially dangerous events.

REFERENCES

Baryakh A.A. (2011). Fedoseev A.K.: Sinkhole formation mechanism. *Journal of Mining Science*, Vol. 47, 2011/4.

Blachowski J. and Walerysiak, N. (2025). Geodatabase of Sinkholes in the Post Mining Area of the Brown Coal Mine "Babina" (W Poland). EGU General Assembly 2025, Abstract EGU25 18583.

Chudek M. (2010). Rock mass mechanics with basics of environment management in mining and post-mining areas. Wydawnictwo Politechniki Śląskiej, Gliwice.

Chudek, M., Janusz, W. and Zych, J. (1988). Studium dotyczące stanu rozpoznania tworzenia się i prognozowania deformacji nieciągłych pod wpływem podziemnej eksploatacji złóż. Zeszyty Naukowe Politechniki Śląskiej, seria Górnictwo;141, Gliwice.

Chudek M. and Staroń T. (1969). Współczynnik rozluzowania skał towrzyszących pokładom węgla w świetle badań laboratoryjnych. Zeszyty Naukowe Politechniki Ślaskiej. Seria Górnictwo z. 36. Gliwice.

Fajklewicz Z., Piwowarski W., Radomiński J., Stewarski E. and Tajduś A. (2004). *Badanie deformacji w górotworze w celu odtwarzania wartości budowlanej terenów pogórniczych*. Wyd. AGH, Kraków.

Gutierrez F., *et al.* (2008). A review on sinkhole hazards in Spain: distribution, causes and preventive measures, *Geomorphology*, 108(3–4), 172–190.

Knothe St. (1984). *Prognozowanie wpływów eksploatacji górniczej*. Wydawnictwo "Śląsk", Katowice.

Kotyrba A. (2005). Zagrożenie i ryzyko zapadliskowe terenów Górnośląskiego Zagłębia Węglowego. *Wiadomości Górnicze.*

Kowalski A. (2015). Deformacje powierzchni w Górnośląskim Zagłębiu Węglowym. Wydawnictwo GIG, Katowice.

Lei M., *et al.* (2022). Mapping and monitoring sinkholes in China: an overview, *Engineering Geology*, 301, 106608.

Paris R. (2020). Sinkhole disasters: a global overview of causes and consequences, *Natural Hazards*, 100(3), 1109–1129.

Popiołek E. (2009). *Ochrona terenów górniczych*. Kraków, Wyd. AGH.

Praca zbiorowa (red.: Popiołek E., Pilecki Z.) (2005). Ocena przydatności do zabudowy terenów zagrożonych deformacjami nieciągłymi za pomocą metod geofizycznych. Instytut Gospodarki Surowcami Mineralnymi i Energią PAN. Studia, rozprawy monografie. 184. Kraków.

Ścigała R. (2013). The identification of parameters of theories used for prognoses of post mining deformations by means of present software. *Archives of Mining Sciences*, Polska Akademia Nauk - Instytut Mechaniki Gorotworu, vol. 58, nr 4, p.. 1347–1357, DOI:10.2478/amsc-2013-0093.

Strzałkowski P. and Litwa P. (2021). Environmental protection problems in the areas of former mines with emphasis on sinkholes: selected examples, *International Journal of Environmental Science and Technology*, vol. 18, nr 3, p. 771–780, DOI:10.1007/s13762-020-02860-4.

Strzałkowski P., Ścigała R. and Szafulera K. (2019). Categories of mining areas of closed mines due to limitations of land use for building purposes on selected examples, In: 5th International Scientific Conference on Civil Engineering-Infrastructure-Mining, Kraków, Poland, January 17–18, 2019 [online] / Pilecka E., Tatara T., Kogut J. P. (red.), E3S Web of Conferences, vol. 106, EDP Sciences, (plik pdf) 1–8, DOI:10.1051/e3sconf/201910601010.

Strzałkowski P. (2021). Duration of the final phase of mining area deformation process in the conditions of Upper Silesia (Poland), *Bulletin of Engineering Geology and the Environment*, vol. 80, p. 5769–5780, DOI:10.1007/s10064-021-02272-9.

Strzałkowski P. (2019). Sinkhole formation hazard assessment, *Environmental Earth Sciences*, Springer Verlag, vol. 78, numberr 1, Article number: 9, p. 1–6, DOI:10.1007/s12665-018-8002-5.

Strzałkowski P. (2021). The influence of selected mining and natural factors on the sinkhole creation hazard based on the case study, *Environmental Earth Sciences*, Springer Verlag, vol. 80, number 3, Article number: 117, p. 1–12, DOI:10.1007/s12665-021-09403-1.

Tihansky A.B. Sinkholes, West-Central Florida. A link between surface water and ground water. https://pubs.usgs.gov/circ/circ1182/pdf/15WCFlorida.pdf.

Waltham T., Bell F. and Culshaw M. (2005). *Sinkholes and Subsidence: Karst and Cavernous Rocks in Engineering and Construction*, Springer, Dordrecht.

Mine Surveying and Mining Area Protection – Sokoła-Szewioła and Jarczyk (Eds)
© 2025 The Author(s), ISBN: 978-1-041-10531-2

Integrated diagnostics of a Heritage Palace: A multi-modal sensor fusion approach with machine learning for structural assessment

K. Karsznia & K. Książek
Warsaw University of Technology, Poland

ABSTRACT: This paper presents an intelligent diagnostics framework for a heritage palace, focusing on a multi-modal sensor fusion approach to ensure cyber-secure structural assessment. The article proposes a diagnostic model that is a key part of the facility information audit that combines structural monitoring data based on the surveying and geotechnical readings captured by the geodetic and fiber-optic sensors, all managed by a robust IT infrastructure. The core of this framework lies in the application of machine learning algorithms to process and interpret the vast, multi-source dataset. This approach enables advanced anomaly detection and predictive analysis, moving beyond simple data collection to create a comprehensive knowledge base of the palace's structural health. Furthermore, the authors address the critical issue of cybersecurity, analyzing potential threats to data integrity and system reliability within the monitoring network. By integrating technical diagnostics with cyber protection, this study highlights the need for a holistic, secure approach to preserving cultural heritage in the digital age.

1 INTRODUCTION

1.1 *Problem definition*

The diagnostic assessment of heritage buildings is a cornerstone of national heritage preservation. This is particularly vital for structures located in high-risk areas, such as post-mining sites, densely urbanized environments, or zones exposed to natural and anthropogenic phenomena. Due to their historical significance, diagnostic studies must be as non-invasive as possible. This has traditionally led to a reliance on remote, non-contact geodetic methods like photogrammetry and laser scanning to assess geometric deformations.

However, a comprehensive understanding of structural threats demands more than just geometric data. Modern monitoring systems integrate dedicated physical instruments, such as geotechnical and fiber-optic sensors, to measure specific parameters like strain, displacement, and tilt. This creates a multi-modal dataset that offers a holistic view of a structure's health. The rapid advancement of these technologies, combined with the proliferation of the Internet of Things (IoT), has led to the widespread use of sophisticated IT infrastructure for real-time data acquisition, transmission, and archival (Lorusso and Guida 2022). This digital transformation allows for the creation of rich knowledge bases, which are essential for long-term structural health monitoring and predictive maintenance.

The integration of such diverse datasets, from both remote and physical sensors, presents a unique challenge and opportunity. The sheer volume and complexity of the data make traditional analytical methods insufficient. This has spurred the adoption of advanced computational techniques, particularly machine learning algorithms. By training models on historical and real-time data, we can identify complex patterns, detect subtle anomalies, and predict future structural behavior with a level of accuracy previously unattainable. These

DOI: 10.1201/9781003655558-10

models can also help quantify the often-elusive measurement uncertainty, providing a more robust basis for decision-making.

Despite technological progress, the diagnostic process remains complex. It requires not only a command of technical skills but also critical thinking and sound judgment. Furthermore, the increasing reliance on interconnected digital systems introduces significant cybersecurity risks. As monitoring becomes more automated and data centralized, the systems become vulnerable to attacks that could compromise data integrity, leading to flawed diagnoses and potentially jeopardizing the safety of heritage structures. This critical aspect of cybersecurity in heritage monitoring is an underexplored area in scientific literature, warranting detailed analysis. Regardless of data security aspects, a key challenge is the proper design of a monitoring system based on the integration of data from multiple sources. This requires the collaboration of various measurement sensors, their integration with classic geodetic methods, and the inclusion of archival, historical, and spatial planning data, as well as information from other available resources.

This approach goes beyond a purely geodetic view, which often reduces the problem to solving a specific set of equations to determine the displacement of individual control points, a process known as adjustment computations. While this method is fundamental to geodetic data processing, it becomes insufficient when dealing with multi-source and diverse data (both quantitative and qualitative).

Therefore, the problem demands the development of more complex methodologies. A significant aid here is the application of artificial intelligence, specifically machine learning, which enables the construction of an integrated knowledge base about the object. Such a cloud-based instance allows for advanced analysis, interpretation, and drawing conclusions from complex relationships.

This perspective, combined with the dynamic progress in these methods worldwide, prompted the authors to develop a methodological framework for two heritage buildings that are a core part of the Coal Basin Museum in Będzin, Poland. Although these objects differ in character and historical era, the implemented solutions allowed for the creation of a sophisticated knowledge base about their technical condition, enabling in-depth interpretative analyses.

1.2 *Reference to the subject literature*

Studying the technical condition of historical buildings is a key aspect of Cultural Heritage. For many years, efforts have focused on combining achievements from technical sciences with historical and conservation knowledge. In this context, monitoring systems that integrate multi-source measurement data play a crucial role in assessing the technical condition of structures (Markiewicz *et al.* 2019; Thirugnanam *et al.* 2022). Research on monitoring systems for engineering objects is a significant branch of science and technology within civil engineering, surveying, and related disciplines. Both conceptual work and specific implementations have been described in numerous publications. It is particularly worth highlighting the development of monitoring systems, especially in modern sensors and data communication technologies (Wielandt 2023).

Most published research (Bassier *et al.* 2018; Chetverikov *et al.* 2024; Lasaponara and Masini 2011) focuses on remote methods of spatial data capture, often without considering metrological (geotechnical) sensors. While these methods allow for detecting external changes in historical structures, a more integral approach requires exploring the entire object. Such an integrated approach necessitates the introduction of multiple additional technologies and, importantly, their integration into a coherent, database-driven examination system. Therefore, it is essential to consider diverse and direct measurement methods. Wireless systems, particularly those based on the Internet of Things (IoT), are becoming increasingly important (Thirugnanam *et al.* 2022). The widespread availability of these modern solutions also makes low-cost technologies noteworthy (Ćmielewski *et al.* 2020;

Lăpădat *et al.* 2021; Wielandt *et al.* 2023). Their continued development enables broader application across various objects, potentially reducing costs.

The integration of measurements can be achieved in multiple ways, from establishing analytical functions to using specific data for advanced numerical methods or those based on Artificial Intelligence (AI) and its subfield, Machine Learning (ML). These solutions are now the backbone of contemporary scientific research, given their ability to process large datasets quickly and reliably and to formulate well-founded conclusions (Aslanyan 2023).

Studying the technical condition of structures, especially those located in high-risk areas, requires a wide range of advanced techniques for spatial data acquisition, processing, and analysis (Karsznia *et al.* 2023; Osada *et al.* 2023). While a vast body of academic literature is available, it does not fully exhaust the topic. In addition to scientific publications, there is an extensive, yet difficult to estimate, volume of technical reports, industry studies, and expert opinions that are not easily accessible. It is also essential to consider the unique nature of virtually every structure under investigation. Even if some characteristics are repeatable, they differ in the specific factors that determine the behavior of the structure. Therefore, each object must be approached individually, referencing the most current state of inter-disciplinary knowledge on assessing the condition of civil engineering structures (Bednarz 2023).

1.3 *Research site*

The research area is located in Będzin, in the Silesian Voivodeship, home to the Coal Basin Museum, which operates under the auspices of the Będzin City Council. The subject of this study is one of the museum's structures: the Mieroszewski Palace, whose history dates back to 1702 (https://muzeumzaglebia.pl/pałac-mieroszewskich - accessed on 7.09.2025). For many years, this structure has been the subject of research conducted by scientists from the Warsaw University of Technology as part of a cooperation agreement.

Będzin is a city with a rich history, and its unique landmarks are a significant historical, cultural, and tourist attraction. The location of the research site within the country is shown in Figure 1.

Figure 1. Location of the research site – Będzin, Silesian Voivodeship (geographical coordinates: 50° 19' 09" N; 19° 08' 04" E) – indication in the contour map generated by ChatGPT.

More information about the history and socio-economic conditions of the region can be obtained from the website of the Będzin District (https://powiat.bedzin.pl/powiat/uklad-urbanistyczny-miasta-wraz-ze-wzgorzem-zamkowym - accessed on 7.09.2025).

The research area is characterized by a distinctive geological structure. Situated in the northeastern part of the Upper Silesian Coal Basin, its geology is dominated by Carboniferous deposits. Specifically, the geological sequence consists of interbedded layers of sandstones, mudstones, claystones, and hard coal seams. Older Devonian and

Carboniferous formations are present in the substratum. At the same time, the upper layers are composed of younger Neogene and Quaternary deposits, including sands, clays, and silts, often of glacial origin (Antoszczyszyn-Szpicka and Anotszczyszyn 2016).

This geological diversity contributes to ground instability, particularly in areas affected by historical mining activity, which necessitates continuous geotechnical monitoring. The extensive mining history in this region has left behind numerous traces that directly impact current ground conditions and building stability. Moreover, the study area is subject to ongoing urban development, involving intensive reconstruction of the road network and changes in land use (Local Spatial Development Plan of the City of Będzin 2018). Nevertheless, as stated in this local legal document, no evidence of landslide phenomena or areas at risk of mass movements was found in the analyzed area. Consequently, the observed problems, such as geometric changes in buildings—especially historical ones—are more closely related to geotechnical and structural factors. Detailed information on these specific conditions and an analysis of case studies are presented in the publication (Oleniacz *et al.* 2023).

Based on the above considerations, the authors aim to contribute to the existing state-of-the-art by integrating multisource surveys and standardizing them within one coherent information system and applying ML inference techniques to establish potential warning models useful in structural monitoring. Hence, the paper describes the investigation of applying ML to diagnose the condition of historic structures and develop a reliable monitoring system capable of integrating diverse measurement data. The study structure was chosen due to its location in a high-risk area. Regarding that, numerous factors influencing stability can be observed, including:

- post-mining activity: these highly urbanized and industrialized areas have undergone significant transformations over time,
- climate change: dynamic shifts in temperature and water relations may cause unpredictable phenomena,
- combined effects: these factors accelerate geotechnical and rheological processes, leading to unforeseen consequences when combined with the aging of historic buildings.

For the purposes of this article, the focus is on the main building of the Coal Basin Museum: the aforementioned Mieroszewski Palace, which has undergone various modifications and reconstructions since its construction (Figure 2).

Figure 2. View of the Mieroszewski Palace in Będzin, Poland.

The two palace bastions, visible as annexes on the left and right, were added more than 150 years after the main body of the building was erected (www.muzeumzaglebia.pl). Due to its location on a slope (Figure 3), the left bastion is subject to increased displacement, which has led to the formation of cracks and dilatations both inside and outside the building (Figure 4). The authors have been conducting condition monitoring of the building since 2022 (Świerczyńska *et al.* 2024). The two-and-a-half-year observation period has allowed for drawing preliminary conclusions regarding the behavior of the historic structure under investigation.

Figure 3. Lateral view of the palace with the visible slope in the foreground.

Figure 4. Visible damage to the façade and walls of the palace.

To perform this task, the authors use existing Machine Learning (ML)-based solutions that adapt to the specific conditions and environment. These studies are unique because they are applied to structures not previously monitored, apart from sporadic measurements taken many years ago. The collaboration established between the Warsaw University of Technology and the Coal Basin Museum has created an opportunity to apply scientific advancements to the study of the technical condition of accessible historic buildings, including both their interiors and exteriors.

2 MATERIALS AND METHODS

2.1 *Integrated structural monitoring*

Site monitoring was designed to consider the multi-scale and multi-source nature of the data to be acquired. Geodetic methods were used to observe geometric changes in the palace structure:

- precise leveling of benchmarks (performed quarterly or after a significant reason) with an accuracy of ±0.3 mm/km,
- total-station measurements with the use of an electronic total station (two campaigns so far), with an accuracy of ±1.0 mm,
- laser scanning of the palace interiors and façade (two campaigns so far) with an accuracy of ±2-3 mm, as well as geotechnical surveys:
- crack measurements using crack gauges (nine campaigns so far) with an accuracy of approximately ±0.1 mm and
- continuous structural monitoring of the left bastion of the Mieroszewski Palace using the Axis precision hydrostatic inclinometer (Ornoch *et al.* 2021) with an accuracy of approximately ±5 μm/1m.

The data being acquired, therefore, has different characteristics, precision, accuracy, and frequency of sensor readings. Consequently, building a coherent system based on them is a real challenge. It should be added that, except for these values, the temperature and humidity of the palace interiors have also been measured. Undoubtedly, it is possible to conduct a comprehensive analysis by assessing the individual results separately and applying expert knowledge. Nevertheless, the task of the monitoring system is to warn users of possible undesirable events on an ongoing basis, which would be impossible in this approach.

A fundamental problem is, therefore, the consistency of the data and the frequency with which it is collected. The answer to such a problem may be linked to using various learning models.

2.2 *Facility information audit*

Based on the work carried out, a facility information audit plan was developed. This plan forms the core of the expert system, which is supported by ML solutions. The system represents an original, proprietary solution developed within the project, dedicated to solving a specific scientific problem. The presented workflow is called an information audit. Its flowchart is presented in Figure 5. To perform the relevant input, the pertinent data was prepared and grouped according to the structure shown in Table 1.

Table 1. Summary of sensors providing input data.

Sensor	Degrees of freedom	Description
Crack meters	2	Linear values 2D
Hydrostatic inclinometer	3	Angular values, temperature, vessel heights
Leveling	1	precise leveling of benchmarks
total station	3	measuring control points
TLS	3	point clouds (3D, reflectivity, RGB)
UAV photogrammetry	3	point clouds, RGB
Meteo data	1	temperature, moisture (indoor, outdoor)

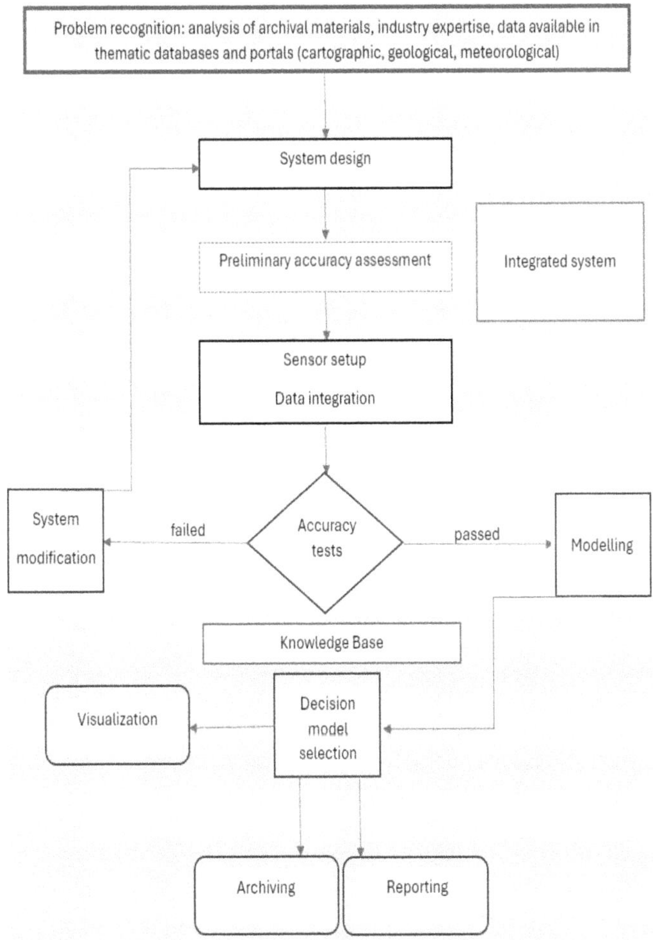

Figure 5. Conceptual scheme of the designed expert system (K. Karsznia).

The variety of data acquired necessitates its integration and modeling. To this point, ML algorithms were used.

2.3 *ML-driven modelling*

The research was divided into two stages - aimed at detecting displacements inside the building (prediction of displacements based on cracks in the museum rooms) and outside (identification of cracks in the palace façade). The diversity of the measurement data determines the need for proper integration and inference. To achieve this goal, various machine learning (ML) models were explored. The calculations were executed in RapidMiner Studio v. 10.3 (Rapid Miner 2025).

The modelling process used data from crack gauges stabilized on three floors of the palace in the location of the most extensive observed cracks, the results of leveling of control benchmarks located outside the building on the ground floor, as well as the results of continuous measurements with a hydrostatic inclinometer situated in the left bastion, in the attic (Figure 6).

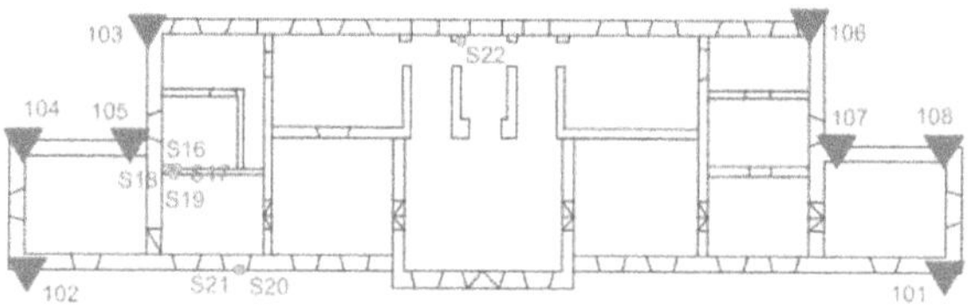

a) ground floor (crack meters marked red and leveling benchmarks marked green)

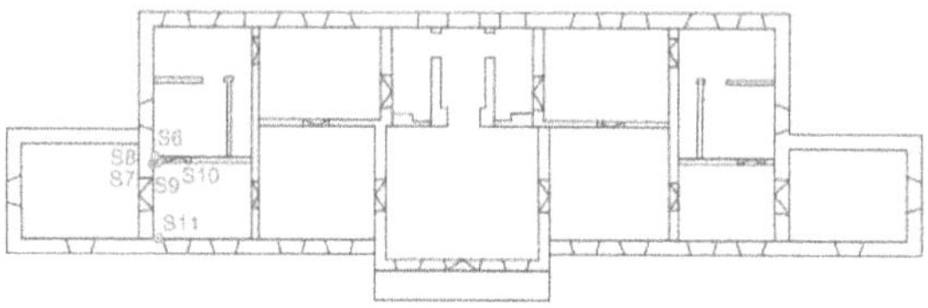

b) 1st floor (crack meters marked red)

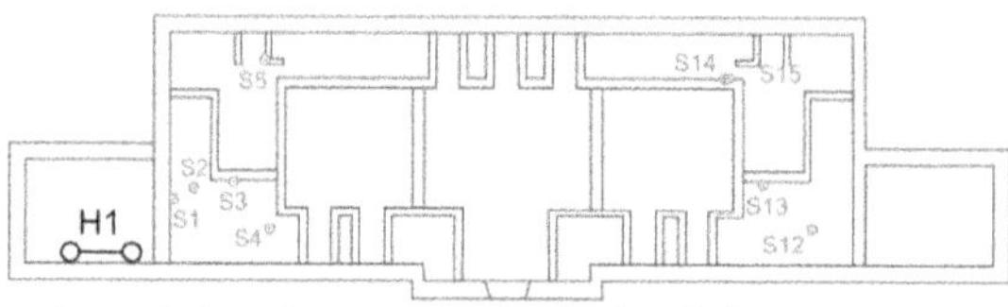

c) attic (crack meters marked red and tiltmeter H1 marked black)

Figure 6. Location plan of measuring sensors (a – ground floor, b – the first floor, and c – the attic) – (E. Świerczyńska).

a) ground floor (crack meters marked red and leveling benchmarks marked green)
b) 1st floor (crack meters marked red)
c) attic (crack meters marked red and tiltmeter H1 marked black)

In addition, temperature and humidity readings from these locations were used in the computing process. These values were obtained from the ongoing meteorological monitoring conducted by the museum as part of its standard operations. A view of the measuring sensors is shown in Figures 7 and 8.

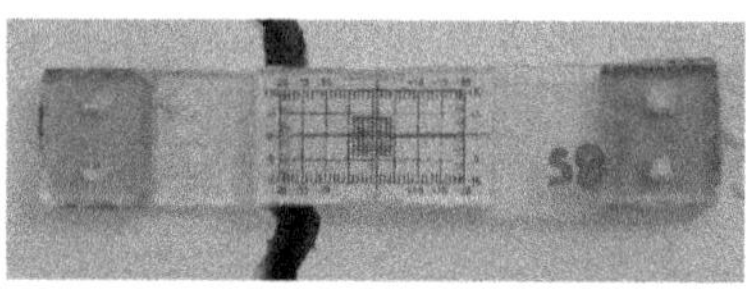

Figure 7. View of a crack meter installed inside the palace.

Figure 8. View of the hydrostatic tiltmeter installed in the attic.

Where accessible, classical crack meters were installed to provide precise geometric data. The reading precision for the simplest type of crack gauge is 0.1 mm. A hydrostatic inclinometer is a high-precision sensor that monitors changes in the movements of building segments relative to the facade. Although valuable, the data from geotechnical sensors is inherently localized, providing information only about the specific section of the structure where a sensor is installed. A unique methodology is required to accurately determine the displacement parameters of individual segments affected by cracks. One approach is to create a mathematical model of the structure, considering the separate segments formed by the cracks. This task can only be performed after thoroughly analyzing the building's structure and any material losses. Another key element is documenting all crack gauges' positions and spatial orientation.

An alternative solution can be proposed for situations where collecting all necessary information is difficult or impossible. This approach uses an AI-based model capable of establishing a relationship between a set of local sensor readings and the overall behavior of the object. The measurement results serve as control data that can be used to train the model.

The data was processed using three ML variants: two based on neural networks and one based on a decision tree (DT). The first variant uses an artificial neural network (ANN), specifically a feed-forward neural network trained by a backpropagation algorithm. The second variant relies on deep learning (DL) and consists of a multi-layer feed-forward artificial neural network trained with stochastic gradient descent using backpropagation (RapidMiner 2025). The procedure workflow is shown in Figure 9. It corresponds to the configuration of the project in the program and contains three basic task blocks: 1) preparation of training data, 2) determination of training and modeling algorithms, and 3) definition of the process of validation of the obtained results.

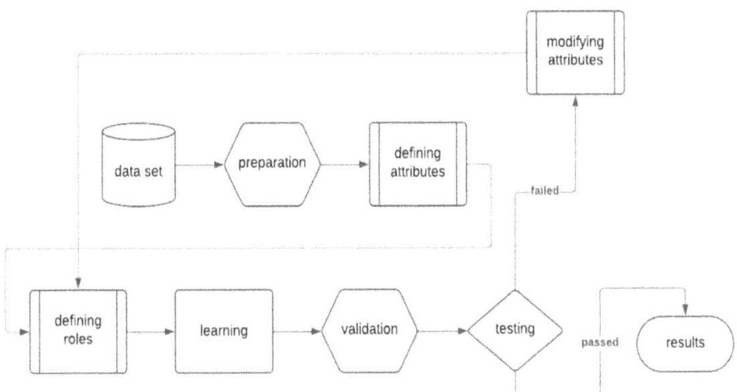

Figure 9. Workflow chart of the ML modeling process (K.Karsznia).

All processes used 33 features for which attributes were defined: 22 crack gauges, two hydrostatic inclinometer vessels, seven control benchmarks (due to destruction, benchmark 104 was eliminated from processing), and two developed fracture scans located on the ground and first floor. The data set comprised the difference between two survey periods - the reference epoch (the moment of installation of the hydrostatic inclinometer in May 2023) and the most recent survey epoch of December 2024. The attributes of the above-mentioned features were the differential values of the 2D displacement vector, the increase in the attic tilt angle, the change in the height of the leveling benchmark characteristic for a specific location of the crack meters (a particular wing of the palace), the difference in temperature and humidity inside the building (different for individual stories), as well as the average width of the cracks identified from the scan. As stated, the calculations were done for three variants: 1) using ANN, 2) DL, and 3) DT. A diagram of the modeling process is shown in Figure 10.

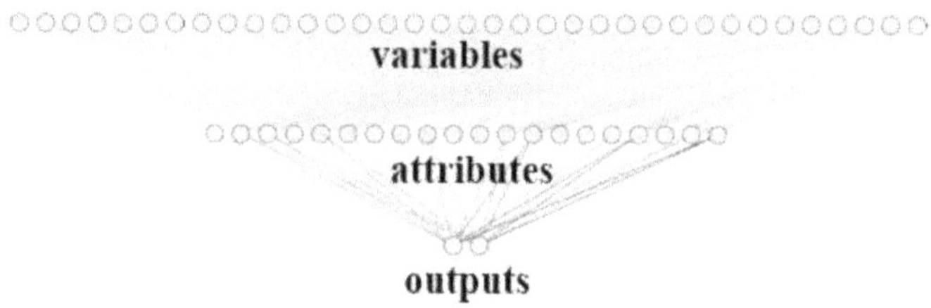

Figure 10. Diagram of the modeling process using neural networks (Rapid Miner).

As a result of the tests, the system identified the attribute relating to the crack gauge readings (V0) as the key variable in determining the deformation that occurred. A graph showing the deformation qualification is presented in Figure 11 a.

(a) graph of linear displacement validation generated by the modeling process (V0 - crack gauge readings, 1 - displacement, 0 - no displacement)

(b) angular displacement validation graph generated by the AI modeling process (dAlfa - angle of inclination, 1 - displacement, 0 - no displacement)

Figure 11. Displacement graphs obtained within the modelling process (Rapid Miner).

The resulting displacement qualification was 0.281 mm ($\sim$ 0.3 mm) for the 2D vector. The resulting accuracy rate was 100%, which, according to the authors, is indicative of the relatively small set of data analyzed. This set should, therefore, be successively supplemented, and, as far as possible, the number of variables should be expanded by adding more measurement points, especially at newly appearing cracks in the rooms of the object. In the next step, the change in the angle of inclination of the structure (dAlpha) of the left palace bastion should is considered (Figure 11 b).

The given limit for the difference in the angle of inclination (Alpha) is 45.213 seconds, which, according to the manufacturer, corresponds to a displacement of 252 micrometers per one meter ($1'' = 4.85\ \mu m/m$) (https://ultraxis.eu – accessed on 7.09.2025). This means that the eligibility limit for displacement is 0.25 mm for each running meter of the structure tested (2.5 mm for 10 m). In this case, however, we are dealing with a relatively low level of accuracy of the prediction made (66.7%). It demonstrates the high variability of the monitored processes and the excessive resolution of the device's operation. Definitely, this implies the need for further measurements and the expansion of the test network at the site.

3 CONCLUSIONS

Developing a monitoring system for engineering structures, especially historic buildings, demands a comprehensive, multi-modal, and integrated approach to both measurement and analysis. A critical aspect indicative of the solution's scientific rigor is the thoughtful definition of safety thresholds that trigger appropriate notifications. In the context of monuments, measurement, and condition monitoring systems must prioritize non-invasive technologies. This does not preclude using traditional sensors in critical locations where structural integrity is at risk and stabilization measures are necessary. A system designed this way leverages multi-source data with varying characteristics, accuracy, and frequency. The processing of such data sets and their integration with TLS point clouds or photogrammetric imagery is not easy and leaves a large margin for interpretation uncertainty. AI and ML solutions come to the rescue, allowing multithreaded and consistent data processing and reliable definitions of alarm thresholds. In this paper, the concept of such integration and, as a result, building a coherent expert system for effective risk management is presented. This is particularly important in the case of monuments, for which the design of reliable diagnostic systems is currently one of the key research trends in heritage safety.

ACKNOWLEDGMENTS

This article summarizes studies conducted as part of scientific projects funded by the Scientific Council of the Discipline of Civil Engineering, Geodesy, and Transport at the Warsaw University of Technology (project titled: "Evaluation and modeling of geometric changes in the structures of historical buildings located in high-risk areas" was governed by Dr. Krzysztof Karsznia). The authors wish to thank Prof. Izabela Karsznia from the University of Warsaw for her consultations and substantive support during the revision of this paper. Her expertise in applied Artificial Intelligence (AI) was invaluable.

REFERENCES

Antoszczyszyn-Szpicka, P. and Antoszczyszyn, T. (2016). Role of preliminary assessment of the suitability of building conditions for the spatial planning – example from Będzin city. *Prz. Geol.*, 64: 177–182.

Aslanyan T.K., The Machine Learning Fundamentals Handbook, freeCodeCamp Press. (2023). https://www.freecodecamp.org/news/machine-learning-handbook (accessed on 7.09.2025).

Bassier M., Vincke S., de Lima Hernandez R., Vergauwen M. (2018). An Overview of Innovative Heritage Deliverables Based on Remote Sensing Techniques, *Remote Sens.* 10:1607; DOI: 10.3390/rs10101607

Bednarz Ł. J. (2023). *Monitoring diagnostyczny obiektów historycznych*, Oficyna Wydawnicza ATUT – Wrocławskie Wydawnictwo Oświatowe, Wrocław: 17–27.

Chetverikov B., Hlotov V., Bakuła K. (2024). Clarification of the Boundaries of Lands of Historical and Cultural Heritage and Determination of Their Protection Zones by Remote Sensing Methods, 13: 923L; DOI: 10.3390/land13070923.

Ćmielewski K., Karsznia K., Kuchmister J., Gołuch P., Wilczyńska I. (2020). Accuracy and functional assessment of an original low-cost fibre-based inclinometer designed for structural monitoring. *Open Geosciences*, 12(1): 1052–1059.

Karsznia, K., Osada, E., Muszyński, Z. (2023), Real-Time Adjustment and Spatial Data Integration Algorithms Combining Total Station and GNSS Surveys with an Earth Gravity Model, *Appl. Sci.* 13: 9380. https://doi.org/10.3390/app13169380.

Lasaponara R., Masini N. (2011). Satellite remote sensing in archaeology: past, present and future perspectives, *Journal of Archaeological Science*, 38: 995–2002.

Lăpădat A.M., Tiberius C.C., Teunissen P.J. (2021). Experimental evaluation of smartphone accelerometer and low-cost dual frequency GNSS sensors for deformation monitoring, *Sensors*, 21(23): 7946.

Local Spatial Development Plan of the City of Będzin for the area located in the Śródmieście District, ["Miejscowy Plan Zagospodarowania Przestrzennego Miasta Będzina dla terenu położonego w dzielnicy Śródmieście"]. (2018). https://www.bedzin.pl/strona-761-konsultacje_projektu_miejscowego_planu.html (accessed on 7.09.2025).

Lorusso, A., Guida, D. (2022). IoT System for Structural Monitoring. In: Karabegović, I., Kovačević, A., Mandžuka, S. (eds) *New Technologies, Development and Application V. NT 2022. Lecture Notes in Networks and Systems*, vol 472. Springer, Cham. https://doi.org/10.1007/978-3-031-05230-9_72.

Markiewicz J., Tobiasz A., Kot P., Muradov M., Shaw A., Al-Shamma. (2019). Review of Surveying Devices for Structural Health Monitoring of Cultural Heritage Buildings, *Developments in eSystems Engineering*, DOI: 10.1109/DeSE.2019.00113.

Oleniacz, G., Kokoszka, W., Ożóg, D., Sikora, A., Słowik, M., Przesmycka, N., Gosk, W. (2023). Geoenvironmental assessment of site suitability for construction purposes, *Materiały Budowlane*, vol. 12: 47–50, DOI:10.15199/33.2023.12.10 (accessed on 7.09.2025).

Ornoch L., Popielski P., Olszewski A., Kasprzak A. (2021). Ultrasonic sensors enabling early detection of emergency trends and analysis of structure inclination and stability by means of highly accurate level measurements, *Sensors*, 21(5):1789.

Osada, E.; Owczarek-Wesołowska, M.; Karsznia, K.; Bęcek, K.; Muszyński, Z. (2023). A Method for the Precise Coordinate Determination of an Inaccessible Location. *Sensors*, 23: 8199. https://doi.org/10.3390/s23198199.

RapidMinerofficial website: https://altair.com/altair-rapidminer (accessed on 7.09.2025).

Świerczyńska E., Karsznia K., Książek K., Odziemczyk W. (2024). Investigating diverse photogrammetric techniques in the hazard assessment of historical sites of the Museum of the Coal Basin Area in Będzin, Poland, *Reports on Geodesy and Geoinformatics*, 118:70–81, DOI:10.2478/rgg-2024-0016.

Thirugnanam H., Uhlemann S., Reghunadh R., Ramesh M.V., Rangan V.P. (2022). Review of landslide monitoring techniques with IoT integration opportunities, *IEEE Journal of Selected Topics in Applied Earth Observations and Remote Sensing*, 2022, 15: 5317–5338.

Wielandt S., Uhlemann S., Fiolleau S., Dafflon B. (2023). TDD LoRa and Delta Encoding in Low-Power Networks of Environmental Sensor Arrays for Temperature and Deformation Monitoring, *Journal of Signal Processing Systems*: 1–13.

Mine Surveying and Mining Area Protection – Sokoła-Szewioła and Jarczyk (Eds)
© 2025 The Author(s), ISBN: 978-1-041-10531-2

An integrated approach to assessing mine shaft deviation: Surveying measurements versus theoretical calculations

M. Kruczkowski & G. Dyduch
Silesian University of Technology, Gliwice, Poland

ABSTRACT: This paper presents a comprehensive assessment of the impact of long-term underground mining activities on the geometric deformations of mine shaft linings, based on data collected over a period exceeding two decades of systematic surveying measurements. The analysis involved identifying both the directions and magnitudes of shaft lining deviations, which were subsequently correlated with the geometry of extraction workings, the intensity of mining operations, as well as local geological conditions and rock mass tectonics. The results enabled the determination of relationships between shaft lining displacements and the orientation and extent of mining activities. The results have significant practical implications for forecasting the impact of underground mining on subsurface infrastructure, particularly vertical structures, and may serve as a foundation for the design and optimization of structural reinforcement systems for shafts located in areas of active mining operations.

1 INTRODUCTION

The operation of an underground mining facility is primarily based on access workings, which provide entry to the deposit and typically remain functional throughout the entire life of the mine. In Polish coal mines, shafts serve as the principal means of ventilation and transportation. Assessing the impact of rock mass behavior on shafts constitutes a significant component of contemporary research, employing various techniques and solutions (Białek *et al.* 2016, Han *et al.* 2019, 2019a, Hejmanowski *et al.* 2014, Jendryś 2018, Kleta 2013, 2017, Strzałkowski 2021).

Monitoring the geometry of mine shaft linings is one of the key elements in ensuring the safe and efficient operation of underground mines. Deformations of shaft linings resulting from mining activities may lead to damage to technical infrastructure, a loss of shaft functionality and increased natural hazards (Bobek *et al.* 2016). Therefore, systematic observations of shaft lining geometry changes using modern measurement technologies and verification of theoretical deformation models are indispensable. Observation and modeling of these processes are becoming fundamental components of geotechnical risk management.

The objective of this study is to compare the results of theoretical calculations of shaft lining deflection, performed according to the W. Budryk – S. Knothe theory, with actual geodetic measurements conducted between 1998 and 2019.

2 CHARACTERISTICS OF THE SHAFT

2.1 *Technical characteristics of the shaft*

The shaft under analysis was sunk in the mid-1960s to a depth of approximately 500 m. In the 0–78 m section, the shaft was constructed with brick masonry lining of thickness ranging

DOI: 10.1201/9781003655558-11

from 0.38 m to 0.51 m, whereas the remaining section was lined with concrete masonry of 0.36 m thickness, with local thickening of the lining to 0.56 m and 0.74 m. Currently, the shaft functions as an intake shaft for materials and personnel transport. The cross-section of the shaft barrel is circular, with an internal diameter of 6.0 m. The shaft is equipped with a hoisting system and a ladder compartment. Additionally, pipelines and signal, electrical, and telecommunication cables have been installed within the shaft. In order to improve resistance of the lining to the impact of underground mining operations, nine levels of wooden yielding inserts, 0.3–0.5 m thick, were installed in the shaft.

2.2 *Geological conditions and rock mass tectonics in the shaft's surroundings*

The rock mass surrounding the shaft barrel consists of Quaternary and Carboniferous formations. The Quaternary formations, occurring within the 0–78 m depth range, are developed as fine sands and sandy or silty clays containing limestone and granite pebbles. In the 78–500 m section, the shaft was sunk in Carboniferous formations, predominantly of clay facies, with interbeds of clay-sandy and sandy sediments. A gradual increase in the content of fine- and medium-grained sandstones is observed with depth.

The stratification of layers in the shaft's vicinity is oriented NW–SE, with a dip ranging from 6° to 10° towards the SW. No tectonic disturbances were identified in the shaft profile.

Hydrogeological conditions in the shaft area are considered stabilized. Static reserves of natural water in both Quaternary and Carboniferous formations have been depleted due to long-term mining activity in this part of the deposit. The Quaternary formations are recharged by precipitation, and their infiltration capacity depends on rainfall intensity. In the Carboniferous formations, two aquifer levels can be distinguished, located at depths of approximately 90 m and 200 m, associated with water-bearing sandstones. Inflows from these aquifer levels range from 10 to 15 dm^3/min.

Long-term mining activity conducted in the shaft's surroundings has caused localized damage to the shaft lining, which has been repaired on an ongoing basis.

3 MINING OPERATIONS IN THE SHAFT'S SURROUNDINGS

Mining operations in the vicinity of the shaft have been carried out in the Ruda and Saddle beds with high intensity since the shaft's construction. This included both mining outside the shaft's protective pillar and mining that encroached upon it. The characteristics of the mining operations conducted are presented in Table 1 and illustrated in Figure 1.

Table 1. Characteristics of mining operations in the vicinity of the shaft.

Coal seam	Period of exploitation	Range of average exploitation depth, H m	Range of average seam thickness, g m	Roof control method	Location relative to shaft
405/2	1961 – 67	~335	~3,5	hydraulic backfilling	SW
407/2	1980 – 82	~385	2,0 – 2,2	caving	W and SW
409	1963 – 72, 1982 – 83	250 – 410	1,4 – 2,4	hydraulic backfilling and caving	W, S and E
410	1987 – 88	~440	~2,3	caving	W
413/2	1996 – 98, 2007 – 10	450 – 580	2,1 – 3,0	caving and caving with sealing	N, W and SE

(*continued*)

Table 1. Continued

Coal seam	Period of exploitation	Range of average exploitation depth, H m	Range of average seam thickness, g m	Roof control method	Location relative to shaft
415/1	1999, 2005 - 10	535 – 610	1,7 – 3,5	caving and caving with sealing	S, SE and E
416	1990 – 2016	490 – 640	1,5 – 4,0	hydraulic backfilling, caving andcaving with sealing	W, S, E and N
416/3	2000 – 02	510	~1,5	caving	W
418/1	2003 – 08, 2015 – 16	640 – 700	2,6 – 2,7	caving and caving with sealing	W and S
502	1979 – 85, 1994 - 2009	645 – 715	2,0 – 2,9	hydraulic backfilling	W, S, E and N
504	1984 – 2003	575 – 800	1,6 – 2,6	caving and caving with sealing	W, S, E and N
506	1970 – 86	700 – 870	1,3 – 2,0	caving	W, S, E and N
507	1973 – 89	635 – 885	2,4 – 3,0	caving	W, S, E and N
510	1979 – 2005	620 – 860	2,0 – 3,0	hydraulic backfilling, caving and caving with sealing	W, S, E and N

Figure 1. Extent of mining operations in the shaft area.

3.1 *Impact of mining operations based on theoretical calculations*

The impact of mining operations on the shaft was calculated using the W. Budryk – S. Knothe theory (Awierszyn 1954, Budryk 1954, Knothe 1953), supported by software developed by R. Ścigała (2013, 2013a).

The following parameters were adopted for the calculations:

- coefficient of roof control for caving of roof strata: a = 0.8,
- coefficient of roof control for sealed caving: a = 0.6,
- coefficient of roof control for hydraulic backfilling: a = 0.2,
- main influence angle parameter: tgβ = 1.9,
- exploitation offset: d = 0,
- parameter B = 0.32 r.

A crucial element of the calculations is the theoretical determination of the shaft barrel deflection based on the extent of mining influence in the rock mass (Kowalski 1985, Strzałkowski *et al.* 2017). Although Strzałkowski *et al.* (2017) presented several solutions, they could not be verified using geodetic measurements. Based on preliminary calculations, the following model was adopted for determining the shaft axis deviation:

1) Variability of the radius of main influence in the rock mass r(z):

$$r(z) = r\left(\frac{z}{H}\right)^n \tag{1}$$

where z – vertical distance in the rock mass from the seam roof [m], r – value of the main influence radius at the surface [m], H – mining depth [m],n – exponent, assumed as n = 0.7 (Drzęźla 1989).

2) Variability of parameter B(z):

$$B(z) = B\left(\frac{z}{H}\right)^m \tag{2}$$

where m – exponent, assumed as m = 3.

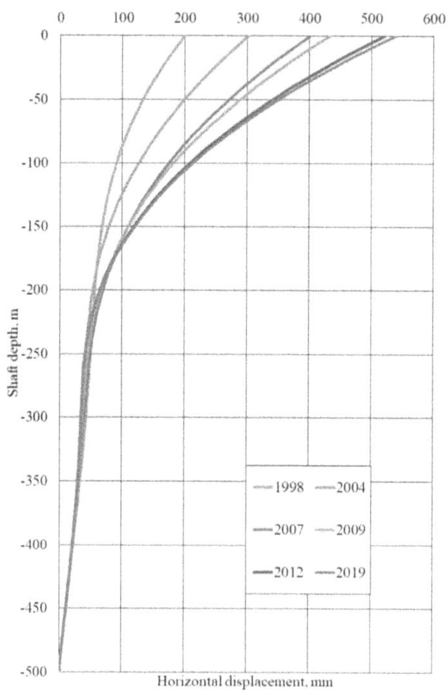

Figure 2. Resultant deviation of the shaft axis based on theoretical calculations for the period 1998–2019.

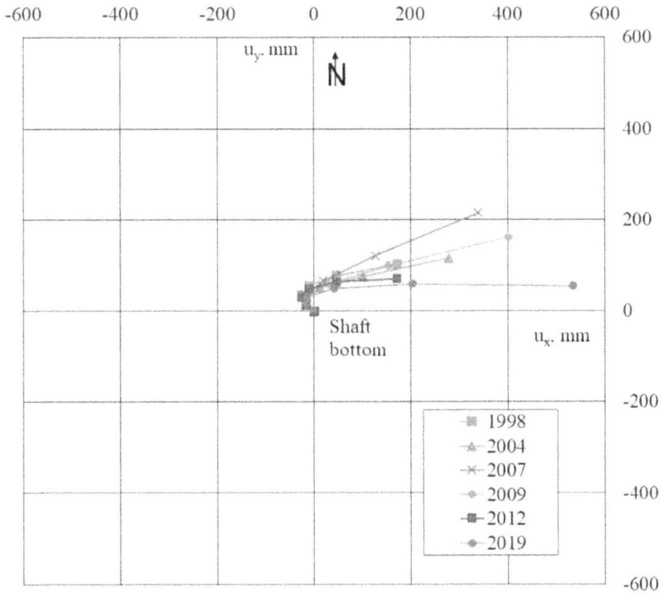

Figure 3. Theoretically calculated shaft center deviations at computational horizons spaced every 100 m.

The time horizons were selected in accordance with the deflection measurements carried out in the shaft.

Based on theoretical calculations, it can be concluded that the mining operations conducted to date have caused a progressive deflection of the shaft barrel over time. The greatest values of deflection are expected to occur in the upper section of the shaft (Figures 2–3). For the shaft collar, the calculated resultant deflection was 200 mm in 1998 and increased to 540 mm in 2019, as a result of mining activities carried out in its vicinity.

Figure 3 presents a cumulative graph of shaft axis deviation for selected computational horizons: 0 m (shaft collar), -100 m, -200 m, -300 m, -400 m, and -500 m (shaft sump). The developed forecast indicates a tendency for the shaft barrel to deflect in the ENE direction, with greater deviation increases occurring towards the east and smaller ones towards the north. The results of the calculations, with reference to the latest measurement period, indicate a trend of shaft barrel displacement towards the south, which was related to the intensified mining operations carried out in that direction in recent years.

The results of the predictive calculations were compared with the outcomes of geodetic measurements performed in the shaft.

4 VERTICALITY MEASUREMENTS OF THE SHAFT LINING

Between 1998 and 2019, cyclic verticality measurements of the mine shaft lining were conducted, enabling the monitoring of changes in shaft geometry over time. Such measurements are typically carried out using four freely suspended plumb lines inside the shaft (so-called mechanical verticality surveying), or with modern laser systems utilizing shaft laser plummets (Jóźwik 1999, Milewski 1991, Pielok 2011). Both techniques were applied in the verticality measurements conducted for this shaft.

The locations of the plumb lines were determined at the surface using the polar method from a surface control point near the shaft. The installation depths of shaft girders and measurement horizons along the shaft perimeter were established using a shaft tape. Measurements to the shaft girders and lining were taken in horizontal cross-sections of the shaft—at each girder level—using the coordinates and offsets method along with face position measurements. In the case of laser plummets, special scaled targets were used.

For the mechanical verticality method, the mean error of determining the plummet's position at the surface—assuming four repetitions of distance measurement using a steel tape with millimetre graduations ($m_l \leq \pm 0{,}7$ mm) and an angular measurement error not exceeding $\pm 10^{cc}$ amounts to approximately ± 0.8 mm. The error in determining the rest position of a freely suspended plumb line, resulting from air currents in the shaft and the uncertainty of reading the position of the oscillating plummet, is estimated at ± 1.4 mm (Milewski 1991). However, this is a systematic error; therefore, differences in offsets determined at successive measurement horizons are unaffected by it. Consequently, the mean error in determining the position of a measurement point on the shaft lining by the mechanical verticality method, also accounting for the accuracy of offset measurements, does not exceed ± 2.0 mm (Pielok 2011).

In the case of laser verticality surveying, significant error components include the irregularity of the surface of the shaft guides to which the scaled target is attached estimated at $\pm(1\text{–}2)$ mm and the uncertainty in determining the position of the centre of the laser spot $\pm(1\text{–}3)$ mm. As a result, the error in determining the position of the laser plummet amounts to $\pm(2\text{–}5)$ mm. When additional errors from orthogonal offset measurements taken from the guide corners to the structural details of the shaft wall are considered, the mean error in determining the position of such measurement points on the shaft lining falls within $\pm(3.8\text{–}7.4)$ mm (Pielok 2011).

Measurement points were marked at the level of each shaft girder, i.e., at vertical intervals of approximately 3.0 m, which corresponds to the structural spacing of shaft girders in the shaft under study. For each horizontal cross-section, measurements were taken to four survey points evenly distributed around the shaft circumference, forming a system of vertical profile lines (Figure 4). Points were marked directly on the shaft masonry in the area of the shaft girder sockets, due to the impossibility of their permanent stabilization on the shaft lining. This limitation is a consequence of ongoing contamination of the shaft lining over time by deposits and accretions formed by the precipitation of mineral components from water in flowing into the shaft.

The measurements made it possible to determine the coordinates of the shaft center at each measurement horizon, thereby enabling determination of the shaft axis alignment and its deviation from the vertical at individual horizons. The analysis of shaft lining deflection was carried out assuming that the geometric axis of the shaft passes through the center of the circular cross-section of the shaft barrel determined at the last measurement horizon, i.e., at a depth of -496.85 m or -487.86 m (2019), for which complete measurement data were available. Figures 5–6 present the development of the resultant shaft axis deflection over the analysed periods.

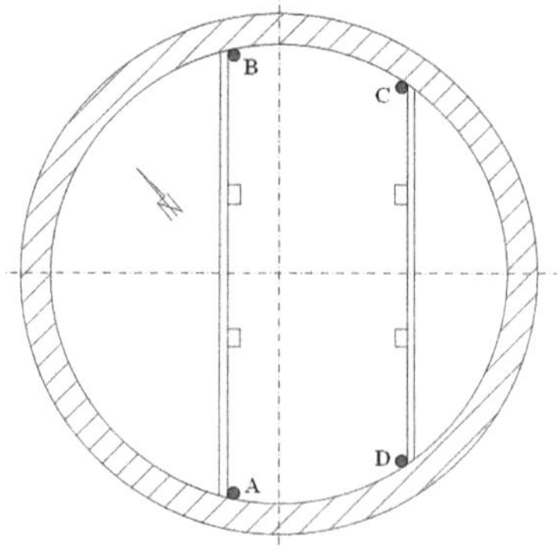

Figure 4. Diagram of the shaft cross-section with marked measurement points along the shaft lining perimeter.

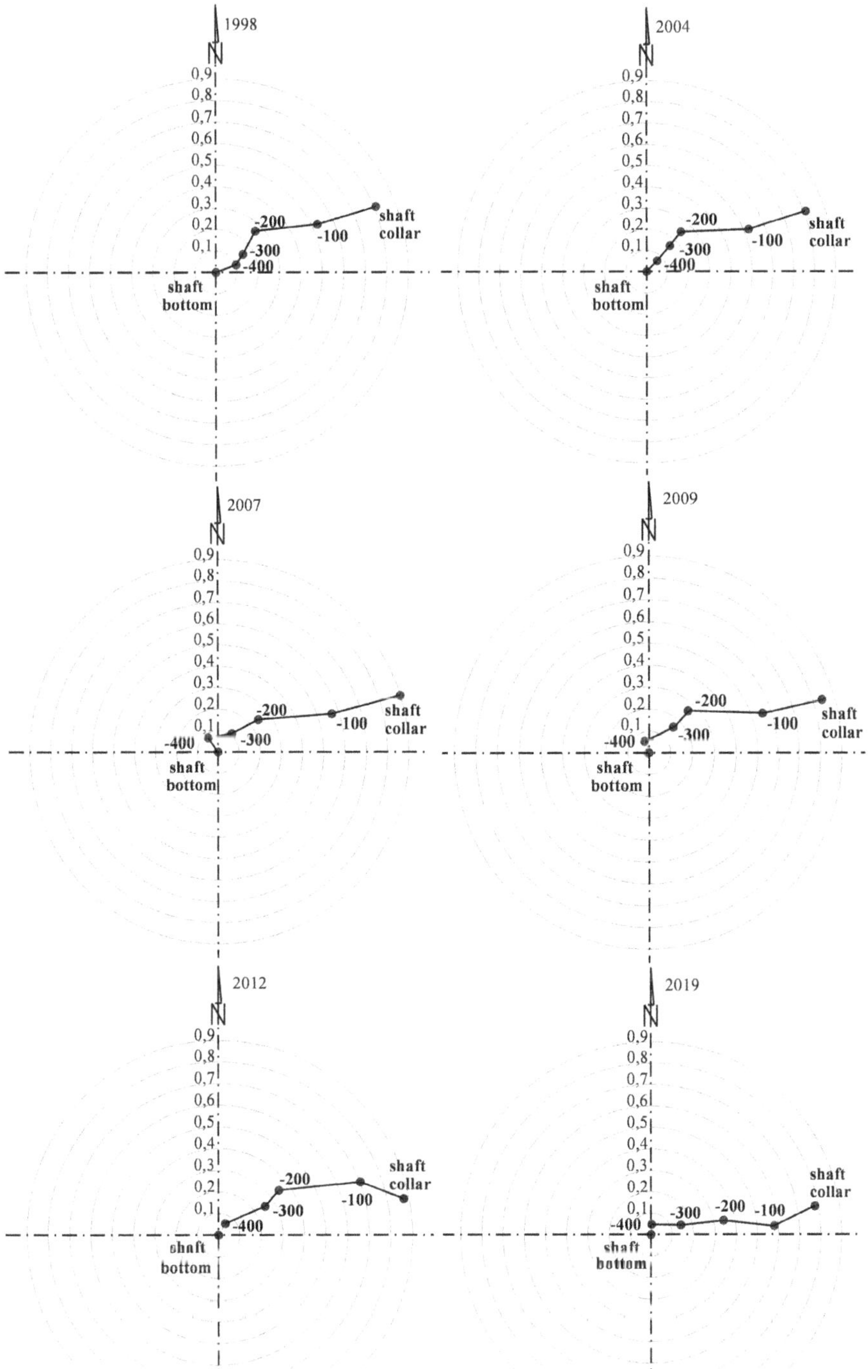

Figure 5. Diagram of shaft center deflection at designated measurement horizons spaced at 100 m intervals, based on measurement results from the years 1998–2019. Numerical values in the figures are given in meters.

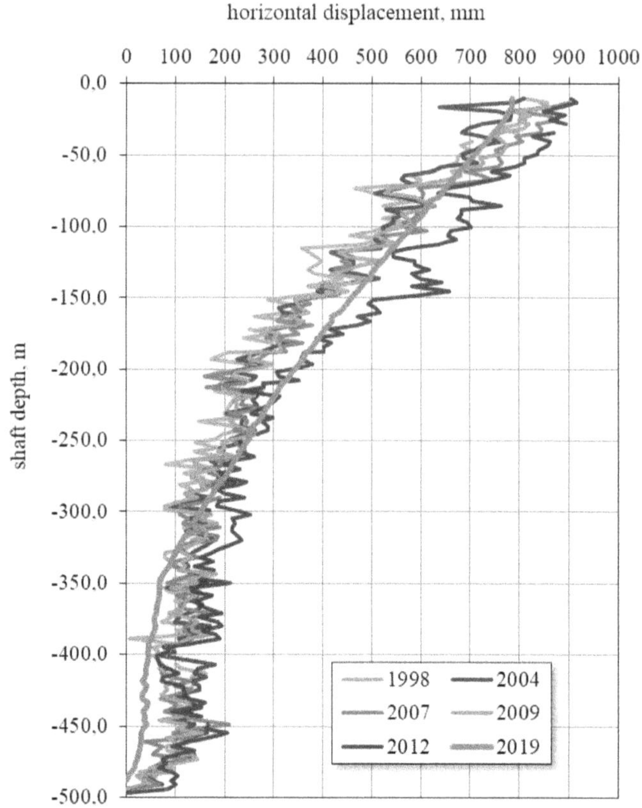

Figure 6. Development of shaft axis deflection in the ENE direction based on shaft lining deflection measurements from 1998 to 2019.

The results of the cyclic measurements indicate a tendency for the shaft lining to deflect in the ENE direction, i.e., toward the area of highest mining concentration, which correlates with the theoretical calculation outcomes. It can be observed that the greatest deviation of the actual shaft axis from the geometric axis—defined at the depth of the last measurement horizon—occurs near the shaft collar and remains relatively consistent over the analysed period. The variability in shaft axis deviation values at individual measurement horizons across the analysed time span results from the precision of shaft lining construction, the influence of mining operations, and the fact that measurements were taken to points that were not permanently stabilized on the shaft masonry.

Based on the measurement results, it can be concluded that the shaft lining deflection had stabilized before the measurement cycle began, and despite the continued intensive mining around the protective pillar, no significant change occurred over the subsequent 21 years (Figure 6). However, the direction of the shaft axis deflection did change during the analysed period, showing a systematic trend of displacement toward the south (Figure 5).

Due to the impossibility of stabilizing permanent measurement points on the shaft lining, the results of such measurements should be interpreted primarily in qualitative terms. Although the exact magnitude of shaft lining deflection cannot be determined with high precision, it is still possible to identify its direction and forecast the continued trend with sufficient accuracy. Long-term, cyclic verticality measurements of the shaft lining provide valuable information not only about the shaft's geometry but also allow for assessing the impact of mining operations on rock mass displacements in the area of the protective pillar.

5 CONCLUSIONS

The integrated analysis of shaft deflection, combining theoretical modelling results and measurement data, enabled a comprehensive assessment of the impact of mining operations on shaft geometry. The results of the study and analyses confirm a high degree of consistency between theoretical deformation models and actual geodetic measurements. The identified shaft axis deflection was slow, gradual, and spread over time, indicating the relative stability of deformation processes.

Both the theoretical analysis and geodetic measurements demonstrated that the dominant direction of shaft axis deflection is toward the ENE, which is correlated with the historical intensification of mining activities in this section of the protective pillar. The measurements also showed that no abrupt anomalies or sudden displacements were recorded over the 21-year period, confirming the effectiveness and reliability of the adopted measurement methodology despite limitations arising from the lack of permanent stabilization of measurement points.

The results highlight the importance of systematic and long-term shaft monitoring as a tool supporting deformation forecasting and geotechnical risk management. The combination of numerical modelling with empirical data constitutes a valuable engineering approach for assessing rock mass displacements in the vicinity of protective pillars. The findings of this analysis may be applicable in engineering practice, particularly in the design and monitoring of shafts subjected to the influence of intensive mining operations.

REFERENCES

Awierszyn S.G. (1954). Gornyje raboty pod sooruženiami i wodojemami. *Ugletiechizdat.* Moskwa

Budryk W. (1953). Wyznaczanie wielkości poziomych odkształceń terenu. *Archiwum Górnictwa i Hutnictwa t. 1. z. 1.*

Białek J., Mielimąka R. and Chowaniec A. (2016). Projekt eksploatacji w filarze szybów głównych uwzględniający ochronę zakładu przeróbczego i szybu. Ochrona środowiska na terenach górniczych i pogórniczych kopalń w subregionie zachodnim województwa śląskiego : XXIII Konferencja naukowo-techniczna, Rybnik, październik 2016 / Duży Stanisław (red.) Rybnik, Stowarzyszenie Inżynierów i Techników Górnictwa. Oddział Rybnik, s.16–25

Bobek, R., Śledź, T., Twardokęs, J., Ratajczak, A. and Głuch, P. (2016). Problems of shaft lining stability in the light of experience Knurów–Szczygłowice coal mine, *Zeszyty Naukowe Instytutu Gospodarki Surowcami Mineralnymi i Energią Polskiej Akademii Nauk vol. 94*: 41–52. Kraków: Instytut Gospodarki Surowcami Mineralnymi i Energią PAN

Drzęźla B. (1989). Opis programów prognozowania deformacji górotworu pod wpływem eksploatacji górniczej. Aktualny stan oprogramowania. *Zeszyty Naukowe Politechniki Śląskiej, Seria: Górnictwo z. 165.* Gliwice: Wydawnictwo Politechniki Śląskiej.

Hejmanowski R., Malinowska A., d'Obyrn K., Ulmaniec P. and Kwinta A. (2014). Badania i prognoza deformacji górotworu w rejonie szybu Św. Kingi w Kopalni Soli „Wieliczka". *Przegląd Górniczy 8/2014.* 150–156.

Han J., Jiuqun Zou J., Hu Ch. and Yang W. (2019). Study on Size Design of Shaft Protection Rock/Coal Pillars in Thick Soil and Thin Rock Strata. *Energies 2019*, 12, 2553; doi:10.3390/en12132553

Han J., Jiuqun Zou J., Hu Ch. and Yang W. (2019a). Deflection Mechanism and Safety Analysis of Coal Mine Shaft in Deep Soil Strata. *Hindawi. Mathematical Problems in Engineering, Volume 2019*, Article ID 9461742, 11 pages, https://doi.org/10.1155/2019/9461742

Jendryś M. (2018), Analysis of stress state in mine shaft lining, taking into account superficial defects. *IOP Conference Series: Earth and Environmental Science*, Volume 261, Mining of Sustainable Development 28 November 2018, Gliwice, DOI 10.1088/1755-1315/261/1/012016

Jóźwik, M. (1999). Geodezyjna obsługa budowy i eksploatacji szybów górniczych, *Zeszyty Naukowe Politechniki Śląskiej Seria. Górnictwo z. 239*: 145–152. Gliwice: Wydawnictwo Politechniki Śląskiej.

Kleta H. (2013) *Zasady oceny bezpieczeństwa szybów i ich odporność na oddziaływania górnicze. Gliwice*, Wydawnictwo Politechniki Śląskiej. Monografia nr 468.

Kleta H., Mielimąka R. and Jendryś M. (2017). Uwarunkowania i możliwości prowadzenia eksploatacji górniczej w filarze szybów głównych w aspekcie ich aktualnego stanu technicznego., W: Ochrona powierzchni na terenach górniczych i pogórniczych kopalń w Subregionie Zachodnim Województwa Śląskiego : XXIV Konferencja naukowo-techniczna, Rybnik, październik 2017 / Duży Stanisław (red.), 2017, Rybnik, Stowarzyszenie Inżynierów i Techników Górnictwa. Oddział Rybnik, s.48-57

Knothe S. (1953). *Równanie profilu ostatecznie wykształconej niecki osiadania.* Archiwum Górnictwa i Hutnictwa, t. l, z. 1.

Kowalski A. (1985). *Zmienność parametru zasięgu wpływów głównych w górotworze.* Ochrona Terenów Górniczych, nr 72/2.

Milewski, M. (1991). *Geodezja górnicza. Cześć II.* Kraków: Wydawnictwa AGH.

Pielok, J. (2011). *Geodezja górnicza.* Kraków: Wydawnictwa AGH.

Ścigała, R. (2013). Software system for identification of parameters needed for forecast of deformations caused by underground extraction. *Exploration Geophysics, Remote Sensing and Environment*: 52–61.

Ścigała R. (2013a). Mining extraction system with advancing longwall as a method of protecting-technical objects and an element of rational deposit management. *Gospodarka Surowcami Mineralnymi-Mineral Resources Management.* DOI: 10.2478/gospo-2013-0032

Strzałkowski P., Ścigała R., Szafulera K. and Kruczkowski M. (2017). Wpływ zmienności parametru B w górotworze na przebieg odkształceń. *Budownictwo Górnicze i Tunelowe 2/2017*: 12–19.

Strzałkowski P. (2021). Sinkhole Hazard Caused by Inactive Mining Shafts as Illustrated by a Selected Example. *Pure Appl. Geophys. 178 (2021)*, 1697–1707 2021 The Author(s) https://doi.org/10.1007/s00024-021-02716-z

Mine Surveying and Mining Area Protection – Sokoła-Szewioła and Jarczyk (Eds)
© 2025 The Author(s), ISBN: 978-1-041-10531-2

Uncertainty of land subsidence model parameter values and its impact on the risk assessment of buildings

A. Kwinta
University of Agriculture, Krakow, Poland

R. Hejmanowski & A. Malinowska
AGH University of Krakow, Poland

Hua-yang Dai
China University of Mining and Technology, Beijing, China

ABSTRACT: This study examines the detrimental effects of underground mining, particularly focusing on surface displacements and deformations. It highlights the consequences of these transformations in urban areas where buildings and critical infrastructure are at risk. The research investigates a stochastic model used for predicting land subsidence, specifically scrutinizing the accuracy related to the uncertainty of model parameters. The concept of the probability influence function is elucidated, describing its mathematical formulation and relevance to subsidence prediction. The parameters involved in the model are detailed, emphasizing their significance and dependency on geological and technical conditions. The statistical analysis assesses prediction accuracy concerning model parameters, discussing the Monte Carlo method and the law of error propagation to estimate the mean errors of deformation indicators due to inaccuracies in model parameters. Two practical examples are presented. The first involves determining the probability of exceeding harmful influence ranges in coal mining using calculated deformation indicators and statistical analyses. The second example examines coal exploitation in a residential area, assessing the resistance of buildings to deformation and the probability of exceeding strain limit values.

1 INTRODUCTION

Human activities within the Earth's crust transform the environment, affecting all its elements (Darling 2011; Faunt *et al.* 2016; Vasco *et al.* 2022). Surface displacements and deformations are among the most significant transformations, with land subsidence being a prevalent issue worldwide (Bagheri-Gavkosh *et al.* 2021; Dinar *et al.* 2021; Saeidi, *et al.* 2021). Groundwater pumping, underground extraction of gas, oil, and solid minerals like metal ores or coal primarily cause land subsidence. Surface deformations frequently result in structural damage to buildings and loss of functionality (Can *et al.* 2012; Kratzsch 1983; Merad *et al.* 2004; Peng 2020; Sheorey *et al.* 2000; Weitao *et al.* 2022; Whittaker and Reddish 1989; Zhu *et al.* 2015). Therefore, any human activity impacting the environment requires an assessment of its consequences. Numerous publications detail deformation prediction methods and their modifications. Accurate prediction of displacement and deformation caused by underground activities is critical, especially in urbanized areas (Gazzola *et al.* 2023). Various theoretical solutions for modeling deformations induced by porous rock compaction or post exploitation void convergence have been developed (Bahuguna *et al.* 1991; Gambolatti and Teatini 2015; Luo and Qiu 2012). These models include numerical models, stochastic models, and AI-based models (Yoginder *et al.* 1989). The practical utility and reliability of these theoretical solutions have been verified (Cai *et al.* 2023). Stochastic models using appropriate transition functions are among the

DOI: 10.1201/9781003655558-12

most widely used computational methods globally. Various transition functions are described in the literature (Fokker and Osinga 2018; Ren *et al.* 1987; Sann 1949; Schmitz 1924; Sheorey *et al.* 2000; Saedi *et al.* 2021). Accurate geological data and parameterization are crucial in modeling. High-accuracy observation of underground activity effects on the ground surface is possible, and estimating model parameters a posteriori from geodetic monitoring data is feasible (Lian *et al.* 2023). However, determining a priori parameter values in the absence of monitoring data is critical (Hoffmann *et al.* 2003; Janna *et al.* 2012). Therefore, defining model parameters suitable for the geological and technical conditions of planned operations is essential (Bang *et al.* 2023; Hejmanowski and Kwinta 2011). The monitoring of land subsidence, whether induced by anthropogenic or natural factors, has advanced significantly, particularly since the European Space Agency (ESA) began providing free radar data (Aschbacher 2017; Ilieva *et al.* 2019). Over the past decade, the investigation of surface displacement through satellite radar interferometry has progressed at an unprecedented rate (Cigna and Tapete 2020; Qiao *et al.* 2023). Innovative methods for radar data analysis are continually being developed, and the resolution of new satellite sensors is consistently improving (Bernardino *et al.* 2002). These enhancements in displacement measurement accuracy facilitate a deeper understanding of rock mass deformation phenomena. The research on the application of photogrammetric sensors from low-altitude UAV's also expands knowledge about the monitored displacement fields (Puniach *et al.* 2021). Despite these advancements, predicting land subsidence remains a complex challenge. A critical issue when evaluating modeled deformation values is their reliability.

1.1 *Research objectives*

In this article, we aim to address this issue by striving to fill the gap in research on the accuracy of deformation modeling. We present the principles of determining displacements and deformation, along with statistical analyses related to the parameterization of a theoretical model. Using a selected influence function as an example, we conduct a statistical analysis based on a pseudo-random number generator and the law of average error propagation. The solutions provided are illustrated with two practical examples.

2 THEORY AND METHODS

2.1 *Ground movement modelling – basic assumptions*

Among the models available for simulating the effects of mineral extraction from the Earth's crust, a stochastic model based on the so-called influence function was selected for further consideration. This model involves the concept of transitioning the volume change of an element to surface subsidence through the influence function (Figure 1).

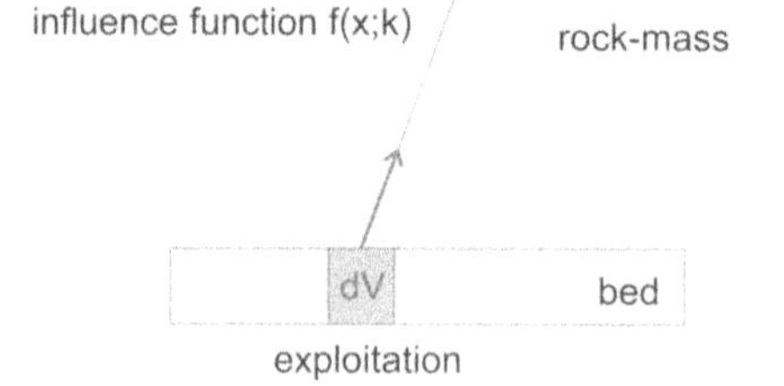

Figure 1. The idea of the influence function.

Exploitation results in the formation of a void in the rock mass, which migrates to the surface and causes the displacement of rock masses towards the extraction area. The exploitation of a mining panel with finite dimensions can be divided into a set of elements with identical dimensions. Each element corresponds to an elementary surface subsidence. The overall impact of the entire exploitation field is obtained by superimposing the influences from these elementary extraction parts (Figure 1). It is assumed that the exploitation of an elementary volume of the deposit, dV, with a surface area of dP, results in an elementary subsidence at point A within the range of influence of this exploitation:

$$dS_A = f(x;k)dP \tag{1}$$

where: dS_A– elementary subsidence at point A, $f(x;k)$ – influence function, where x is a set of data and k is a set of model parameters, dP – surface of the exploitation element.

Applying the principle of superposition, it is assumed that the vertical displacement dS at point A is the sum of the elementary subsidence contributions from all elementary volumes within the exploitation field. This is expressed as an integral over the exploitation area using the influence function:

$$S_A = \iint_P f(x;k)dP \tag{2}$$

The shape of the subsidence basin varies depending on the form of the influence function used. Each influence function is characterized by a set of input data, x, and a set of parameters, k. The most significant parameters are typically related to the dimension and position of the elementary void within the rock masses. The strength of the rock masses, which determines the displacement distribution, is usually represented by a parameter known as the angle of the influence function distribution (Table 1).

Table 1. Examples of Influence functions historically used in modeling rock mass displacements.

Author	Function $f^*(x,k)$
Bals1932	$\frac{1}{H^2}\cos^2\delta$
Sann 1949	$\frac{2.256}{r}\exp(-4r^2)$
Knothe 1950 (1953)	$\exp\left(-\pi\frac{x^2}{r^2}\right)$
Erhard, Sauer 1961	$\sqrt{\frac{-\ln(0.01)}{\pi}}\cdot\frac{1}{r}\exp\left(\ln(0.01)\frac{x^2}{r^2}\right)$
Kochmański 1955 (1956)	$\exp\left[-\left(\frac{x}{r_0}\right)^b\right]$
Ren *et al.* 1987	$\frac{1}{r^2}\exp\left(-\pi\frac{d^2}{r^2}\right)$
Sheorey *et al.* 2000	$\frac{0.5352}{r^2}\left(1+\cos\frac{\pi d}{r}\right)$

where: δ, r, r_0, b - model parameters, H, x, d - geometric data

The influence function proposed by Knothe has been specifically selected for further analysis. Although its impact on the overall body of presented research may not be paramount, its selection is instrumental due to the proprietary software embedded within this model. Subsequent analyses of other theoretical solutions outlined in Table 1 will adopt a similar procedural approach.

The Knothe model (1953) has undergone continuous modification over approximately 70 years, establishing itself as a cornerstone theoretical framework for predicting deformations in Poland. A closely analogous model has been applied in Germany (Ehrhardt and Sauer 1961; Kratzsch 1983), with adaptations extending its utility to China, Australia, and other mining-centric nations (Luo and Cheng 2009; Marian *et al.* 2017; Zhou *et al.* 2015). This widespread adoption can be attributed to its commendable alignment between computational outcomes and geodetic measurements, as well as its notable simplicity in application. Notably, this model is instrumental in predicting deformations resulting from the underground mining of diverse solid minerals (Hejmanowski *et al.* 2016; Hejmanowski *et al.* 2018; Taherynia *et al.* 2017).

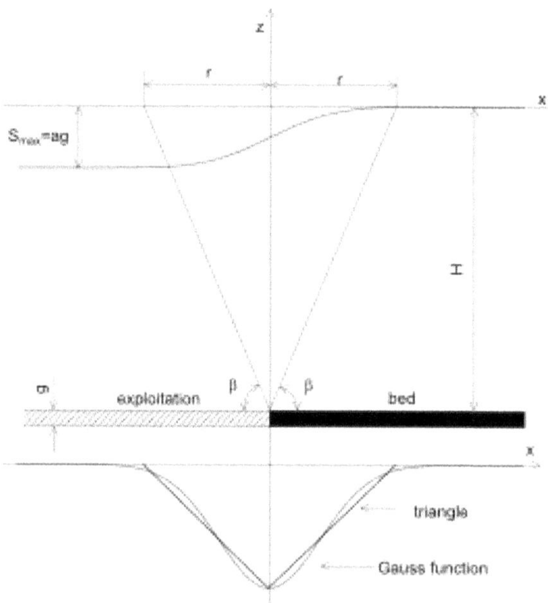

Figure 2. Parameter of the influences' dispersion r and an angle of the main influences' range β (Knothe 1953).

The profile of the subsidence trough was described using a parameterized and generalized normal distribution function adapted to a triangular shape. This approach facilitated the determination of the influence function:

$$f(x;k) = S_{\max}\frac{1}{r}\exp\left(-\pi\frac{x^2}{r^2}\right)_{\max} \tag{3}$$

where: $f(x;k)$ - the Knothe's influence function,

$$S_{\max} = a \cdot g \tag{4}$$

$S_{\max}$- maximum subsidence [m], a - model parameter, known as the exploitation coefficient, dependent on the method of filling the post-mining void [/], g - geological layer thickness of the exploited deposit [m],

$$r = \frac{H}{\tan\beta} \tag{5}$$

r – radius of exploitation influences for a surface [m], H - depth of the exploited deposit [m], β - angle of the main influence zone according to Knothe's model (see Figure 2), x - horizontal distance of the computational point from the exploitation site.

In the outlined Knothe model, two fundamental parameters are defined, corresponding to the conditions of underground mining k: $k = \{a, tg\beta\}$, and information regarding the geometric layout and dimensions of the underground operation concerning the calculation point X:

$$X = \{x, g, H\} \tag{6}$$

The angle of the main influence zone, denoted as 'β', is related to the physico-mechanical properties of the rock mass and provides a simplified description of the complex structure and properties of the rock mass. In geomechanical (numerical) models, this role is fulfilled by the Young's modulus (E) and Poisson's ratio (ν), which are estimated and scaled for each geological layer of the overburden. The value of the parameter 'a' depends on the method of post-mining void closure and the adopted exploitation system. It is essential to emphasize that determining both parameters for specific exploitation conditions is best achieved based on empirical data. This can involve data from monitored exploitation measured by geodetic or satellite methods. This is achievable, for example, through the application of inverse problem-solving techniques (Chai *et al.* 2023; Hoffman *et al.* 2003; Janna *et al.* 2012; Lian *et al.* 2023).

2.2 *Statistical analysis in ground movement modelling*

The statistical analysis of safe exploitation possibilities in protected area regions involves determining the probability that the predicted theoretical values of deformation indicators will exceed permissible boundary values. Discrepancies between theoretical deformation indicators and geodetic measurements can arise from various causes (Malinowska and Hejmanowski 2016). These discrepancies can be either random or systematic in nature. In this study, the authors aim to examine the impact of errors in computational model parameters on the accuracy of deformation modeling. Specifically, the focus will be on determining the probability of exceeding permissible threshold values of deformation indicators due to inaccurately assumed model parameter values.

Analyzing the formulas in Table 1 and the Knothe model reveals that the equations directly involve the ranges of influence and maximum subsidence. These two variables integrate model parameters with mining-geological data and significantly impact the prediction accuracy. Therefore, detailed statistical analysis should consider both the accuracy of the adopted parameter values and the accuracy of the mining exploitation data used for displacement prediction (Gruszczyński *et al.* 2018).

In cases where there are discrepancies between predicted deformation indicators and actual deformation measurement results, the initial step should be to analyze whether the executed mining operation aligns with the original design. Subsequently, a thorough analysis of the adopted model parameters should be conducted to verify the model's suitability for the specific operating conditions. This publication focuses solely on prediction errors arising from erroneous assumptions in the parameters of the computational model.

In computations, it is essential to consider the mean (expected value) of the assumed computational model parameters, along with their average error values m_a, $m_{tg\beta}$ (standard deviation):

$$\begin{cases} a \pm m_a \\ tg\beta \pm m_{tg\beta} \end{cases} \tag{7}$$

2.2.1 *Determining probability of observing normal distribution variable values*

Given that most natural and physical phenomena typically follow a normal distribution, this distribution was specifically adopted for the statistical analysis. Under this assumption, the probability of observing a value equal to or greater than a certain value x_l for a variable following a normal distribution is represented by the area under the Gaussian curve from that value to infinity:

$$P(x_1 \leq x) = \frac{1}{\sigma\sqrt{2\pi}} \int_{x_1}^{\infty} \exp\left[\frac{-(x-\mu)^2}{2\sigma^2}\right] dx \tag{8}$$

Where: P – probability, x - random variable, σ - Standard deviation (its estimator is the mean error), μ - Expected value of the variable (its estimator is the mean value).

For computational purposes, standardizing a random variable based on its expected value and standard deviation is a viable approach. This process allows for the consideration of variable error. The standardized variable is formulated as follows:

$$z = \frac{|x-\mu|}{\sigma} \tag{9}$$

Consequently, Equation (8) is transformed into Equation (11) under this scenario.

$$P\left(z_1 \leq \frac{x-\mu}{\sigma}\right) = \frac{1}{\sqrt{2\pi}} \int_{z_1}^{\infty} \exp\left[\frac{-z^2}{2}\right] dz \tag{11}$$

In specific computations, employing the error function (ERF) represented by:

$$ERF(z) = \frac{2}{\sqrt{\pi}} \int_{0}^{z} \exp(-t^2) dt \tag{12}$$

Utilizing this function allows for the determination of probability as defined:

$$P(D_{li} \leq D) = 50 \begin{Bmatrix} 1 - ERF(t) & t \geq 0 \\ 1 + ERF(-t) & t < 0 \end{Bmatrix} \quad [\%] \tag{13}$$

Where: D_{li} -n critical value of the selected indicator, D - predicted value of the indicator, σ_D - average error of the determined indicator.

Before conducting statistical computations, it is necessary to determine the predicted values of deformation indicators along with their mean errors (standard deviations).

2.2.2 *Mean errors of deformation indicators*

Monte Carlo method

The average errors of deformation indicators can be determined using two methods. In the first scenario, the Monte Carlo method is employed, utilizing a pseudo-random number generator with a normal distribution. A predetermined number of cases is generated, producing pairs of predicting model parameters. Subsequently, histograms of the determined parameters are created to verify whether the parameter sets were correctly generated according to the assumed normal distribution. Next, deformation indicators are determined for each parameter pair. After the predictive calculations, the average value and the average error of this value are calculated for each deformation index. However, this method's drawback lies in its computational inefficiency, as deformation indicators must be calculated separately for each parameter pair. The computation duration depends on the number of generated parameters (Busch *et al.* 2015, 2017; Kwinta and Gradka 2020). For the correct application of the Monte Carlo method, it is necessary to determine the number of iterations required to achieve a specified maximum percentage error at the assumed confidence level.

Numerous algorithms have been described in the literature for generating sets of pseudo-random numbers with predefined probability distributions. For determining parameter sets of prediction models with a normal distribution, the Box-Muller generator (1958), also known as the transformation method, can be applied. In this case, two independent random variables, U_1 and U_2, are generated with a uniform distribution in the range (0,1]:

$$U_1 \in (0,1] \wedge U_2 \in (0,1] \tag{14}$$

Subsequently, auxiliary coordinates are determined in a polar coordinate system:

$$\begin{cases} \varphi = 2\pi U_2 \\ r = \sqrt{-2\ln(U_1)} \end{cases} \tag{15}$$

This transformation yields two independent coordinates, which are then used to compute random variables with a normal distribution N(0,1):

$$\begin{cases} Z_1 = r \cdot \cos(\varphi) \\ Z_2 = r \cdot \sin(\varphi) \end{cases} \tag{16}$$

To obtain variables with a normal distribution satisfying condition (9), a transformation must be performed:

$$\begin{cases} a = a_0 + m_a Z_1 \\ tg\beta = tg\beta_0 + m_{tg\beta} Z_2 \end{cases} \tag{17}$$

It is crucial to note that the Monte Carlo method yields satisfactory results with a very high number of repetitions, ideally over 1000 (Hassan Z. *et al.* 2020). This ensures that the distribution of generated parameters approximates a normal distribution. Before proceeding with further computations, it is essential to verify the correct generation of parameter sets according to the assumed normal distribution. This involves creating histograms of the parameters, comparing them with the normal distribution, and generating a 2D parameter plot. Figure 3 presents the histograms and corresponding plot for exemplary prediction model parameter values.

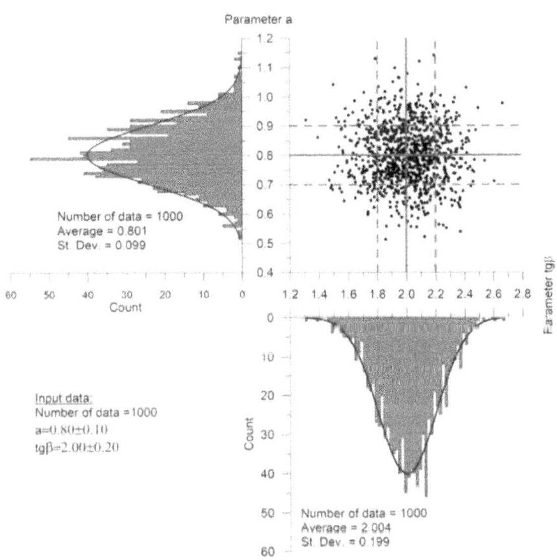

Figure 3. Checking the correctness of generated parameter values.

As observed in the obtained set of parameter pairs (Figure 3), the distribution is satisfactory and can be utilized for deformation prediction. The subsequent step involves calculating the deformation values for each parameter pair, followed by determining their mean values and standard deviations.

Law of Error Propagation

The second approach for statistical analysis involves using the law of mean error propagation for independent observations. The initial assumptions for applying this law are as follows:

- Model parameters are independent quantities.
- Mining exploitation plots are independent of each other.
- Elementary parts of a given plot are mutually independent.

In this context, the general law of propagation of average errors for a given deformation indicator D can be expressed as follows:

$$m_D^2 = \left(\frac{\partial D}{\partial a}\right)^2 m_a^2 + \left(\frac{\partial D}{\partial tg\beta}\right)^2 m_{tg\beta}^2 \tag{18}$$

For the adopted model and the assumptions presented above, the entire issue boils down to determining the appropriate derivatives of deformation indicators concerning specific model parameters.

For the Knothe model and vertical displacements, the formula can be derived as follows:

the subsidence of any point A due to n exploitation plots

$$S_A = \sum_{i=1}^{n} S_i \tag{19}$$

in accordance with the assumptions and formulas (18) and (19), we can express:

$$m_S^2 = \left(\sum_{i=1}^{n} \frac{\partial S_i}{\partial a} m_a\right)^2 + \left(\sum_{i=1}^{n} \frac{\partial S_i}{\partial tg\beta} m_{tg\beta}\right)^2 \tag{20}$$

the subsidence S_i resulting from the exploitation of a single mining plot is obtained by substituting relation (4) into (3):

$$S_i = \iint_{P_i} S_{max} \frac{1}{r_i^2} exp\left(-\pi \frac{x^2+y^2}{r_i^2}\right)_{imax} \tag{21}$$

applying (20) to (21), we determine the respective derivatives:

$$\frac{\partial S_i}{\partial a} = \frac{\partial S_i}{\partial S_{imax} \frac{\partial S_{imax}}{\partial a}} \tag{22}$$

thus, based on (3)

$$\frac{\partial S_i}{\partial a} = g_i \iint_{P_i} \frac{1}{r_i^2} \exp\left(-\pi \frac{x^2+y^2}{r_i^2}\right) dP = \frac{S_i}{a} \tag{23}$$

$$\frac{\partial S_i}{\partial tg\beta} = \frac{\partial S_i}{\partial r_i} \frac{\partial r_i}{\partial tg\beta} \tag{24}$$

further based on (3)

$$\frac{\partial S_i}{\partial tg\beta} = \frac{1}{tg\beta} \iint_{P_i} \frac{2S_{imax}}{r_i^2} \left(1 - \pi \frac{x^2+y^2}{r_i^2}\right) \exp\left(-\pi \frac{x^2+y^2}{r_i^2}\right) dP \tag{25}$$

Consequently, the final formula for the error in vertical displacements due to the acceptance of model parameters can be described according to (20) with the use of (23) and (25).

For the purposes of this study, derived formulas for subsidence S, maximum basin tilt T_{max}, and extreme values of horizontal strains ε_{max} were utilized. The correctness of these derived formulas was verified through a test example by comparing them with the results of computations based on 1000 generated pseudo-random pairs of model parameters. The test calculations demonstrated very good agreement, confirming the correctness of the derived formulas. It is important to note that differences between theoretical and measured deformation values arise from the following sources:

- Prediction inaccuracies (model, parameters).
- The random distribution of the phenomenon.
- Measurement errors.

3 EMPIRICAL APPLICATIONS AND DISSCUSSION

The stochastic ground deformation model is utilized to demonstrate two distinct examples. The first case study focuses on investigating the probability of exceeding predefined safety factor limits within the mining impact area. The second case study interprets the potential hazards and assesses the potential risk to buildings due to uncertainties in model parameters

3.1 *Example 1 – Influence Range*

The first example presented concerns determining the probability of exceeding harmful influence ranges in the context of coal mining using a longwall method with roof collapse. For this case study, a subset of data encompassing four mining plots and 656 calculation points distributed within a grid of 100x100-meter squares was utilized. Technological data on mining are provided in Table 2, while Figure 4 illustrates the layout of the mining plots and the extent of the calculation grid.

Table 2. Geometrical data of mining exploitation.

Longwall plot N	Depth H [m]	Thickness g [m]
158	932	2.05
180	985	1.60
181	973	1.60
182	974	1.50

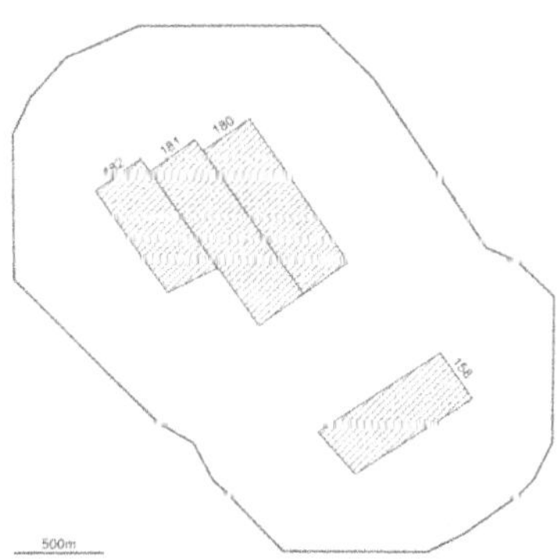

Figure 4. Scheme of mining exploitation and boundary of calculation grid.

The theoretical calculations of deformation indicators utilized the Knothe model parameters as shown in Figure 3. To determine the harmful influence ranges and the probability of exceeding threshold values, the following limits for deformation indicators were adopted based on technical guidelines used in Polish mining and construction (Instruction No. 12 GIG):

- Subsidence: $S_{\text{limit}} = 0.010$ m,
- Tilt: $T_{\text{limit}} = 0.50$ mm/m,
- Horizontal strain: $\varepsilon_{\text{limit}} = 0.30$ mm/m.

Figure 5 presents the predicted values of deformation indicators alongside the probability of exceeding their threshold values within the mining impact area.

Based on the analyses carried out, probability bands of exceedance were obtained for the three key deformation indicators. The spatial probability distributions of exceeding limit values for individual deformation indicators demonstrate significant similarity in shape across the displacement field created around the case study mining area (Figure 5).

Particularly in the center of the subsidence trough, there is a substantial probability of exceeding the displacement S limit values. This logically aligns, as the adopted limit values corresponded more to the range of harmful influences rather than the regions of maximal displacement values. The values of both considered model parameters, namely the exploitation coefficient "a" and the influence dispersion parameter "β" determine the range of harmful impacts on the surface depending on the planned mining depth and the deposit thickness.

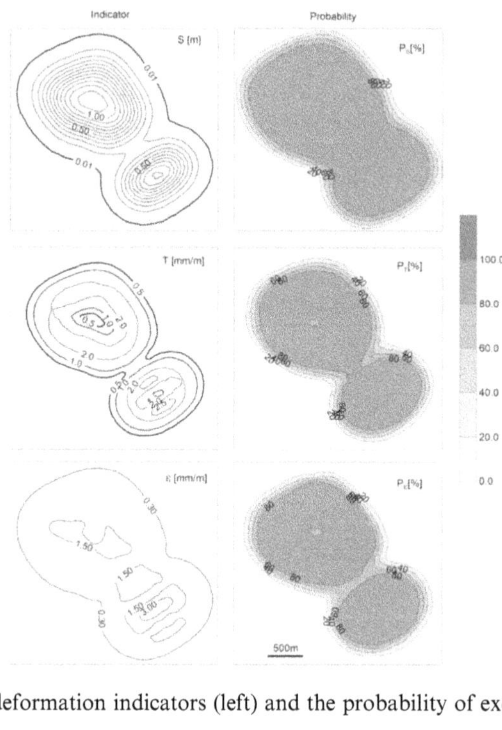

Figure 5. Calculated deformation indicators (left) and the probability of exceeding the limit values (right).

The range of damaging impacts determined for vertical displacements may be exceeded with a probability significantly less than 10% (see Figure 5). A similar trend is observed for trough slopes and horizontal strains. The presented method of investigating the probability of exceeding the limits set for a given indicator can therefore be used to assess the safety of objects on the surface. For instance, for a classic building, it could be assumed that buildings located in the 20% probability zone may not suffer any damage relevant to the comfort of their use. However, objects that are more sensitive to ground deformation (e.g., hydro-technical structures, energy facilities) should be moved further away from the contours of the mining operation to a probability zone of less than 5%. This method can also be used to indicate a zone that is not at risk of deformation at all.

3.2 *Example 2 – exploitation under a housing estate*

One of the criteria for permitting exploitation is predicting its impact on surface structures (pipelines, railroads, highways, bridges). To analyze the impact of mining on structures and technical infrastructure, it is crucial to determine the allowable horizontal strains for specific structures. A viable approach involves engineering experts assessing structural strength. Accepted threshold values for different building types are considered as limits that should not be exceeded. The hazard assessment compares estimated extreme deformation indicators against these threshold values, highlighting at-risk structures. In such cases, understanding the probability of exceeding deformation limits, as defined by Equation (14), is beneficial.

The case study presented concerns coal exploitation in a residential area (Figure 6). A total of 111 buildings were investigated. Thirteen mining walls across four coal layers, ranging in depth from 450 m to 670 m, were chosen for computations. The mining method employed is the longwall system with roof caving. For this region and mining method, specific model parameter values have been adopted as follows:

$$\begin{cases} a = 0.7 \pm 0.07 \\ tg\beta = 1.9 \pm 0.19 \end{cases} \tag{26}$$

Prior to commencing new operations, a survey of existing buildings was conducted due to previous mining-related damage. The regulations in force in Poland for the protection of buildings in areas affected by deformation require an assessment of the resilience of these structures in terms of their vulnerability to ground deformation. Such assessments are typically carried out by building specialists and involve analyzing various attributes of a building (e.g., soil homogeneity in the subsoil, type of foundation, building shape, type of masonry material, etc.) to determine their ability to withstand deformation.

During the assessment, each attribute is assigned a certain number of points indicating whether it is rated low or high in terms of resistance. The sum of these points corresponds to

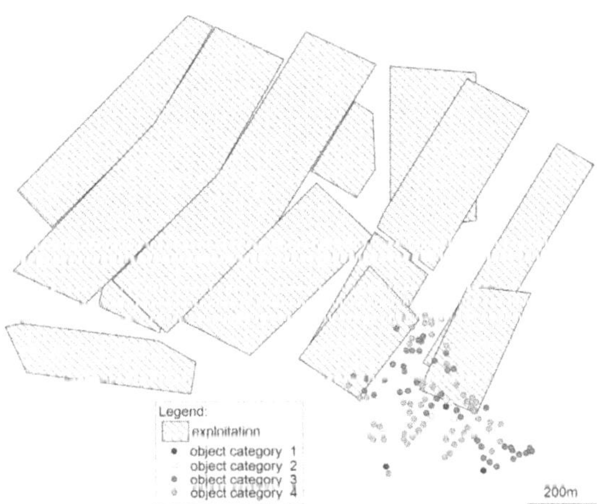

Figure 6. Designed mining operation and residential area on the surface.

the building's resistance category (Dzegniuk *et al.* 1997; Malinowska and Hejmanowski 2010). This classification reflects the building's sensitivity to horizontal strain. Each resistance category corresponds to a specific range of horizontal strain values and is marked on the map with the corresponding color (Figure 6). The most vulnerable buildings are marked in blue, while the most resilient are marked in green.

For the example analyzed here, the individual buildings located in the area at risk from mining deformation are summarized in Table 3

Table 3. Number of categorized buildings based on their resistance.

Building resistance category	Horizontal Strain ε [mm/m]	Colour (Figures 6, 7)	Count
0	(0.0;0.5⟩	gray	0
1	(0.5;1.5⟩	blue	3
2	(1.5;3.0⟩	yellow	7
3	(3.0;6.0⟩	red	42
4	(6.0;9.0⟩	green	59

Most buildings fall within resistance categories 3 and 4, although categories 1 and 2 are also represented. Using theoretical solutions, deformation indicators were computed and subjected to statistical analysis. The upper strain limits associated with each resistance category (Table 3) were considered for this statistical analysis. The results are summarized in Table 4 and depicted graphically in Figure 7.

Table 4. Number of buildings with calculated probability of exceeding horizontal strain limits.

Probability [%]		Number of buildings in resistance category			
from	to	1	2	3	4
0	10	0	0	31	59
10	20	0	0	1	0
20	30	0	0	0	0
30	40	0	0	0	0
40	50	0	0	1	0
50	60	0	0	3	0
60	70	0	0	4	0
70	80	0	0	1	0
80	90	0	2	1	0
90	100	3	5	0	0

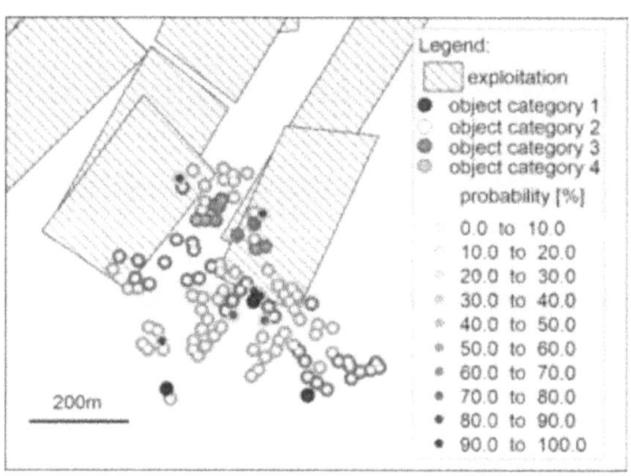

Figure 7. Calculated probability of exceeding horizontal strain limits.

The analysis suggests that most buildings will not suffer significant damage due to mining operations. Of the 111 surveyed buildings, around 20 exhibit a probability of exceeding the horizontal strain limit values by over 40%. All Category 1 buildings are at risk of exceeding horizontal deformation limits with a probability of more than 90%. For Category 2 buildings, two have a probability exceeding 80%, and five have a probability exceeding 90%. Most Category 3 buildings are not at risk, but for nine of them, the probability of exceeding the horizontal strain limits is greater than 50%. For two of these, the probability is very significant, exceeding 70%. Therefore, damage due to developing surface deformation is almost certain for these buildings. All 59 buildings included in the highest resilience category remain unthreatened. For buildings with a very high probability of exceeding the threshold values, appropriate prophylactic measures (e.g., building safeguards) should be recommended prior to the start of mining operations.

4 SUMMARY

The presented study investigates the random distribution of modeled ground deformations caused by underground mining and their influence on the safety of buildings and engineering structures.

The research focuses on assessing the impact of errors in the basic parameters of the stochastic deformation model on the results of the predicted surface deformation distribution. Based on the estimated surface deformation probability, the likelihood of exceeding safety limits for any surface object was estimated. These limits refer to the absence of significant hazards in the flanking regions of the displacement field.

Using the stochastic ground deformation model, two diverse examples were presented to interpret the potential hazards to buildings and the likelihood of exceeding predefined safety factor limit values. The first example assesses the probability of harmful effects during coal mining exploitation in a longwall system. The issue of ground deformation exceeding the determined safety factor value was investigated. By merging the deformation distribution with the probability distribution, the probability of limit value exceedance can be established. Prediction certainty can be estimated based on the importance of the objects or areas subjected to ground deformation. This solution supports the determination of harmful influence ranges and the probability of exceeding threshold values.

Statistical analysis and deformation calculations enabled the determination of the likelihood of exceeding the limits of horizontal strain, which has a harmful effect on building and engineering infrastructure.

The paper emphasizes the role of statistical analyses in improving the reliability of deformation modeling. The presented statistical analyses support theoretical simulations of ground deformation by incorporating the randomized character of model parameters into hazard estimation.

The article outlines fundamental statistical analysis for prediction models involving error propagation laws and the Monte Carlo method. The study reveals that statistical analysis provides additional information about the randomness of the deformation process and that knowledge about the uncertainty of the prediction results can increase awareness of the risk of damage to fragile objects. The investigation is not limited to the probability influence function of ground deformation. The assumption about the randomized character of modeling parameters can be applied in any geo-mechanical or stochastic solution for ground deformation estimation.

The presented methodology demonstrates that integrating statistical analyses with theoretical predictions can enhance safety in mining terrains.

This research was supported by the AGH University of Krakow under subsidy No. 16.16.150.545.

REFERENCES

Aschbacher, J. (2017): ESA's Earth Observation Strategy and Copernicus. In: Onoda, M., Young, O. (eds) *Satellite Earth Observations and Their Impact on Society and Policy*. Springer, Singapore. https://doi.org/10.1007/978-981-10-3713-9_5, pp. 81–86.

Bagheri-Gavkosh, Mehdi and Hosseini, Seiyed Mossa and Ataie-Ashtiani, Behzad and Sohani, Yasamin and Ebrahimian, Homa and Morovat, Faezeh and Ashrafi, Shervin (2021): Land subsidence: A global challenge. *Sci. Total Environ.*, vol. 778. https://doi.org/10.1016/j.scitotenv.2021.146193

Bahuguna P.P., Srivastava A.M.C. and Saxena N.C. (1991): A critical review of mine subsidence prediction methods. *Mining Science and Technology*, Volume 13, Issue 3, 1991, Pp 369–382,

Bals R. (1931/32): *Beitrag zur Frage der Vorasusberechnung bergbaulicher Senkungen*. Mitt. Aus dem Markscheidewesen, Stuttgart

Bang, Z., Yueguan, Y. and Jianrong, K. (2023): Dynamic Prediction Model for Progressive Surface Subsidence Based on MMF Time Function. *Appl. Sci.*, 13(14), 8066; https://doi.org/10.3390/app13148066

Bernardino, P., Fornaro, G., Lanari, R. and Sansosti, E.: A New Algorithm for Surface Deformation Monitoring Based on Small Baseline Differential SAR Interferograms. *IEEE Transactions on Geoscience and Remote Sensing*. 40(11):2375–2383. DOI: 10.1109/TGRS.2002.803792

Busch W., Walter D., Coldewey W.G. and Hejmanowski R. (2015): Bergwerk Auguste Victoria der RAG AG. Analyse von Senkungserscheinungen außerhalb des prognostizierten Einwirkungsbereiches. Gutachten im Auftrag der Bezirksregierung Arnsberg. *Institut für Geotechnik und Markscheidewesen*, TU Clausthal, Clausthal-Zellerfeld. (in German)

Busch W., Walter D., Yin X., Coldewey W.G. and Hejmanowski R. (2017): Bergwerk Lippe der RAG AG. Analyse von Senkungserscheinungen außerhalb des prognostizierten Einwirkungsbereiches. Gutachten im Auftrag der Bezirksregierung Arnsberg. *Institut für Geotechnik und Markscheidewesen*, TU Clausthal, Clausthal-Zellerfeld. (in German)

Cai, Y., Jin, Y., Wang, Z. *et al.* (2023): A review of monitoring, calculation, and simulation methods for ground subsidence induced by coal mining. *Int J Coal Sci Technol* 10, 32. https://doi.org/10.1007/s40789-023-00595-4.

Can E., Kuşcu S., and Kartal M.E. (2012): Effects of mining subsidence on masonry buildings in Zonguldak hard coal region in Turkey. *Environmental Earth Sciences*, 66(8): 2503–2518.

Chai, H.; Xu, M.; Guan, P.; Ding, Y.; Xu, H.; Zhao, Y. (2023): Research on Mining Subsidence Prediction Parameter Inversion Based on Improved Modular Vector Method. *Appl. Sci.* 2023, 13, 13272. https://doi.org/10.3390/app132413272

Cigna F., Tapete, D. (2020): Mapping land subsidence in urban areas using ESA's G-POD and the P-SBAS InSAR technique: examples in Asia, South America and North Africa. *2020 Mediterranean and Middle-East Geoscience and Remote Sensing Symposium (M2GARSS)*. DOI: 10.1109/M2GARSS47143.2020.9105169

Darling P. (2011): *SME Mining Engineering Handbook*, Third Edition. Society for Mining, Metallurgy, and Exploration (U.S.)

Dinar, Ariel and Esteban, Encarna and Calvo, Elena and Herrera, Gerardo and Teatini, Pietro and Tomás, Roberto and Li, Yang and Ezquerro, Pablo and Albiac, Jose (2021): We lose ground: Global assessment of land subsidence impact extent. *Science of The Total Environment*. 2021, vol.786, pp.147415, https://doi.org/10.1016/j.scitotenv.2021.147415

Dzegniuk, B., Hejmanowski, R., Sroka, A. (1997): Evaluation of the damage hazard to building objects on the mining areas considering the deformation course in time. *Proceedings of Xth international congress of the international society for mine surveying*, Perth/Australia, 1997, pp. 2–6.

Ehrhardt W., Sauer A. (1961): Die Vorausberechnung von Senkung, Schieflage und Krümmung über dem Abbau in flacher Lagerung. *Bergbauwissenschaften*. Vol. 8. pp. 415–428 (in German)

Faunt, C.C., Sneed, M., Traum, J. and Brandt, J.T. (2016): Water availability and land subsidence in the Central Valley, California, USA. *Hydrogeol. J.* 24, 675–684.

Fokker, P.A., Osinga, S. (2018): On the Use of Influence Functions for Subsidence Evaluation. *ARMA*. 18–613.

Gambolati, G., and P. Teatini (2015): Geomechanics of subsurface water withdrawal and injection, *Water Resour. Res.*, 51, 3922–3955, doi:10.1002/2014WR016841.

Gazzola, L., Ferronato, Massimiliano and Teatini, Pietro and Zoccarato, Claudia and Corradi, Anna and Dacome, Maria Carolina and Mantica, Stefano (2023): Reducing uncertainty on land subsidence modeling prediction by a sequential data-integration approach. Application to the Arlua off-shore reservoir in Italy. *Geomechanics for Energy and the Environment.* 2023, vol.33, pp. 100434, https://doi.org/10.1016/j.gete.2023.100434

Gruszczyński, W.; Niedojadło, Z.; Mrocheń, D. (2018): Influence of model parameter uncertainties on forecasted subsidence. *Acta Geodyn. Geomater.* 2018, 15, 211–228
Hassan Z. Al Garni and Anjali Awasthi (2020): in *Handbook of Probabilistic Models*, 2020
Hejmanowski R., Malinowska A., Kwinta A. and Patykowski G. (2016): Prediction of land subsidence and deformations at copper ore underground mining site: experiences and verification based on KGHM mines in Poland. *Proceedings XVI International Congress for Mine Surveying*. Brisbane 2016. str.183–186,
Hejmanowski R., Malinowska A., Kwinta A. and Ulmaniec P. (2018):Modeling of land subsidence caused by salt cavern convergence applying Knothe's theory. Markscheidewesen 125, No.2, str. 40–45,
Hoffmann, J., Galloway, D.L. and Zebker, H.A. (2003): Inverse modeling of interbed storage parameters using land subsidence observations, Antelope Valley, California. *Water Ressources Research*. VOL. 39, NO. 2, 1031
Ilieva M, Polanin P, Borkowski A, Gruchlik P, Smolak K, Kowalski A and Rohm W (2019): Mining deformation life cycle in the light of InSAR and deformation models. *Remote Sens* 11(7):745
Janna, C., Castelletto, N., Ferronato, M., Gambolati, G. and Teatini, P. (2012): A geomechanical transversely isotropic model of the Po River basin using PSInSAR derived horizontal displacement. *International Journal of Rock Mechanics and Mining Sciences*, 2012, vol.51, pp.105–118. https://doi.org/10.1016/j.ijrmms.2012.01.015
Knothe S. (1953): Równanie profilu ostatecznie wykształconej niecki osiadania. *Archiwum Górnictwa i Hutnictwa*, t.1, z.1, Warszawa
Kochmański T. (1956): Obliczanie ruchów punktów górotworu pod wpływem eksploatacji górniczej. *Państwowe Wydawnictwo Naukowe*, Warszawa,
Kratzsch H. (1983): *Mining Subsidence Engineering*. Springer-Verlag Berlin Heidelberg New York.
Kwinta A. (2011): Application of the least squares method in determination of the Knothe deformation prediction theory parameters. *Archives of Mining Sciences*, Vol. 56, Issue 2, Kraków str. 319–329
Kwinta A. (2010): Assessment of Outliers in Deformation Calculated from Geodesy Observation on the Mining Areas. *Journal of Mining Science*. Vol 46, no. 4, Wyd. Springer, str. 375–383
Kwinta, A. and Gradka, R. (2020): Analysis of the damage influence range generated by underground mining. *International Journal of Rock Mechanics and Mining Sciences*, vol. 128,
Lian Sue, Sue Zhao, Hu Li, Jie Zheng, Xiuqiang Lei and Sue Gong (2023): Genetic algorithm-based parameter inversion and pipeline subsidence prediction. *Journal of Applied Geophysics*, Jul. 2023
Luo Y and Cheng JW (2009): An influence function method based subsidence prediction program for longwall mining operations in inclined coal seams. *J Min Sci Technol* 19(9):592–598
Luo, Y. and Qiu, B. (2012): Enhanced subsurface subsidence prediction model that considers overburden stratification. *Min Eng* 64(10):78–84
Malinowska, A. and Hejmanowski, R. (2010): Building damage risk assessment on mining terrains in Poland with GIS application. *International Journal of Rock Mechanics and Mining Sciences* 47 (2), 238–245
Malinowska, A. and Hejmanowski, R. (2016): Significance of the uncertainty level for the modeling of ground deformation ranges. *International Journal of Rock Mechanics & Mining Sciences* 83, 140–148, doi: 10.1016/j.ijrmms.2015.12.019
Marian, D.P., Onica, I., Marian, R.R., and Floarea, D. (2017): Surface Subsidence Prognosis Using the Influence Function Method in the Case of Livezeni Mine. *Mining Revue*, 23(1), 9–14.
Merad, M.M., T. Verdel, B. Roy, and S. Kouniali (2004): "Use of Multi-Criteria Decision-Aids for Risk Zoning and Management of Large Area Subjected to Mining-Induced Hazards." *Tunnelling and Underground Space Technology* 19 (2): 125–138
Peng SS (2020): *Surface subsidence engineering: theory and practice*. CSIRO Publishing, Clayton South.
Puniach, E., Gruszczynski, W., Cwiakala, P. and Matwij, W. (2021): Application of UAV-based orthomosaics for determination of horizontal displacement caused by underground mining. *ISPRS Journal of Photogrammetry and Remote Sensing*; ISSN 0924-2716. 2021 vol. 174, s. 282–303.
Qiao, Xiaojun, Chu, Tianxing, Tissot, Philippe, Holland and Seneca: (2023): Sentinel-1 InSAR-derived land subsidence assessment along the Texas Gulf Coast. *International Journal of Applied Earth Observation and Geoinformation*. Vol. 125. https://doi.org/10.1016/j.jag.2023.103544
Ren, G., D.J. Reddish, and B.N. Whittaker (1987): Mining Subsidence and Displacement Prediction Using Influence Function Methods. *Mining Science and Technology* 5 (1): 89–104.
Saeidi A., Deck O., Seifaddini M., Marwan A.L. Heib and Verdel T. (2021): An improved methodology for applying the influence function for subsidence hazard prediction, *Georisk: Assessment and Management of Risk for Engineered Systems and Geohazards*
Sann B. (1949): Betrachtungen zur Vorausberechnung von Bodensenkungen infolge Kohlenabbaues. Bergbau-Rundschau 1 t.7

Schmitz A. (1924) Bodenbewegungsvorgänge im Bergbau. Mitt. Aus dem Markscheidewesen, Stuttgart.
Sheorey P.R, Loui J.P, Singh K.B and Singh S.K., (2000). Ground subsidence observations and a modified influence function method for complete subsidence prediction. *International Journal of Rock Mechanics and Mining Sciences* vol 37 no.5, pages 801–818. DOI: 10.1016/S1365-1609(00)00023-X
Taherynia M.H., Aghda S.M., Ghazifard A. and Moradi E. (2017): Prediction of Subsidence Over Oil and Gas Fields with Use of Influence Function (Case study: South Pars Gas Filed, Iran). *Iranian Journal of Science and Technology*, 2 (41), 375.
Vasco, D.W., Kim, K.H., Farr, T.G., Reager, J.T., Bekaert, D., Sangha, S.S., Rutqvist, J. and Beaudoing, H. K (2022): Using Sentinel-1 and GRACE satellite data to monitor the hydrological variations within the Tulare Basin, California. *Nature, Scientific reports*, 12:3867.
Weitao Yan, Junting Guo, Junjie Chen, Yi Tan, Shaoge Yan and Yueguan Yan (2022): Severe Damage Law on the Ground Surface Induced by High-Strength Mining: A Case Study From the Shendong Coal Field in China. *Front. Earth Sci.*, 29 April 2022 Sec. Geohazards and Georisks Volume 10 – 2022. https://doi.org/10.3389/feart.2022.827826
Whittaker B.N. and Reddish D.J. (1989): *Subsidence, Occurrence, Prediction and Control.* Elsevier.
Yoginder P. Chung Qing-Wang Hao Fu-Sheng Zhu (1989) State of the art. *In mine subsidence prediction.* Land Subsidence 1989" Publ. Balkema.
Zhou, DW., Wu, K., Cheng, GL. (2015): Mechanism of mining subsidence in coal mining area with thick alluvium soil in China. *Arab J Geosci* 8, 1855–1867.
Zhu L, Gong H, Li X, Wang R, Chen B, Dai Z and Teatini P (2015): Land subsidence due to groundwater withdrawal in the northern Beijing plain, China. *Engineering Geology* 193:243–255, DOI: https://doi.org/10.1016/j.enggeo.2015.04.020.

Mine Surveying and Mining Area Protection – Sokoła-Szewioła and Jarczyk (Eds)
© 2025 The Author(s), ISBN: 978-1-041-10531-2

Multivariate analysis of the accuracy of the basic geodetic control network for guiding the axis of the tunnel built using a TBM machine on the LK104 railway line

T. Lipecki & W. Jaśkowski
AGH University of Krakow, Poland

ABSTRACT: Precise guidance of the TBM (Tunnel Boring Machine) machine requires ensuring the accuracy of the drilling direction location at the level of +/- 1 cm for each kilometer of the tunnel with a length of almost 4 km. This means the need to determine and control not only the points of the implementation control network but above all the basic control network, the accuracy of which will be at the above-mentioned level. The article presents analyses that allow determining the technology of measuring this basic control network, in which an essential element was the connection of angular-linear measurements with gyro-azimuth measurements. This analysis also indicated the need to provide geodetic service using the highest accuracy instruments, which will enable the implementation of the design assumptions in the form of a transverse axis error at the end of the tunnel. The article also presents analyses of GNSS measurements confirming the determination of the accuracy of the reference points near the starting and ending chambers of the tunnel.

1 INTRODUCTION

After the social transformation in the 1990s, Poland also began to transform its road and rail transport network. In the first decades, the network of expressways and motorways was developed, expanding them to the current total of over 5,100 km compared to about 500 km in the 1990s. Every year, vehicle traffic increases by about 10-15% per year, which further determines the development of expressways. The transformation of rail routes looks slightly different, as they have been significantly limited and operators minimized the crisis by eliminating connections, but the strategic need to return to the development of this type of transport was still perceived. When Poland joined the European Union, the reconstruction and modernization of rail routes began, adapting them to modern trains traveling at higher speeds. Currently, an airport construction program is being implemented, called the Central Communication Port (CPK), between Warsaw and Łódź, for which construction of the so-called spokes of the High Speed Railway (HSR), which are to provide a radial connection of all major Polish cities with the CPK. Both types of expanded transport networks (expressways and railway routes) will significantly interfere with the existing development network, often passing through foothill and mountainous areas. In connection with this, tunnel construction is starting to play an important role, carried out using classic mining methods but also using tunnel boring machines (TBM), which require carrying out the investment to the so-called break-in point, located in the middle or end of the tunnel (Camus and Moubarak 2015). Due to the simultaneous implementation of the remaining engineering infrastructure on both sides of the tunnels under construction, it is necessary to guide the tunnel axis with very high accuracy, which will allow the tunnel under construction to fit into the already

DOI: 10.1201/9781003655558-13

built engineering structures. A huge role in this task is played by the surveying services, which must prepare the so-called surveying breakthrough project (SBP), containing the tunnel network project (basic and implementation), acting as a georeference for guiding the shield and a control for any correction of its axis.

The article presents multi-variant analyses showing the method of preparing the final form of the SBP as well as indicating the very strict measurement regime that must be maintained to meet the design assumptions of the construction of the tunnel on the LK104 railway line.

2 STUDY AREA AND MATERIAL

2.1 *Study area*

The analysed tunnel facility is 3750 m long and is located on the new, 57 km section of the modernized LK104 railway line, connecting Nowy Sącz with Kraków (Figure 1) and further with CPK, as a supplement to the TEN-T network (Trans-European Transport Network) covering the European transport corridors of the core network. Thanks to this modernization, it will be possible to travel at a speed of up to 160 km/h. On the modernized and newly designated section, it is planned to build 16 tunnels, of which the two longest will be about 4 km long and in addition to the transport tunnels, emergency tunnels will be drilled in parallel.

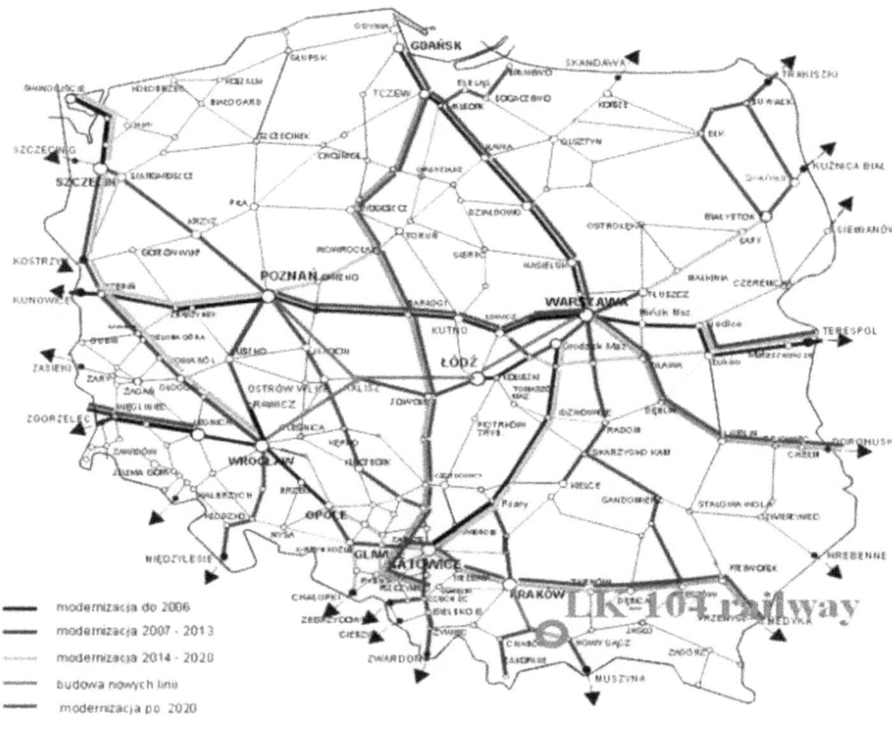

Figure 1. Location of the modernized LK104 railway line in Poland (based on https://mavink.com/post/9F4820272774514FFDFA3720EDFEA6EFEBAM75FC44/pkp-rail-map) access on June 6, 2025.

The use of TBM shields is planned for them (Figure 2), in which the largest one will have a diameter of almost 11 m and a weight of 1,700 tons. Prefabricated housing segments with a length of 2 m, thickness of 0.4 m and a single segment weighing 10,500 kg will be installed after it. A navigation system based on laser setting of the direction in relation to the backward control points and control of this set direction in relation to the design position and orientation in a dedicated computer system in the operator's cabin is planned for guiding the shields (Park *et al.* 2023).

The analysed tunnel is being constructed between the towns of Męcina and Mordarka under the Pisarzowa mountain and will be 3,750 m long. The smaller TBM shield (diameter 4.8 m) is currently drilling an evacuation tunnel, while the larger shield (diameter 10.8 m) is being prepared for drilling. Every 500 m both tunnels will be connected by cross tunnels, allowing evacuation in case of danger.

Figure 2. View of the 11 m diameter TBM under construction in front of the tunnel. (source: https://limanowa.in/aktualnosci/prawie-300-metrow-tunelu-druga-tarcza-tbm-ruszy-w-czerwcu/77077?image=7, left) and: https://www.plk-sa.pl/o-spolce/biuro-prasowe/informacje-prasowe/szczegoly/niemal-300-metrow-podziemnej-podrozy-kingi-jadwiga-w-blokach-startowych-10761 (right), access on June 6, 2025.

The TBM shield is equipped with at least three guidance devices. The basic guidance system consists of a gyroscope and a laser beam emitter (Mao *et al.* 2013, 2015) with a sufficient effective range. The number of markers placed at the back is at least 1 (depending on the radius of curvature) and allows for angular-linear navigation in the horizontal plane but also vertically (Figure 3)

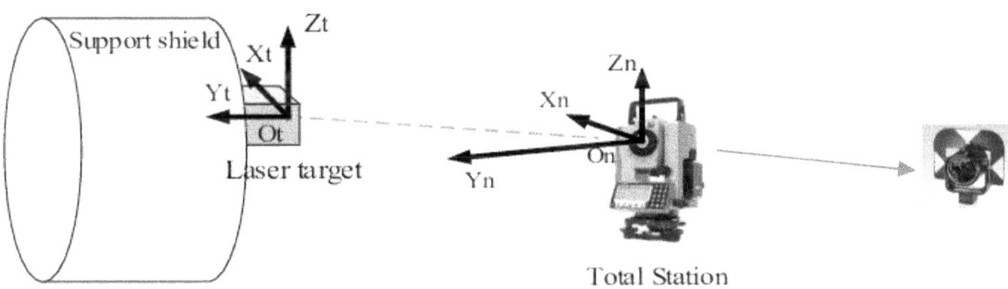

Figure 3. Principle of dial guidance using angular linear measurements (https://www.researchgate.net/figure/The-schematic-for-measuring translation matrix T n t_fig2_356516308).

2.2 *Material*

The requirements for the accuracy of guiding the basic tunnel network are essential for the correct implementation of implementation activities, performed simultaneously on many engineering objects located along the entire length of the modernized section of the railway line. For this reason, the precision of guiding the tunnel boring machine (TBM) is the basic criterion for the quality of work. According to the project, the required verification covers a radius of 0.1 m from the functional radius of the tunnel. The diameter given on the tunnel section plans includes all permissible deviations in the alignment, guidance, implementation and deformation of the segments. The shield shift has a constant value with information on the deviation at the end of each cycle. In other words, the permissible tolerance in relation to the theoretical axis is 10 cm along the radius, i.e. perpendicular to the boring axis. Therefore, it is primarily important to maintain the criterion of not exceeding the transverse boring error. Taking into account the relationship between the probability characterizing the definition of tolerance T (P=100%) and the measurement uncertainty - the mean measurement error M (P=68%), they are described by the value of the coefficient $k=T/M \geq 3$. This means that when characterizing errors in geodetic measurements and determination of network coordinates at the error level M, one should strive to ensure that they do not exceed the value T, taking into account the coefficient $k = 3$. For this reason, with the assumed design tolerance of 10 cm, the value of the so-called transverse error in determining coordinates should not exceed 100 mm / 3 = 33 mm, also at the end after drilling almost 4 km of tunnel. Due to the accumulation of elementary errors of geodetic measurements, including the impact on the measurement of each activity (centering, leveling, accuracy of angle and distance measurement, number of measurement series, weather conditions, measurement techniques), it was necessary to perform a multi-variant analysis of the accuracy of the geodetic control network for guiding the TBM shield, allowing for the selection of the optimal method of measurement and determination of coordinates. This translates into the accuracy of guiding the shield in accordance with the design, within the imposed tolerance. This analysis is based on the so-called adjustment of observations using the method of minimizing squared deviations (Least Square Method - LSM), assuming that the mean unit error of adjustment $m_0 = 1$ (Weiss *et al.* 2022). The general calculation scheme is based on the following formulas:

$$m_x = \sqrt{Q_{xx}}, \quad m_y = \sqrt{Q_{yy}}, \quad m_{xy} = \sqrt{Q_{xy}} \tag{1}$$

$$Q = m_0 \cdot \left[N^{-1} + A^T \cdot P \cdot A\right]^{-1} \tag{2}$$

$$N = \begin{bmatrix} m_{x_0}^2 & m_{xy_0} \\ m_{xy_0} & m_{y_0}^2 \end{bmatrix} \tag{3}$$

where: Q_{xx}, Q_{yy}, Q_{xy} = elements for the matrix point from the variance-covariance matrix Q of the analysed matrix, where the unit error of the matrix $m_0 = 1$; N = error matrix of the reference points; P = error matrix of the mean observations; A = matrix of observation equations.

3 METHODS

The network near the starting and ending portals of the bored tunnels was also designed, by connecting it with the railway network of the route. The location of the starting point G1 and the ending point G2 was designed as the center of gravity of the points surrounding the external network, located no further than 550 m away. The error of the average coordinates of the reference points of the network should not exceed ±3 mm for each of the coordinates. Then, points were planned in the tunnels halfway between the transverse tunnels, i.e. about 240 m long, alternately located under the left and right sidewalls of the tunnel (to reduce the effect of refraction). The scheme of the designed points of the tunnel network is presented in Figure 4. This network, which is a polygon close to a rectilinear one, was planned so that its points in each tunnel could be used for mutual connection after the transverse tunnels were driven through (which will only happen after each tunnel is bored). Unfortunately, such an investment schedule also affects the measurement technique, making it impossible to strengthen the network by chain-linking the networks in parallel tunnels. Therefore, the only possibility, apart from the network geometry, is to perform angular-linear measurements in the appropriate measurement regime with high-precision equipment, or to add gyro-azimuths (Figure 5), which allow to minimize the growth of the reference azimuth errors at the very beginning of the tunnel and the accumulation of transverse errors from angular measurements (Pielok *et al.* 2011). Based on the previously presented analyses, it was assumed that the average error of the situational coordinates of the tunnel network should not exceed ±10 mm/km of the tunnel and ±2 mm/km of the tunnel in the vertical direction.

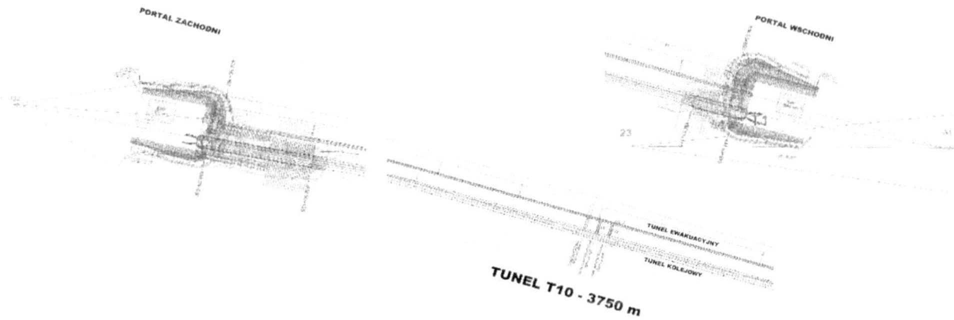

Figure 4. View of the geodetic network at the tunnel portals (west - left and east - right) and inside the tunnel (middle).

As part of the multi-variant analysis, it was determined that modeling of the formation of the mean error of the coordinates of the control network (especially the transverse error) will be performed, and additionally an assessment of the error in determining the azimuth will be performed, which is in accordance with Polish regulations on conducting surveying works in mining plants. These also include tunneling works using TBM.

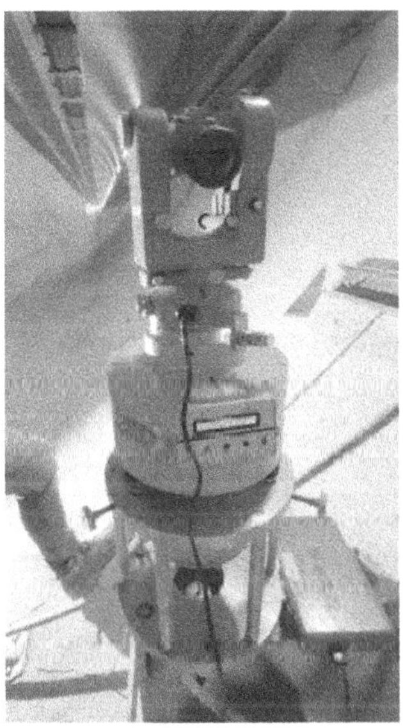

Figure 5. Gyrotheodolite Gyromat 2000 during measurement in the tunnel (own photo).

Eight variants were prepared, the parameters of which are presented in Table 1:

Table 1. Parameters assigned to calculation variants.

Variants	Angle error [mgon]	Distance error [mm+ppm]	Reference coord error m_x i m_y_[mm]	number of gyro-azimuths	giroazim. error [mgon]
I	0.5	0.8+1ppm	2	0	—
II	0.5	0.8+1ppm	2	3	1.2
III	0.5	0.8+1ppm	2	4	1.2
IV	0.3	0.8+1ppm	2	0	—
V	0.3	0.8+1ppm	2	3	1.0
VI	0.3	0.8+1ppm	2	4	1.0
VII	0.3	0.8+1ppm	2	4	1.2
VIII	0.3	0.8+1ppm	3	3	1.0

The angular-linear measurements declared in the calculations will be carried out at each point of the network and the gyro-azimuth measurements (Lipecki and Jaśkowski 2011, Szafarczyk *et al.* 2019) will be carried out in schemes – 3 stations (0, 1, 2 and 2.8 km of the tunnel) or 4 stations (0, 1, 2 and 3 km of the tunnel). As mentioned, the analysis was based on LSM allowing for minimization of the sum of squares of the assumed observation errors, which are given in Table 1, assuming that their balancing factor m_0 is 1. The Geodezja B. Cisło program was used for modeling, from which the calculation values are obtained in the form of tabulations and statistical summaries describing the data sets. The most important element used in the analysis are the obtained azimuth errors of the network sides, mean errors along the X and Y axes of the coordinate system. Due to the different orientation of the tunnel to the cardinal directions (other than XY), it is crucial to analyse the parameters of the error ellipse, which in the discussed case of a rectilinear, unilaterally connected line is identified with its semi-axes A (transverse error) and B (longitudinal error). The relationship between the reference system 2000 in which the XY coordinates are calculated and the axes of the error ellipses A, B and the angle φ of their inclination are presented in the following formulas:

$$\tan 2\varphi = \frac{2m_{xy}}{m_x^2 - m_y^2} \tag{4}$$

$$A^2 = \frac{1}{2}\left(m_x^2 + m_y^2 + \sqrt{\left(m_x^2 - m_y^2\right) + 4m_{xy}^2}\right) \tag{5}$$

$$B^2 = \frac{1}{2}\left(m_x^2 + m_y^2 - \sqrt{\left(m_x^2 - m_y^2\right) + 4m_{xy}^2}\right) \tag{6}$$

Where: m_x, m_y, m_{xy} = errors of the coordinates of the collision points and the covariance of the coordinates of the last point, A, B = semi-axes of the error ellipse, φ = slope of the error ellipse with respect to the XY axes.

4 RESULTS

Based on the assumed analysis criteria, the results were obtained, which are presented in Table 2. It should be noted that none of the calculation variants without gyroazimuth measurement (variants I and IV) meet the tolerance criterion (the transverse error of the final side should be less than 33 mm). Also, one of the variants with gyroazimuth measurement (variant II) does not meet this criterion. The reason is too few gyroazimuth measurement stations with lower accuracy of determining this azimuth. The remaining variants meet the project assumptions and it can be stated that the presented regimes of angular and linear measurements together with gyroazimuth measurement will guarantee the correctness of the execution of the drilling using TBM shields, with simultaneous laying of the final support.

Table 2. Accuracy characteristics of the tunnel network end points (3750m).

Variant	Angle error [mgon]	Number of gyro-azimuth	Gyro-azimuth error [mgon]	Error of last side azimuth [mgon]	Transverse error of end side [mm]*
I	0.5	0	—	1.5	58
II	0.5	3	1.2	1.05	36.4
III	0.5	4	1.2	0.9	32.5
IV	0.3	0	—	0.92	37.9
V	0.3	3	1.0	0.71	27.2
VI	0.3	4	1.0	0.68	24.9
VII	0.3	4	1.2	0.72	27.3
VIII	0.3	3	1.0	0.73	29.3

* This error corresponds to the semi-axis of the error ellipse A and should not be greater than ±33 mm in order to maintain the design tolerance of the tunnel axis position at the end of boring.

Referring to the values of the obtained azimuths of the last sides of the tunnel network, it can be assumed that each variant allows for a decrease in the valuesof the topographic azimuths of the sides after the application of gyro-azimuth measurements, stabilizing at the level of ±0.7 mgon (7^{cc}). This means that there is no need to increase the number of gyro-azimuths and it is enough to perform measurements at three stations, while maintaining the accuracy of angular measurements at the level of ±0.3 mgon (3^{cc}) to obtain the desired design effect. For this reason, variant V seems to be the optimal. However, there is an alternative (variant III), which indicates the possibility of realizing the assumptions even when the angle measurements are burdened with an error of ±0.5 gon (5^{cc}). However, it is then necessary to determine 4 gyroazimuths, even if they have an error of ±1.2 mgon (12^{cc}).

The results described in the table are also presented in Figure 6, and the distribution of coordinate errors in the form of their ellipses is presented for the optimal variant (V) in Figure 7.

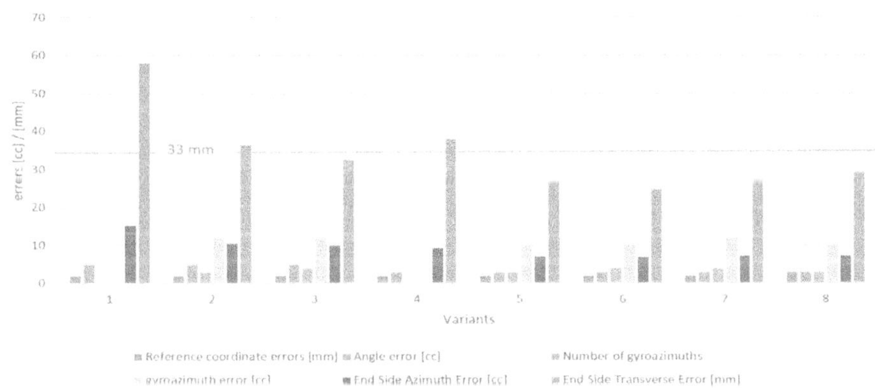

Figure 6. Propagation of lateral errors of the last point of the tunnel network (3750 m) in relation to the maximum error and the adopted calculation variants.

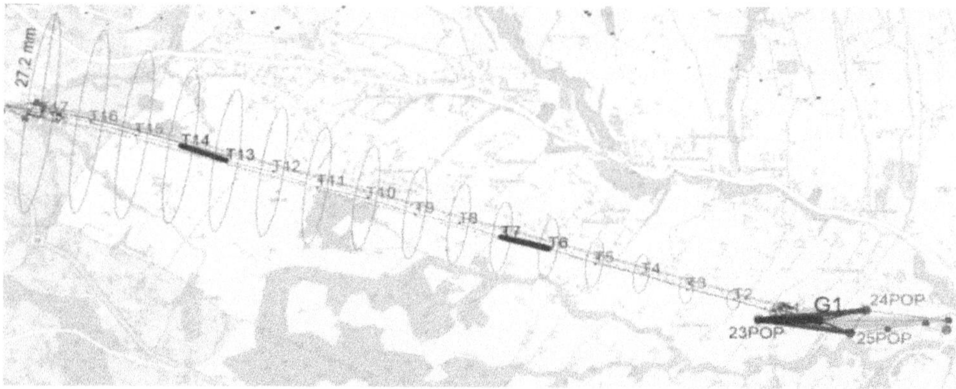

Figure 7. Variant V optimal - distribution of coordinate errors of points of the basic tunnel network in the form of error ellipses on the background of the map and the railway network. Bold lines indicate the sides of the determined gyroazimuths at the beginning and in the middle of the tunnel (0, 1.2, 2.8 km).

The influence of increasing the errors of the reference coordinates (variant VIII) was also analysed, in which the same assumptions were adopted as in variant V (optimal), however, the reference errors mx and my were assumed to be ±3 mm each. As can be seen in the table, increasing these errors did not in principle change the accuracy of determining the transverse error at the end of the tunnel (±29.3 mm compared to ±27.2 mm in variant V). The correctness of adopting the reference error value at the level of ±2 mm for both coordinates X and Y is evidenced by the GNSS campaign conducted using the static method, carried out twice independently on the reference points of the tunnel network and selected points on the ground surface. The observations were linked to the nearest reference stations of the Polish state service ASG-EUPOS, which allowed for determining the coordinates and assessing their accuracy. Vector postprocessing and coordinate alignment were performed independently using Trimble Business Control, Topcon Office, and the automatic calculation module of the ASG-EUPOS system (Kadaj and Świętoń 2009). The obtained results differed slightly, up to 6 mm along the Y axis and 2 mm along the X axis of the 2000/7 system (Kadaj 2021), with the accuracy of coordinate determination at the level of ±1.5 mm. Therefore, it can be assumed that the declared accuracy of mx and my = ±2 mm was justified.

5 CONCLUSION

The article presents an issue that seems to be a typical engineering problem, concerning the optimization of measurement methods in the aspect of maintaining the accuracy required by the excavation project. However, the challenge was to perform such an analysis for the specific conditions of the tunnel construction project. It is the first, currently implemented and carried out undertaking in Poland, which for the needs of railway construction assumes the excavation of the currently longest tunnel using TBM with a length of almost 4 km. What is specific is that typical excavation tolerances known from the construction of other railway tunnels (e.g. metro) must be maintained also in this case, despite the lack of indirect connections with the surface, allowing for performing geodetic control measurements and strengthening the tunnel network. Another problem was the lack of possibility to connect both tunnels under construction during their excavation using a crossbeam, which also complicated the issue of selecting measurement techniques and the designed basic network in the tunnel. Thus, using the experience of mining surveying carried out in underground mines, gyro-azimuth measurements were successfully used to strengthen the network (Pielok *et al.*

2011). To sum up, based on the multi-variant analysis of the accuracy of the designed network, variant V was selected as the optimal one, with the following parameters: the accuracy of the external tunnel network connection points was assumed to be (mx = ± 2 mm and my = ± 2 mm), for linear-angle measurements, an instrument with an angle measurement accuracy of ±3cc and (e.g. Trimble S9 instrument) should be used. The length measurement error of ± (0.8 mm + 1 ppm) will be obtained by forced centring of the instruments and measurement signals. The control network will be reinforced by determining 3 topographic azimuths, based on gyro-azimuth measurements with a direction determination accuracy of ± 1 mgon (10^{cc}), on the side of the tunnel control network connection (before the starting portal) and at distances from the tunnel starting portal: 1.2 km and 2.8 km (respectively, station points: G1, T6 and T13). The preliminary analysis of this variant presented above allows determining the point of the basic tunnel control network (POT) on the exit (western) portal of the tunnel with an accuracy of $Mp = \pm 27.4$ mm in the horizontal plane (transverse error at the level of ± 27.2 mm). In addition, during the analysis it was found that obtaining the accuracy criterion in the vertical plane is possible on the basis of precise levelling with an accuracy of no worse than ± 1 mm/km, using the standard procedure related to such measurement, for this reason such analysis was not discussed in the article.

REFERENCES

Camus, T., and Moubarak, S. (2015). Maintenance robotics in TBM tunnelling. In *ISARC. Proceedings of the International Symposium on Automation and Robotics in Construction* (Vol. 32, p. 1). IAARC Publications.

Kadaj, R., and Świętoń, T. (2009). Algorytm i oprogramowanie modułu automatycznego postprocessingu (APPS) w polskim systemie satelitarnych stacji referencyjnych (ASG-EUPOS). *Zeszyty Naukowe Politechniki Rzeszowskiej. Budownictwo i Inżynieria Środowiska*, 37-56.

Kadaj, R. (2021). Conformal and empirical transformation between the PL-ETRF89 and PL-ETRF2000 reference frames using the new adjustment of the former Polish I class triangulation network. *Geodesy and Cartography*, *70*(2).

Lipecki, T., and Jaśkowski, W. (2011). The accuracy of determination of the gyroscopic azimuth by means of the GYROM AT 2000 instrument, in conditions of Polish underground mines. Dokładność wyznaczenia azymutu giroskopowego przyrządem GYROMAT 2000, w warunkach polskich kopalń podziemnych. *Przegląd Górniczy*, *67*.

Mao. S., Shen, X., Lu M.,, and Wu X., (2013), Real-time tablet-based virtual reality implementation to facilitate tunnel boring machine steering control in tunnel construction https://itc.scix.net/pdfs/convr-2013-23.pdf

Mao, S., Shen, X., and Lu, M. (2015). Virtual laser target board for alignment control and machine guidance in tunnel-boring operations. *Journal of Intelligent and Robotic Systems*, 79, 385–400.

Park, B., Choi, S.W., Lee, C., Kang, T.H., Do, S., Lee, W.Y., and Chang, S.H. (2023). A Geometric Model for a Shield TBM Steering Simulator. *Applied Sciences*, *13*(18), 10087.

Pielok, J., Cisło, B., Gawałkiewicz, R., Jaśkowski, W., Jura, J., Lipecki, T., ... and Szewczyk, J. (2011). Geodezja górnicza. *Wydawnictwa AGH, Kraków.*

Szafarczyk, A., Skaba, A., and Sokalla, K. (2019). Implementation of gyroscope measurements in underground mines; focus on the mine of ruch (unit)„Borynia" in the Jastrzębie Coal Company. *Geoinformatica Polonica*, *18*, 113–120.

Weiss, G., Labant, S., Gasinec, J., Stankova, H., Cernota, P., Weiss, E., and Weiss, R. (2022). Establishment of local geodetic networks based on least-squares adjustments of GNSS baseline vectors. *Advances in Geodesy and Geoinformation*, *71*(1).

https://limanowa.in/aktualnosci/prawie-300-metrow-tunelu-druga-tarcza-tbm-ruszy-w-czerwcu/77077?image=7, access on June 6, 2025.

https://www.plk-sa.pl/o-spolce/biuro-prasowe/informacje-prasowe/szczegoly/niemal-300-metrow-podziemnej-podrozy-kingi-jadwiga-w-blokach-startowych-10761, ss on June 6, 2025

https://www.researchgate.net/figure/The-schematic-for-measuring-translation-matrix-T-n-t fig2 356516308 access on June 6, 2025.

Mine Surveying and Mining Area Protection – Sokoła-Szewioła and Jarczyk (Eds)
© 2025 The Author(s), ISBN: 978-1-041-10531-2

A multi-criteria framework for assessing groundwater retention potential

S. Łabojko & A. Malinowska
AGH University of Krakow, Poland

B. Stypiński
Lubelski Węgiel "Bogdanka" S.A., Poland
AGH University of Krakow, Poland

ABSTRACT: Groundwater retention is a critical component of regional water resource management, particularly in areas affected by low terrain variability. This study presents a multi-criteria framework integrating Analytic Hierarchy Process (AHP) and Geographic Information Systems (GIS) to evaluate groundwater retention potential over a 130 km^2 flat study area. Five key geographical determinants—geologic layer permeability, distance to streams, Topographic Position Index (TPI), slope, and land cover—were incorporated into the analysis. High-resolution LiDAR-derived elevation data enabled detailed characterization of microtopographic features, while precise geological maps provided robust estimates of subsurface permeability. The proposed framework allows systematic integration of multiple thematic layers into a unified decision-making model. Pairwise comparison and weighting identified geologic layer permeability as the dominant controlling factor (45.92%), with distance to streams and TPI contributing equally (19.81% each). Slope and land cover exhibited comparatively minor influence (10.39% and 4.07%, respectively), reflecting the homogeneity of the study terrain. The resulting groundwater retention potential map demonstrates the capacity of combining high-accuracy topographic and geological datasets with multi-criteria decision analysis to produce reliable spatial predictions. This approach provides a practical framework for groundwater management, land-use planning, and hydrological assessment in regions characterized by flat landscapes and low terrain variability.

1 INTRODUCTION

Groundwater is a vital natural resource that plays a central role in both the hydrological cycle and human activities. Its sustainable management is essential, as disruptions to the water balance can lead to significant environmental consequences (IPCC 2022). Over the past decades, interest in surface and groundwater retention has grown steadily, extending beyond scientific and engineering communities into broader societal discussions (Ayazi *et al.* 2010; Gao *et al.* 2018). Despite this trend, public awareness of water management practices remains limited, and the importance of natural retention processes is still often underestimated (Kud *et al.* 2024). This gap underscores the need for models that are not only scientifically robust but also simple, interpretable, and suitable for both management and educational purposes.

Geographic Information System (GIS)-based approaches provide powerful tools for visualizing areas with high groundwater retention potential and for delineating protection zones. To support such analyses, multi-criteria decision-making (MCDM) techniques are frequently employed. Among them, the Analytic Hierarchy Process (AHP) has become one

DOI: 10.1201/9781003655558-14

of the most widely applied methods for groundwater retention assessment (Arulbalaji *et al.* 2019; Hosseinzadeh *et al.* 2023; Kudamnya *et al.* 2019; Nampak *et al.* 2014). AHP facilitates the integration of diverse geographical factors into a unified framework but depends heavily on expert judgment, which can introduce subjectivity.

To address this limitation, data-driven approaches have gained increasing attention. Machine learning methods, such as Random Forests and neural networks, offer robust alternatives for integrating complex datasets and reducing reliance on subjective weighting (Arulbalaji *et al.* 2019; Kudamnya *et al.* 2019; Nampak *et al.* 2014). At the same time, simpler statistical techniques such as the Frequency Ratio (FR) remain effective, although they may be sensitive to data reclassification (Kudamnya *et al.* 2019). Groundwater retention potential is typically evaluated using indicators such as groundwater flow rates or well yields (Razandi *et al.* 2015; Sander *et al.* 1996). However, these datasets are often sparse or unavailable. Groundwater levels provide a practical and widely accessible proxy: shallow water tables usually signal higher retention potential (Balamurugan *et al.* 2017). Key geographical factors commonly included in retention modeling encompass geology, land cover, hydrographic networks, distance to streams, and topographic attributes such as slope or Topographic Position Index (TPI) (Arulbalaji *et al.* 2019; Kudamnya *et al.* 2019; Nampak *et al.* 2014; Zhao *et al.* 2024).

Building on these methodological foundations, the present study focuses on a flat terrain region of approximately 130 km^2, where low relief amplifies the importance of subtle geological and hydrological factors in shaping groundwater retention potential. To address these challenges a multi-criteria frameworkwas developed integrating Analytic Hierarchy Process (AHP) with Geographic Information Systems (GIS). The approach combines high-resolution LiDAR-derived elevation data with detailed geological mapping to capture both microtopographic variability and subsurface permeability. By systematically integrating five thematic determinants—geologic layer permeability, distance to streams, Topographic Position Index (TPI), slope, and land cover—into a unified decision-making model, the study provides a robust basis for assessing groundwater retention potential in low-variability landscapes. The following sections describe the study area in detail and outline the methodology applied to construct and validate the groundwater retention model.

2 MATERIALS AND METHODS

2.1 *Study area and data sources*

The study was conducted in the Bogdanka coal mine area (Lubelskie Voivodeship, eastern Poland), covering approximately 130 km^2 (Figure 1).

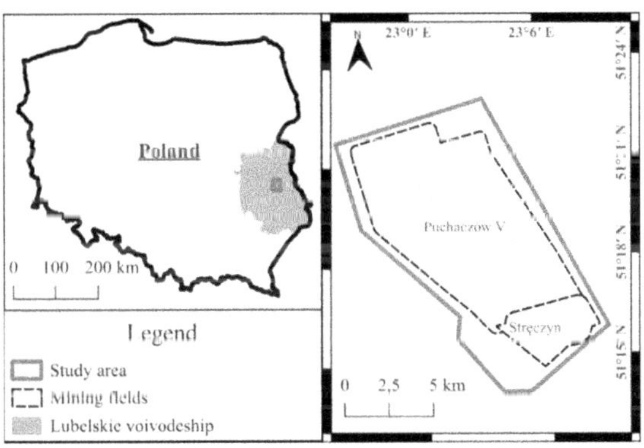

Figure 1. Map of the study area and its location in Poland.

The region is characterized by predominantly flat terrain with low relief variability, making it suitable for analyzing groundwater retention potential under homogeneous surface conditions. Groundwater retention was evaluated using five key geographical determinants: Topographic Position Index (TPI), slope, distance to streams, land cover, and permeability of geological layers, selected based on their recognized influence on groundwater infiltration, storage, and flow dynamics (Arulbalaji *et al.* 2019; Kudamnya *et al.* 2019; Nampak *et al.* 2014; Zhao *et al.* 2024).

A high-resolution Digital Elevation Model (DEM) obtained from a LiDAR survey in 2015 (1 m horizontal resolution, 2 cm vertical accuracy) provided an exceptionally detailed representation of terrain microfeatures, critical for detecting small depressions and subtle slopes that control surface water accumulation and infiltration. This high-precision DEM allowed us to overcome limitations of conventional elevation data, particularly in flat landscapes where standard DEMs often fail to capture meaningful topographic variability.

Land cover and hydrographic datasets were obtained from the 2015 archival database of the Head Office of Geodesy and Cartography (GUGiK). These standardized datasets provide consistent, high-quality information on surface cover types and hydrographic networks, enabling precise calculation of factors such as distance to streams and surface water retention potential.

Geological data were sourced from archival maps of the Polish Geological Institute. By georeferencing and vectorizing these maps in a GIS environment the lithological units were delineated and assigned permeability values to each geological layer. This step was essential to integrate subsurface hydraulic properties into the analysis, allowing the creation of a spatially coherent permeability layer that captures the heterogeneity of subsurface conditions.

2.2 *Data Preprocessing*

2.2.1 *DEM resampling and terrain analysis*

While high-resolution DEMs offer unparalleled detail, they can exaggerate minor surface features that are irrelevant at a regional scale. To optimize the data for large-scale TPI and slope analysis, the DEM was resampled to 10 m resolution using bilinear interpolation. The combination of LiDAR accuracy and careful resampling ensured that microtopographic features affecting groundwater retention were preserved while noise was minimized.

Generally, the terrain slope is low, with the mining waste dump in the central-west region being the only exception, exhibiting slopes up to 40°. Minor slopes also appear in the southeast of the area (Figure 2).

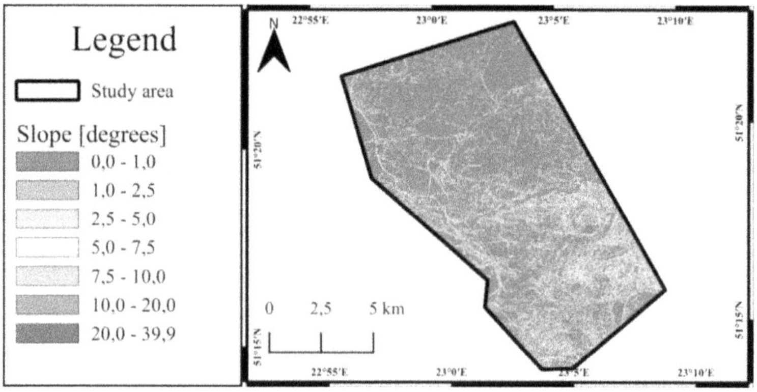

Figure 2. Slope map.

2.2.2 *Topographic position index mapping (TPI)*

The flat terrain results in TPI values ranging predominantly between -1 and +1, indicating that within a 200 m radius, local elevation differences rarely exceed 1 m. This reflects the scarcity of significant depressions that could promote groundwater retention, with minor depressions observed mainly in the northwest and central-east regions. Extreme TPI values are associated with the mining waste dump, which represents the only prominent topographic feature in the study area (Figure 3).

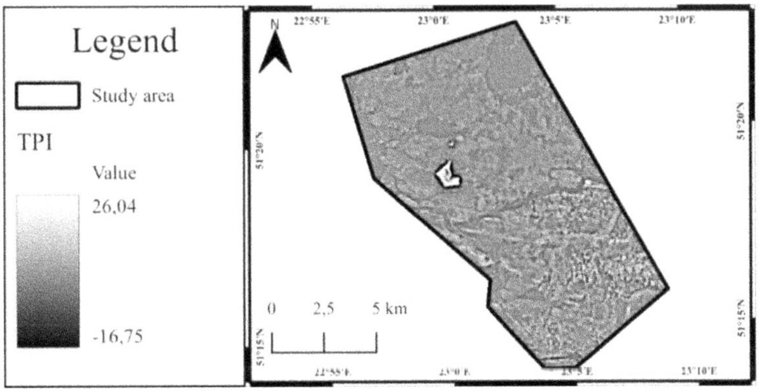

Figure 3. Topographic position index map.

2.2.3 *Hydrographic data*

Distance to streams was derived from the hydrographic network dataset, resulting in a raster representation of local flow paths. The hydrographic network exerts a strong influence on groundwater infiltration. To minimize potential errors associated with inappropriate search radius in hydrographic density calculations. The distance-to-stream raster was applied as a more reliable indicator. Most of the study area is densely intersected by rivers, streams, and drainage ditches, which substantially shape local infiltration Distances to streams greater than 1 km occur only in limited zones, primarily in the southeast (Figure 4).

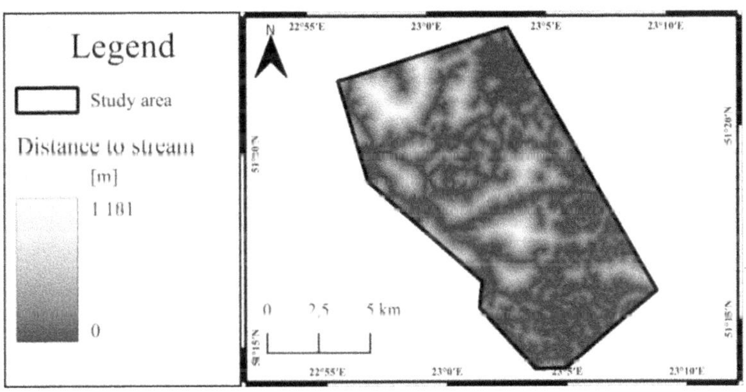

Figure 4. Distance to streams across the study area.

2.2.4 *Land cover data*

Land cover classes were assigned numerical scores reflecting their relative water retention and infiltration potential. Forests and wetlands were assigned the highest values due to dense vegetation, high organic matter, and excellent water storage capacity. Grasslands and meadows were given moderately high scores, while shrublands received medium values depending on vegetation density and soil type. Croplands were assigned lower values reflecting their limited retention due to tillage and management practices. Urban and built-up areas, as well as bare soils, were scored the lowest because impervious surfaces and compaction severely restrict infiltration. Water bodies were considered reservoirs and not included in the infiltration cycle. This scoring system is consistent with previously established hydrological assessments (Arulbalaji *et al.* 2019, Krnáčová *et al.* 2020). Land cover across the study area is largely homogeneous, dominated by croplands, with forests representing the only other substantial class (Figure 5).

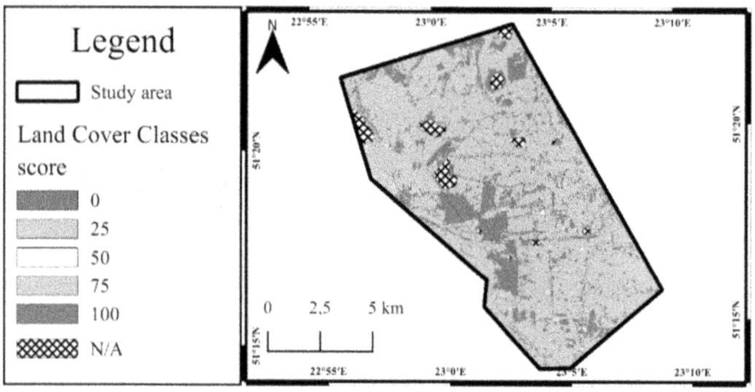

Figure 5. Land cover classes score map.

2.2.5 *Geological data*

Geological layers were classified according to hydraulic conductivity and assigned Darcy Factor scale ranging from 0 to 5, enabling the integration of subsurface properties into a single permeability raster. Coarse-grained materials such as gravel, coarse sand, and fractured limestone or basalt exhibited very high hydraulic conductivity and were assigned the highest Darcy Factor, representing major aquifers. Medium- to fine-grained sands and weathered sandstones displayed high to moderate conductivity, functioning as primary aquifers and recharge layers. Fine sands, cemented sandstones, and sandy silts showed moderate to low conductivity and were interpreted as secondary aquifers. Low-permeability units, including silt, glacial till, and clayey loess, act as aquitards, restricting vertical flow. Finally, clay, shale, and unfractured granite represented nearly impermeable layers, functioning as confining beds (Figure 6)

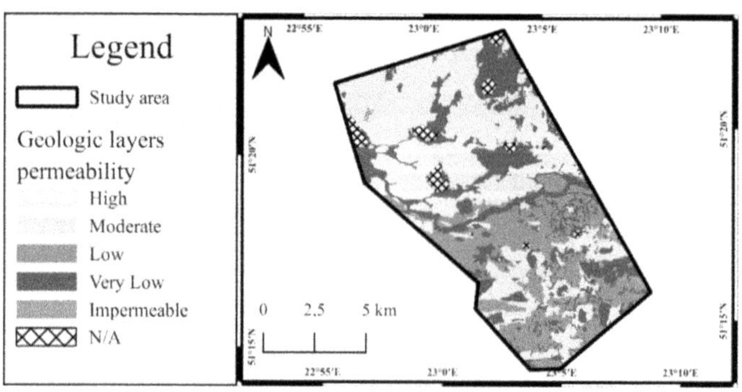

Figure 6. Spatial distribution of geological layers and their permeability.

2.2.6 *Raster normalization*

All raster layers—including TPI, slope, distance to streams, geological permeability, and land cover—were normalized to a 0–100 scale, where higher values indicate more favorable conditions for groundwater retention. This normalization ensured that all datasets contributed proportionally to the subsequent AHP-based multi-criteria analysis, enabling robust integration of continuous and discrete variables into a coherent groundwater retention model.

2.3 *Analytical methods*

The Analytic Hierarchy Process (AHP) was applied to integrate the five geographical factors into a composite groundwater retention potential map (Saaty 1980). A detailed step-by-step workflow was constructed to guide the analysis (Figure 7).

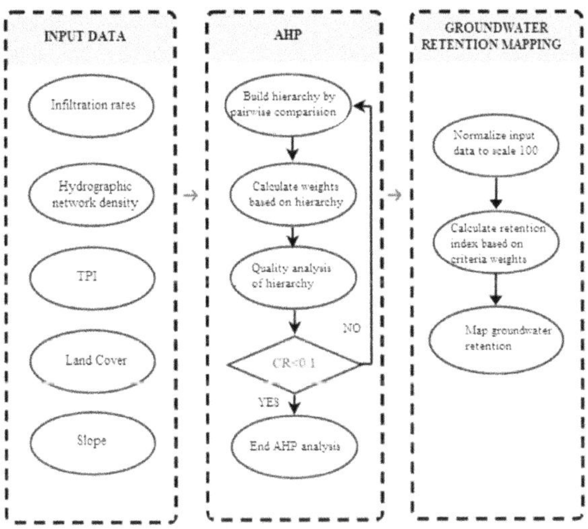

Figure 7. Workflow of AHP-based groundwater retention modeling in the study area.

Factors were compared pairwise using a standardized numerical scale. Results highlighted the dominant influence of geological permeability, with distance to streams and TPI as secondary factors. Slope and land cover were of minor importance due to flat and homogeneous terrain.

Weights derived from normalized pairwise comparisons were:

- Geological layers permeability: 45.92%
- Distance to streams: 19.81%
- TPI: 19.81%
- Slope: 10.39%
- Land cover: 4.07%

Internal consistency was evaluated, yielding a consistency index of 0.063 and a consistency ratio of 0.057, both below the accepted threshold of 0.10, indicating reliable weights for the model.

3 RESULTS

To generate the final groundwater retention potential map, all five thematic layers were first standardized to a common numerical scale to enable direct comparison and integration. Topographic indices (TPI and slope) were rescaled so that areas of minimal elevation variability received the highest scores, while steeper terrain was assigned the lowest (Figure 8).

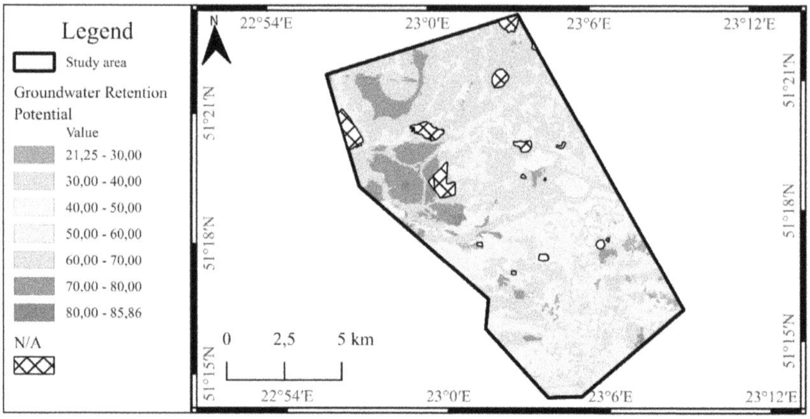

Figure 8. Groundwater retention potential map.

Distance to streams was transformed in the opposite manner, with greater distances corresponding to higher scores. Geologic layers were evaluated according to permeability, with highly permeable formations assigned maximum values and impermeable rocks the lowest. Land cover had already been expressed on a comparable scale and therefore required no further adjustment. Following this harmonization, each raster layer was multiplied by its corresponding weight derived from the AHP analysis, allowing the integration of all factors into a single, weighted model of groundwater retention potential. The resulting map highlights spatial variability in groundwater storage capacity across the study area.

The highest groundwater retention potential is observed in the northern and northwestern regions, where highly permeable Quaternary sands and gravels coincide with areas more distant from surface water features. Moderate retention potential occurs centrally, where geological permeability is mixed, while low retention zones dominate the southern and southeastern portions of the study area, characterized by low-permeability silts, clays, and chalk deposits. Topographic factors, including TPI and slope, have a limited impact on groundwater retention due to the generally flat terrain. Localized exceptions occur, such as mining waste dumps, where minor variations in slope and elevation create small-scale differences in water retention. Land cover contributes minimally, reflecting the dominance of cropland and forest, which show little variation in groundwater storage potential. Hydrographic proximity, while not the dominant factor, has a moderate influence, as the dense network of rivers, streams, and drainage channels affects local infiltration and subsurface water accumulation. Overall, the spatial distribution of groundwater retention potential demonstrates that subsurface geological properties and hydrographic distance are the primary determinants, whereas topography and land cover play secondary roles. This integrated, weighted mapping approach provides a clear and actionable tool for groundwater management, highlighting areas of high retention potential for protection or targeted utilization.

4 DISCUSSION

The integration of high-resolution LiDAR-derived DEM and precise geological datasets into the AHP framework represents a significant methodological advancement in groundwater retention assessment for flat terrains. Conventional DEMs often fail to capture subtle topographic variations in areas of low relief, yet these microfeatures may control surface water accumulation, infiltration, and redistribution. By utilizing a LiDAR DEM with 1 m horizontal resolution and 2 cm vertical accuracy reliably of depressions detections raised. This enhanced resolution proved essential for quantifying slope and TPI in a geomorphologically uniform setting, thereby reducing the uncertainty commonly associated with topographic predictors in groundwater studies. Similarly, the geological vectorization process ensured that lithological boundaries and permeability classifications were represented with a degree of accuracy seldom achieved in groundwater retention mapping. In many studies, geological inputs rely on generalized or interpolated maps, which can mask local heterogeneity. Here, the use of detailed archival geological maps and permeability categorization allowed for a more robust assessment of subsurface hydraulic properties. As a result, the dominant role of geological permeability in determining groundwater retention potential was quantified with high confidence.

Another key consideration is the physiographic context of the study area. The Bogdanka region is characterized by flat topography, low terrain variability, and intensive land-use change associated with mining activity. These features amplify the importance of geological and hydrographic factors, while diminishing the role of topography and land cover. Presented analysis confirmed that permeability accounted for nearly half of the weighting (45.92%), while slope and land cover together contributed less than 15%. Such weighting reflects not only the inherent environmental characteristics but also the necessity of adapting AHP to local terrain conditions. Without adjustment, the method might overemphasize variables of limited relevance in flat landscapes.

From a broader perspective, the study highlights the dual vulnerability and opportunity of flat terrains. The flat and homogeneous terrain in the Bogdanka area naturally limits the landscape's ability to retain water, which can increase the risk of waterlogging. However, the availability of high-resolution datasets, including LiDAR-derived elevation models and detailed geological maps, allows for the development of accurate, site-specific groundwater retention models. By applying the AHP framework, expert knowledge is systematically integrated with these precise data to produce results that are transparent, reproducible, and directly applicable for proactive water management and planning in flat landscapes. with high-quality data to produce transparent, replicable, and decision-oriented results.

5 CONCULSION

This study demonstrates the potential of integrating high-resolution LiDAR-derived elevation data, precise geological information, and multi-criteria decision analysis (AHP) to assess groundwater retention in flat terrains. Applied to the Lubelski Węgiel "Bogdanka" area, the methodology enabled the identification of spatial patterns of retention potential even in landscapes with minimal surface variability. Geological permeability emerged as the dominant factor, accounting for nearly half of the overall influence, while hydrographic distance and the Topographic Position Index also contributed. Slope and land cover had a limited effect due to the homogeneity of the terrain.

LiDAR-derived DEMs proved particularly valuable in capturing subtle microtopographic variations, which are critical for local water storage and infiltration, while detailed geological data provided essential subsurface context for assessing permeability and retention realistically in a mining-affected environment. The study confirms that the AHP method can be effectively applied in low-relief areas when factor weighting is adapted to local conditions. By combining geospatial data with structured decision analysis, the framework offers a reliable and practical tool for groundwater management, land-use planning, and hydrological risk assessment.

Overall, the approach provides a replicable methodology applicable to other regions with similar environmental characteristics, supporting sustainable groundwater use and helping mitigate hydrological risks associated with mining and land transformation.

ACKNOWLEDGEMENTS

The authors thank the Ministry of Education in Poland (DWD/8/0064/2024), Lubelski Węgiel 'Bogdanka' S.A. for well survey and LiDAR data, AGH University of Krakow (subsidy 16.16.150.545), and Ms. Małgorzata Bodziony for geological data support.

REFERENCES

Arulbalaji, P., Padmalal, D., and Sreelash, K. (2019). GIS and AHP techniques based delineation of groundwater potential zones: A case study from Southern Western Ghats, India. *Scientific Reports*, 9: 2082. https://doi.org/10.1038/s41598-019-38567-x

Ayazi, M.H., *et al.* (2010). Disasters and risk reduction in groundwater: Zagros Mountain Southwest Iran using geoinformatics techniques. *Disaster Adv*, 3(1): 51–57.

Balamurugan, G., Seshan, K., and Bera, S. (2017). Frequency ratio model for groundwater potential mapping and its sustainable management in cold desert, India. *Journal of King Saud University*, 29(3). https://doi.org/10.1016/j.jksus.2016.08.003

Darcy, H. (1856). Les fontaines publiques de la ville de Dijon. Victor Dalmont.

Gao, Y., Church, S.P., Peel, S., and Prokopy, L.S. (2018). Public perception towards river and water conservation practices: Opportunities for implementing urban stormwater management practices. *Journal of Environmental Management*, 223: 478–488. https://doi.org/10.1016/j.jenvman.2018.06.059

Hosseinzadeh, A., Behzadian, K., Rossi, P., Karami, M., Ardeshir, A., and Torabi Haghighi, A. (2023). A new multi-criteria framework to identify optimal detention ponds in urban drainage systems. *Journal of Flood Risk Management*, 16(2). https://doi.org/10.1111/jfr3.12890

Intergovernmental Panel on Climate Change (IPCC). (2022). *Climate Change and Land: IPCC Special Report on Climate Change, Desertification, Land Degradation, Sustainable Land Management, Food Security, and Greenhouse Gas Fluxes in Terrestrial Ecosystems.* Cambridge: Cambridge University Press. doi:10.1017/9781009157988.

Krnáčová, Z., Kenderessy, P., Hreško, J., Kubínsky, D., and Dobrovodská, M. (2020). Assessment of landscape retention water capacity and hydrological balance in traditional agricultural landscape (model area Liptovská Teplička settlements, Slovakia). *Water*, 12(12): 3591.

Kud, K., Badora, A., and Woźniak, M. (2024). Sustainable management in river valleys, promoting water retention—The opinion of residents of south-eastern Poland. *Sustainability*, 16(11): 4648. https://doi.org/10.3390/su16114648

Kudamnya, E.A., Sylvanus, R.O., Essien, M.T., Vulegbo, A.A., and Omang, B.O. (2019). A 2D GIS approach for mapping aquiferous zones using remotely sensed data within Obubra, Southeast-Nigeria. *Journal of Water Resource and Protection*, 11(6). doi:10.4236/jwarp.2019.116046

Nampak, H., Biswajeet, P., and Manap, M.A. (2014). Application of GIS-based data-driven evidential belief function model to predict groundwater potential zonation. *Journal of Hydrology*, 513. https://doi.org/10.1016/j.jhydrol.2014.02.053

Saaty, T. (1980). *The Analytic Hierarchy Process: Planning, Priority Setting, Resources Allocation.* New York: McGraw-Hill.

Zhao, R., Fan, C., Arabameri, A., Santosh, M., Mohammad, L., and Mondal, I. (2024). Groundwater spring potential mapping: Assessment of the contribution of hydrogeological factors. *Advances in Space Research*, 74(1). https://doi.org/10.1016/j.asr.2024.03.038

Mine Surveying and Mining Area Protection – Sokoła-Szewioła and Jarczyk (Eds)
© 2025 The Author(s), ISBN: 978-1-041-10531-2

Geographical information systems in terms of available data sources on mining areas

A. Mierzejowska
Silesian University of Technology, Gliwice, Poland

ABSTRACT: For many years, Geographic Information Systems have been increasingly used in many activities carried out by various institutions, including mining companies, both during and after their operations. In order for analyses to be reliable, the data used must be up-to-date, complete and come from reliable sources. In recent years, so-called web services, which allow any user with Internet access to view and download data, have become very popular. The most widely used services are WMS, WMTS, WCS and WFS. Software is an element that supports the functioning of Geographic Information Systems. Thanks to continuous development, it allows for the easy implementation of the basic functions of these systems, i.e. the collection, management, processing, analysis and visualisation of data. The author of the paper indicates possible sources of data (both vector and raster) relevant to supporting change monitoring and decision-making in mining areas. She focused in particular on data that is made available free of charge through web services.

1 INTRODUCTION

Since the beginning of mining activities, negative effects have been observed on the ground surface and in the rock mass. As a result of mining, the prevailing equilibrium in the rock mass is disturbed, leading to displacement of rocks towards the selected space. This results in landform changes and hydrological transformations. They may also be accompanied by mining tremors. Direct impacts of mining exploitation are most often observed, visible on the surface in the form of continuous deformations. These are mild depressions of the terrain, revealing themselves in the form of subsidence troughs. Activation of old goafs is also observed, resulting in so-called discontinuous deformations on the surface, in the form of funnels, fissures, thresholds or landslides. Under current legislation, the detrimental effects of mining work become apparent in the space known as the mining area, and every mining company is obliged to have survey and geological documentation to document all work and assess its effects. The studies forming part of this documentation are produced on the basis of surveying and geological work involving measuring, calculating and cartographic activities.

Examples of such documents are survey documents (calculation forms and, if IT techniques are used, electronic documents containing the results of the calculations) and cartographic documents such as situation and elevation maps of the surface within the boundaries of the mining area and seam maps. As a result of the constant monitoring and recording of changes occurring in the environment (whether on the surface or deep within the rock mass), vast amounts of diverse information are collected. Management of this data is made possible by information systems, which include Geographical Information Systems.

Geographical Information Systems (GIS) are characterised by the presence of spatial information, i.e. information about location, geometric properties and spatial relationships.

DOI: 10.1201/9781003655558-15

The location of any object (the location of a phenomenon) can be unambiguously identified in relation to the Earth. One of the basic elements of this system is a database containing information about objects. The rapid technological development that we have seen over many years gives us enormous possibilities in terms of storing, viewing, processing, analysing and sharing data. Geographical Information Systems allow data from different sources to be easily integrated, so that the data are up-to-date, reliable and complete, and high-level analyses are carried out.

The aim of this paper is to review the available sources of data related to mining activities, with a particular focus on those made available through web services, including WMS, WMTS, WCS and WFS. The author of this paper has focused her attention on services that collect this type of data and that make it available free of charge. This is all the more important as the data obtained in this way constitute an important source of information for various analyses, and the use of, for example, GIS class software makes it possible to quickly and easily acquire, integrate and process many different data models.

2 THE MINING AREA AS A SPACE OF NEGATIVE EFFECTS OF MINING ACTIVITIES STARTED

Under current legislation, a mining area is an area covered by the foreseeable detrimental effects of the mining operations of a mining company (Act of June 9, 2011. Geological and Mining Law). The conduct of mining operations, regardless of the method of mineral extraction, involves interference with the environment to a greater or lesser extent. As a general rule, the effects of mining operations on the surface and the rock mass become apparent successively, during the course of mining operations. As the selected space expands, a subsidence trough forms. As a result of the activation of old, shallow goafs (post-coal or ore seams) caused by mining operations carried out very often many years after the cessation of mining in a given area, so-called discontinuous deformations (sinkholes, sills, fissures, cracks, etc.) occur on the surface. This type of deformation may also be caused by the presence of access pits, e.g. shafts, shafts. Discontinuous deformations, due to the fact that:

- may occur many years after the end of mining activities,
- are not preceded by phenomena indicating that they may occur,
- are characterised by their suddenness,

pose a huge threat to the surface and its development facilities.

In addition to impacts in the form of displacement of rock mass elements, mining operations cause changes in water relations and mining tremors. On the surface, there are hydrological and natural transformations, changes to the terrain and damage to buildings (Kwiatek 1998; Kawulok 2015).

All works carried out by a mining enterprise, including their adverse effects, are subject to documentation. The results of the works are presented, inter alia, in geological documentation, including geological documentation of a mineral deposit excluding a hydrocarbon deposit, geological-investment documentation of a hydrocarbon deposit, hydrogeological, geological-engineering and surveying-geological documentation. Their collection, recording, archiving, protection and provision of access to them shall be the responsibility of the relevant bodies of the geological administration and the State Geological Survey (Act of June 9, 2011. Geological and Mining Law; Regulation of the Minister of the Environment on geological documentation of deposits, excluding hydrocarbon deposits). The documents forming part of the surveying and geological documentation of a decommissioned mine shall be stored at the Higher Mining Authority (Regulation of the Minister of the Environment on surveying and geological documentation). The documents that make up geological or geological survey documentation, in textual, graphical and tabular form, contain huge amounts of data. Their management and integration with data from many different sources is possible thanks to Geographical Information Systems.

3 GEOGRAPHICAL INFORMATION SYSTEMS DATA MODEL

A characteristic feature of Geographical Information Systems is information about the location of real world objects, the spatial relationships between them and the possibility of analysis. The concept of an object is very broad. The definition of an object is given in the ISO 19109 standard, according to which the term object should be understood as an abstraction of phenomena occurring in the real world with characteristic features, the so-called attributes. An attribute of an object has a name, a data type and a set of values assigned to it (PN-EN ISO 19110). There is no clear definition of a GIS. In the work (Gaździski 1990). *J. Gaździcki defined this system as a system for acquiring, processing, verifying, integrating, manipulating, analysing and presenting data that are spatially referenced to the Earth. It typically includes a spatial database and relevant software.*

According to K. Clarke, a GIS is an automated system for collecting, storing, retrieving, analysing and displaying spatial data (Clarke 1990).

At the 5th PTIP Scientific and Technical Conference, GIS was defined as an information system supporting the collection, analysis and sharing of data about uniquely spatially located objects.

Despite the lack of a precise definition, five basic components can be distinguished in any GIS. These are:

- data,
- people,
- hardware,
- software,
- methods of acquiring, securing and sharing data (Figure 1).

Figure 1. Components of Geographic Information Systems (own elaboration on the basis of https://www.geoportal.gov.pl/).

Each element is equally important and influences the main functions of the system, i.e. data collection, data verification and preprocessing, data management, data integration, data processing and analysis, data visualisation.

In Geographical Information Systems, the representation of real world objects is done by means of a data model. Basically, two main data models can be distinguished - vector and raster.

The vector data model (spaghetti or topological model) is object-oriented. Each object in this model is represented by a vector or set of vectors. Real world objects can be represented by spatial objects with a point, linear or surface representation. The type of geometric representation is determined by the substantive assumptions of the specific data specification.

A special case of a vector model is the TIN (Triangulated Irregular Network) model, which allows continuous objects such as relief to be modelled.

In a raster data model, information is stored in the form of regular elements - pixels. Each element of this model is associated with attribute information. The model is used to collect data from scans of map bases, aerial and satellite photographs and remote sensing images. (Głażewski 2006). In the case of a raster model, a distinction can be made between a so-called GRID model (a regular structure in the form of a grid of rectangles, in which each elementary field has plane coordinates and stores an average value such as height) and an image model (e.g. an orthophoto).

4 DATA SOURCES ON MINING ACTIVITIES AND THEIR IMPACTS

As mentioned earlier in GIS, real world objects are represented by a data model. Basically, two main models can be distinguished - vector and raster - with which descriptive information, closely related to the objects of interest in these systems, is combined. The data feeding the databases can come from many different sources. They can be obtained directly (as a result of traditional, photogrammetric, remote sensing or GNSS measurements) or indirectly, by transferring data from existing databases (other information systems). All this contributes to the huge interest in this type of solution in many areas of life. The use of this type of technology to collect, process, share spatial information related to mining activities has also been observed for several years (Sokoła-Szewioła and Żogała 2016; Mierzejowska and Żogała 2018; Mierzejowska 2018, 2021).

The obligation to create spatial data services in the countries of the European Union was introduced by the INSPIRE Directive (Directive 2007/2/WE). In Poland, its implementation is the Act of 4 March 2010 on spatial information infrastructure (Spatial Information Infrastructure Act of March 4, 2010). This law regulates issues related to the establishment and operation of network services for spatial data sets and services. Within the scope of the network services a distinction can be made between:

- discovery, making it possible to search for spatial data sets and services on the basis of the content of the corresponding metadata and to display the content of the metadata,
- view, making it possible, as a minimum, to display, navigate, zoom in and out, pan, or overlay viewable spatial data sets and to display cartoon symbols and the content of metadata,
- download, making it possible to download copies of a file or part of a file and, where practicable, to directly access the file,
- transformation, enabling sets to be transformed with a view to achieving interoperability between spatial data sets and services,
- enabling spatial data services to be invoked (Spatial Information Infrastructure Act of March 4, 2010).

As a result of enormous technological advances (development of information technology, widespread access to the Internet), so-called web services have become extremely important for data acquisition.

Four services, whose specifications have been developed by the OGC (Open Geospatial Consortium), have been widely used. These include:

- WMS (Web Map Service),
- WMTS (Web Map Tile Service)
- WFS (Web Feature Service),
- WCS (Web Coverage Service) (Izdebski 2021).

The WMS service is currently the primary web service for viewing data in the form of dynamically generated map images. Within this service, maps are made available as raster files. The WMTS service is also a data browsing service, with the difference that it is limited to image fragments created from various data in the form of so-called tiles, forming an image pyramid. Such data is prepared for a range of magnifications.

The WFS service is used to retrieve data in vector form. This is done on the basis of user-defined criteria. The form of this data allows the user to process it and carry out any analysis based on it. The WCS service is also a service that allows data to be downloaded by the user to their computer, where they can then be analysed and processed. However, the data is stored in a raster model or other model characterised by matrix coverage, e.g. orthophotomap, numerical terrain model.

In Poland, most of the information made available by means of web services is data contained in the National Geodetic and Cartographic Resource (GIS). They constitute almost 70% of the data in the spatial data infrastructure. (Izdebski 2021). The scope of data available from the GIS is discussed in the work (Mierzejowska and Pomykoł 2024). The data are made available through the national geoportal. The service www.geoportal.gov.pl in the main menu under Services makes data available through WMS, WMTS, WFS, WCS, among others. Some of this data directly relates to mining areas and issues related to the impacts of mining exploitation on the surface. Taking the above into account, in the following part of the article the author reviewed the sources of spatial data, which are made available free of charge by means of the above-mentioned services.

In addition to data made available through WMS, WMTS, WCS and WFS services such as:

- elements of the principal map and zoning,
- address data,
- administrative boundaries,
- topographic and general geographic data,
- topographic, general geographic and thematic maps,
- geodetic warp,
- photogrammetric data,

the geoportal also provides data relevant to changes in the shape of the mining area (depiction of continuous, discontinuous deformations, floodplains, heaps) in the form of WMS, WMTS and WCS services. These are:

- Orthophotomap (Figure 2)
- Numerical Terrain Model (Figure 3),
- Numerical Land Cover Model.

In Poland, the basic numerical terrain model is a 1m x 1m grid model. This model is systematically updated based on airborne laser scanning (ALS) (https://www.geoportal.gov.pl/pl/dane/numeryczny-model-terenu-nmt/).

Figure 2. Excerpt from an orthophoto (own elaboration on the basis of https://www.geoportal.gov.pl/).

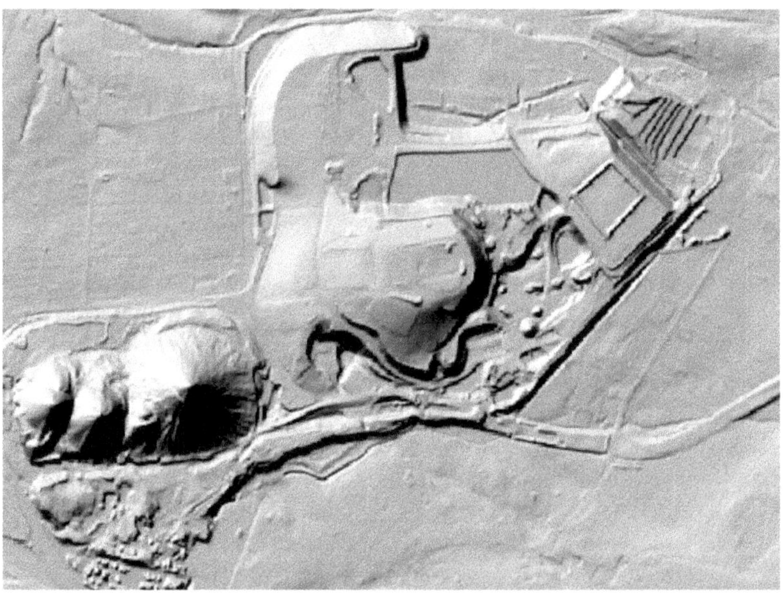

Figure 3. Excerpt from NMT (own elaboration on the basis of https://www.geoportal.gov.pl/).

Spatial data in raster form can be obtained from OpenStreetMap. They provide a package of information with which we can also feed Geographic Information Systems. OpenStreetMap is a project carried out by the Internet community worldwide. It is about creating a single global geographic data resource that is fully free and accessible to all. (https://www.openstreetmap.org/copyright). An example of data extracted from OpenStreetMap with NMT loaded is shown in Figure 4.

Figure 4. Fragment of NMT with spatial data extracted from OpenStreetMap (own elaboration on the basis of https://www.geoportal.gov.pl/).

Many spatial data are also made available by specific state institutions. Environmental data can be downloaded from the website of the General Directorate for Environmental Protection (GDOŚ). Data is available via WMS service, WFS or as shapefiles (.shp) within the GDOŚ INSPIRE Spatial Information Infrastructure node. Spatial data made available by GDOŚ are presented on *crfop.gdos.gov.pl*, *geoserwis.gdos.gov.pl* and www.gov.pl/web/gdos/dostep-do-danych-geoprzestrzennych. Data on individual forms of nature conservation can also be downloaded as a zipped file (.zip) using the links provided (e.g. national parks, landscape parks, nature reserves, protected landscape areas, ecological sites, documentation sites and others. (https://www.gov.pl/web/gdos/dostep-do-danych-geoprzestrzennych).

Another institution that makes spatial data available is the Polish Geological Institute. PGI provides geological and hydrogeological data. The data are available at https://www.pgi.gov.pl/dane-geologiczne/geologiczne-bazy-danych.html#hydrogeologia.

Spatial data from the Polish Geological Institute (PGI-PIB) can be downloaded in several ways, mainly through the Central Geological Database (CBDG). These include the GEOLOGY map viewer, the GeoLOG mobile application, the CBDG GIS (map services) and the CBDG Download Manager. The GEOLOGY map viewer allows access to detailed information from the CBDG, including data from various geological disciplines. It allows data to be viewed in tabular form, sorted, filtered and exported to file. The GeoLOG mobile application offers access to spatial data from the CBDG on mobile devices. CBDG GIS (map services) allows spatial data to be viewed in any geoportal or web services capable software (WMS, WFS). CBDG Download Manager is a tool that allows you to download spatial data from the CBDG in the form of shapefiles (.shp). Data from various geological domains such as mineral resources, hydrogeology, boreholes and others are available. An example of data made available on the PGI website in the Heap Database application is shown in Figure 5.

Figure 5. Example of data made available by PGI (own elaboration on the basis of https://www.geoportal.gov.pl/).

A great help in using web services is the presence on the market of so-called GIS-class software, e.g. QGIS or ArcGIS. This type of software enables data collection, presentation, analysis and processing. Data can be entered in several ways, depending on the data format and purpose. It is possible to add data from the web services discussed above, to load vector layers e.g. as shapefiles, to use maps as raster layers and to input data as CSV files.

A valuable source of data for GIS is LIDAR (Light Detection and Ranging) measurement data from airborne laser scanning (ALS). They represent the terrain in the form of a measurement point cloud with defined XYZ coordinates. In addition, the files contain information about the class of the point and the intensity of the signal reflection. An example of a point cloud file is shown in Figure 6.

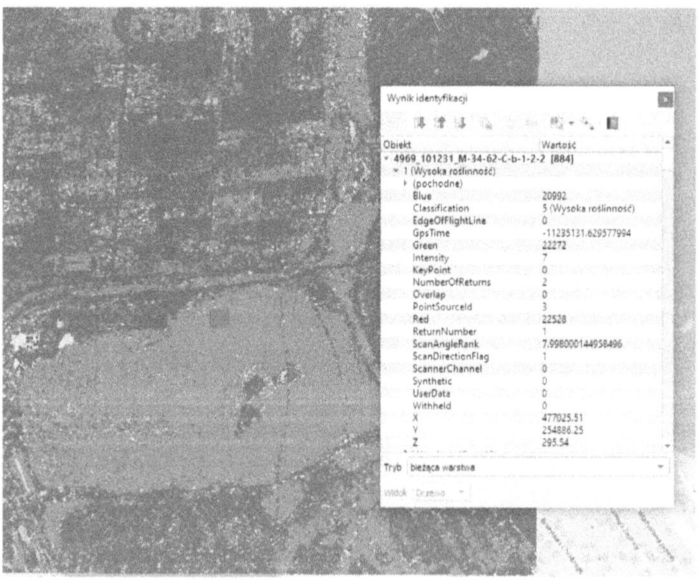

Figure 6. Point cloud layer (own elaboration on the basis of https://www.geoportal.gov.pl/).

A point cloud allows the precise representation of the shapes and locations of objects in 3D space, which is useful in the inventory of buildings, infrastructure or natural objects. This type of data can be used to perform advanced spatial analyses, such as calculating volumes, areas, distances or visibility analyses.

5 SUMMARY AND CONCLUDING REMARKS

A Geographical Information System is an example of an information system that makes it possible to collect, store, process, analyse and also visualise data that has a spatial reference. Undoubtedly, such data are the data collected by a mining company, from the moment it obtains a mining concession until the moment the plant is decommissioned.

It is important that the data feeding this system is adequate, both in terms of reliability, quality and completeness. This is a prerequisite for carrying out analyses at the appropriate level and for producing reliable studies as a result.

These systems enable the management of huge amounts of data from many different sources and in different forms (formats). All data sources can be integrated into GIS, using the available GIS software. It is a tool that allows effective management of a space such as a mining area, supports taking the right decisions based on up-to-date, complete and reliable information and planning future activities. In recent years, web services (WMS, WMTS, WCS, WFS) have become very popular for viewing and downloading data.

The use of data in various formats (raster, vector) and the increase in the availability of a wide range of information at the level of central and local government (thanks to their accessibility via web services) allows for the integration of data from many years ago as well as the latest data, which increases the efficiency and accuracy of decision-making. This is particularly important in the case of analyses of changes occurring in mining areas as a result of mining operations and related to environmental transformations and damage to buildings (the impact of mining on the surface - continuous and discontinuous deformations, change of water relations, mining tremors, post-mining waste, etc.).

REFERENCES

Clarke, K.C. (1990). *Analytical and Computer Cartography*: 195–204. Prentice Hall, Oxford.

Directive 2007/2/ECof the European Parliament and of the Council of 14 March 2007 establishing an Infrastructure for Spatial Information in the European Community (INSPIRE).

Gaździcki, J. (1990). *Systemy informacji przestrzennej*: 182. PPWK, Warszawa.

Głażewski A. (2006). *Modele rzeczywistości geograficznej a modele danych przestrzennych*: 217-225. Polski Przegląd Kartograficzny, tom 38, nr 3.

https://www.geoportal.gov.pl/pl/dane/dane-pomiarowe-lidar-lidar/, access on July 10, 2025

https://www.gov.pl/web/gdos/dostep-do-danych-geoprzestrzennych, access on July 10, 2025

https://www.openstreetmap.org/copyright, access on July 10, 2025

Izdebski W. and Seremet A. (2021). *Praktyczne aspekty infrastruktury danych przestrzennych w Polsce*. Część II: 139. Główny Urząd Geodezji i Kartografii, Warszawa.

Kawulok M. (2015). Szkody górnicze w budownictwie: 340. Instytut Techniki Budowlanej, Warszawa.

Kwiatek J. (1998). Ochrona obiektów budowlanych na terenach górniczych: 726. Wydawnictwo GIG, Katowice.

Mierzejowska A. (2021). *Dokumentacja mierniczo-geologiczna jako źródło informacji dla systemów GIS*: 138-150. Wybrane problemy środowiska przyrodniczego w ujęciu naukowym / Kalbarczyk K., Bujalska B. (*red.*), Wydawnictwo Naukowe TYGIEL Sp. z o. o., s., ISBN 978-83-66489-62-2, Lublin.

Mierzejowska A. and Pomykoł M. (2024). *Dane przestrzenne w systemach informacji geograficznej*: 81-92. Współczesne wyzwania nauk przyrodniczych – analiza i perspektywy rozwoju. Tom 2, Wydawnictwo Naukowe TYGIEL Sp. z o. o., s., ISBN 978-83-67881-62-3, Lublin.

Mierzejowska A. (2018). The use of modern IT tools to acquire information about geographical space from archival maps, IOP Conference Series Earth and Environmental Science, vol. 174 1755–1315.
Mierzejowska A. and Żogała M. (2018). *The characteristics of geographical information systems in terms of their current use:* 101-108. *Jurnal of Water and Land Development*, vol. 39.
PN-EN ISO 19110:(2016) Informacja geograficzna – Metodyka katalogowania obiektów.
Rozporządzenie Ministra Środowiska z dnia 1 lipca 2015r. w sprawie dokumentacji geologicznej złoża kopalin, z wyłączeniem złoża węglowodorów. Dz. U. 2015, poz. 987.
Rozporządzenie Ministra Środowiska z dnia 28 października 2015r. w sprawie dokumentacji mierniczo-geologicznej. Dz. U. 2015, poz. 1941.
Sokoła-Szewioła V. and Żogała M. (2016). The Adoption of the ArcGIS System to Support the Analyses of the Influence of the Mining Tremors on the Building Object: 228–236. *Management Systems in Production Engineering*, vol. 4(24).
Ustawa z dnia 4 marca 2010 *r. o infrastrukturze informacji przestrzennej.*
Ustawa z dnia 9 czerwca 2011r. *Prawo geologiczne i górnicze.* Dz. U. 2011, Nr 163, poz. 981 (z późniejszymi zmianami).

Mine Surveying and Mining Area Protection – Sokoła-Szewioła and Jarczyk (Eds)
 ISBN: 978-1-041-10531-2

Accuracy analysis of the digital elevation model made on the basis of a low-altitude photogrammetry

J. Paszek & J. Mazur
Silesian University of Technology, Gliwice, Poland

ABSTRACT: This article outlines the process of creating a terrain altitude model using data collected from a photogrammetric flight conducted with a drone. The flight was carried out in a village located in the northeastern part of the Silesian Voivodeship, in Poland. A total of 129 aerial photographs were taken, with a degree of coverage ranging from 60% to 80%. The photos were processed in the Agisoft Metashape Professional program, and a point cloud and the terrain altitude model were created based on them. The accuracy of the altitude coordinate of the points forming the Digital Terrain Model was determined based on the heights of 126 points, previously measured using a GPS receiver using the RTN method. The average value of the relative error in the determination of the point height in the Digital Elevation Model was 0.88 m.

1 INTRODUCTION

In geodetic practice, modern measurement technologies based on low-altitude photogrammetry, Simultaneous Localization And Mapping (SLAM) of the terrain surface (Chen *et al.* 2018, Grzechca *et al*, 2020, Keitaanniemi *et al.* 2021, Kim *et al.* 2018, Liu *et al.* 2024, Urban *et al.* 2024, Yarovoi and Cho 2024) and aerial laser scanning of the terrain surface (Light Detection And Ranging – LiDAR) (Chio and Hou 2021, Diara and Roggero 2022, Jia *et al.* 2021, Šafář *et al.* 2021, Sestras al. 2025a,b, Štroner *et al.*, 2023) made from Unmanned Aerial Vehicles (UAV) (Crommelinck *et al.* 2016, Puniach *et al.* 2018, Ramadhani *et al.* 2018, Stöcker *et al.* 2020) are increasingly used. This trend is related to the ubiquitous technological progress, strong competition on the market among companies offering geodetic services, and the growing need to make geodetic measurements of large areas of the terrain surface in a relatively short time and with minimal effort (in urban planning, agriculture, forestry, etc.). Properly processed data obtained using these measurement technologies can be used to create an orthophotomap (OPM) (Janusz *et al.* 2016, Oniga *et al.* 2016, Popescu *et al.* 2024, Sestras *et al.* 2023), a Digital Terrain Model (DTM) (Błaszczak-Bąk *et al.* 2019, Poniewiera 2018, Poniewiera *et al.* 2019) or many studies related to the analysis of the landform, e.g. profiles and cross-sections of a riverbed, the volume of an open pit and a stone stack (Paszek 2024), etc.

The article presents a case study of low-altitude photogrammetry conducted from a drone in 2023 in the Nakło village, located in southern Poland. The acquired data was processed using appropriate software, resulting in the creation of a point cloud with assigned georeferences, followed by the generation of a digital terrain model. The research aim was to determine the value of the average error in the altitude coordinate of the obtained model in relation to the accuracy applicable in this subject in Poland, as specified in current legal regulations. This error was determined by comparing the altitude positions of selected model points with the altitudes of the same points surveyed using the GPS (Global Positioning System) and RTN (Real-Time Network) satellite measurement techniques.

DOI: 10.1201/9781003655558-16

2 A RESEARCH AREA

The research area is situated in southern Poland, in the northeastern part of the Silesian Voivodeship, specifically in the Częstochowa district, within the Lelów commune, in the Nakło.

Nakło is a small village with 476 inhabitants (Główny Urząd Statystyczny, 2021). The centre of the village is built up of single-family houses and public utility buildings (primary school, library, church, fire station). In the central part of the village there is a historic church of St. Nicholas from the 18th century with a belfry from the first half of the 19th century (Figure 1a). In the south-eastern part of the village there is a historic palace and park complex from the 18th century (Figure 1b). The buildings are surrounded by: from the south – the National Road No. 46, from the east – the Zawiercie-Wieszowa railway line and numerous cultivated fields (Figure 1c).

a) b) c)

Figure 1. Development of the village area: a) a church of St. Nicholas, b) a palace and park complex, c) a national road (red line) and a railway line (blue line).

The research area covered an area of approximately 20 ha. From the north, it is bounded by a local road leading to Przyłęk, and from the south by National Road No. 46. Measurements were taken in the centre of a village, among numerous but not very compact buildings. This area was clear for measurement using photogrammetric methods, and at the same time, it was rich in terrain details that were the subject of surveys using the RTN method.

3 MEASUREMENTS

3.1 *Low-altitude photogrammetry*

The flight over the research area was carried out in the autumn of 2023, using the WingtraOne UAV (unmanned aerial vehicle), a drone with a wingspan of 1.25 m and a maximum speed of up to 16 m/s. It has a built-in the QX1 Sony camera with a resolution of 20 MP, a focal length of 20 mm and the APS-C matrix. Thanks to such camera parameters, the created images are characterised by good clarity. In addition, the mechanical shutter of the photogrammetric camera eliminates the rolling shutter effect, i.e. a blurry image caused by the camera or the photographed object moving at a high speed. The UAV also features a VTOL (Vertical Take-Off and Landing) system, enabling vertical take-off and landing capabilities. The initial and final stages of flight are solved in this way, reducing the risk of the drone crashing during these manoeuvres. The WingtraOne can be operated in areas that are difficult to land and measure, including those with rocky ground, heavy urbanisation, dense vegetation, and other challenging conditions. Table 1 presents the values of the basic technical parameters of the WingtraOne drone.

Table 1. The WingtraOne technical parameters.

Parameter	Value
Wingspan	1.25 m
Maximum speed	16 m/s
Camera	Sony QX1
Resolution	20 MP
Focal length	20 mm
Matrix	APS-C
Take-off/landing system	VTOL
Maximum take-off weight	4.8 kg
Battery capacity	2 x 99 Wh
Radio link	10 km (bi-directional)

Before the start of the flight, 7 photo points numbered f4, f5, f6, f11, f12, f13 and f14 were marked in the terrain. They were characteristic terrain details, e.g. the corner of the pavement and markers painted on the road surface in the form of white crosses with arms 20 cm long.

The flight was carried out at an average altitude of 124 m above the ground and an average speed of 4 m/s. During the flight, 129 photos were taken, and the degree of their mutual overlap ranged from 60% to 80%, which ensured sufficient accuracy of calibration and image matching. The direction of the drone's flight relative to the sides of the rectangle intended for research was parallel to its longer side. The values of the basic parameters of the drone flight were presented in Table 2.

Table 2. The drone flight parameters.

Parameter	Value
Average flight altitude	124 m
Average flight speed	4 m/s
Number of photos	129
Degree of photo coverage	60%-80%
Surface of the area	20 ha
Shape of the area	rectangle
Flight direction	parallel to the longer side of the rectangle

3.2 *The RTN method*

The situational and altitude coordinates of photo points and 126 measurement points, which constitute characteristic elements of land development, were measured using the RTN method.

The RTN method involves positioning the GPS receiver in real time and, unlike the RTK (Real-Time Kinematic) method, allows the use of only one measuring device. The appropriate measurement accuracy is achieved thanks to a network of 125 reference stations located throughout the country and in the border zone. These stations receive data on their position from satellites that are part of the GNSS (Global Navigation Satellite System) and then transmit it to a central server. The same server also receives data on the approximate position of the GPS receiver, determined by using the direct method. The server identifies the 3 closest reference stations and creates, near the mobile GPS receiver, a Virtual Reference Station (VRS) for a given measurement period. It then interpolates corrections to the measurement that are proportional to the distance of the VRS from each of the 3 physical reference stations and then sends the result to the mobile receiver. Corrections are determined separately for each measurement epoch. A necessary condition for using this method is to connect the GPS receiver to the Internet in order to establish a connection with the reference stations.

The surveys were taken in the PL-EVRF2007-NH system, which is currently the coordinate system in Poland, in accordance with the Regulation of the Council of Ministers of 15 October 2012 on the national spatial reference system (*Rozporządzenie Rady Ministrów z Dnia 15 Października 2012 r. w Sprawie Państwowego Systemu Odniesień Przestrzennych*, 2012). Altitude measurements made in this system are referred to the mean level of the North Sea determined by the tide-gauge located in Amsterdam (The Netherlands).

During the measurements, the Trimble R8 GPS receiver with a pole (Figure 2a), the Trimble TSC3 controller (Figures 2b) and 3 reference stations: Kłobuck (KLOB), Lelów (LELO) and Tarnowskie Góry (TAR1) (Figure 3a) were used.

a) b)

Figure 2. Measuring equipment: a) the GPS receiver with a pole (Trimble R8), b) a controller (Trimble TSC3).

By paragraph 9, point 5 of the Regulation of the Minister of Development of August 18, 2020, on technical standards for making geodetic situational and altitude measurements, developing and submitting the results of these measurements to the national geodetic and cartographic resource (*Rozporządzenie Ministra Rozwoju z Dnia 18 Sierpnia 2020 r. w Sprawie Standardów Technicznych Wykonywania Geodezyjnych Pomiarów Sytuacyjnych i Wysokościowych Oraz Opracowywania i Przekazywania Wyników Tych Pomiarów Do Państwowego Zasobu Geodezyjnego i Kartograficznego*, 2020), before starting the right surveys, a control measurement was done at the point no. 52220203 (Figure 3b) with the m_H altitude error of ± 0.01 m, which belongs to the basic altitude geodetic network. As a result of the measurement, the H altitude coordinate of the control point was obtained, differing by -0.07 m compared to its initial coordinate. According to the above Regulation, the maximum value of the permissible deviation of the point height (Δ_{Hmax}) is equal to ± 0.09 m. The obtained difference is within the accuracy requirements and allowed for the continuation of further measurements of photo points and terrain pickets.

There was no need to check the heights of the other points in the altitude geodetic network and their differences between the data contained in the documentation and those obtained from the GNSS measurements, because the point no. 52220203 was the least convenient point for measurement, located in the center of a heavily developed village and obscured on the north by a church wall. Such intense horizon obscuration should have resulted in the largest height difference compared to the height differences of the other points, which should have been smaller than the obtained value (-0.07 m).

Measurements of 126 points characterising terrain details were taken in fourteen measurement epochs, while photo points were taken in thirty epochs. Each of the above points was measured twice.

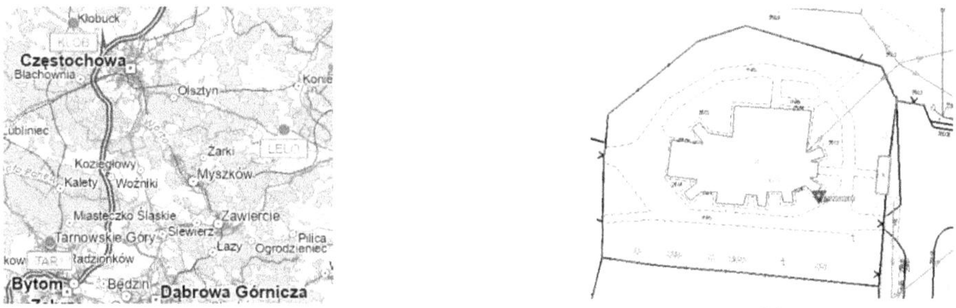

a) b)

Figure 3. Geodetic network: a) the reference stations: KLOB, LELO and TAR1, b) a point of the basic altitude geodetic network (a blue triangle, point no. 52220203).

4 RESULTS AND DISCUSSION

4.1 *Digital terrain model*

The Agisoft Metashape Professional computer program was used to process the low-altitude photogrammetry images. It enables precise photo matching, polygon model generation and texture mapping. This software is used, among others, to create dense point clouds, Digital Terrain Models (DTMs) and orthophotomaps.

The DTM of the Nakło village was created based on 129 imported photos, which were georeferenced in the ETRF2000-PL system, zone 6 (meridian 18°). After entering 7 photo points (f4, f5, f6, f11, f12, f13, f14) as model adjustment points, the program created a sparse point cloud (SPC) consisted of 91,359 points (Figure 4a). The Root Mean Square Error (RMSE) for the H point altitude in the SPC is equal to 0.56 m (RMSE H_{SPC} = 0.56 m). In the next step, the photos were accurately calibrated based on the manual location of photo points in the images where a physical marker labelling an individual photo point was visible. Then, a density map was created, resulting in a dense point cloud. In the next stage, a polygonal mesh model was created, which was covered with texture from the captured photos (Figure 4b). The last stage of the work was to create the DTM together with the Digital Elevation Model (DEM) (Figure 4c). The DEM shows that the terrain of the Nakło village is not complicated and its average altitude is approximately 259 m. The RMSE of the *H* point altitude in the DEM achieved a lower value by over 16% than in the SPC (RMSE H_{DEM} = 0.47 m).

a) b) c)

Figure 4. Stages of creating a Digital Terrain Model: a) a sparse point cloud, b) a polygonal mesh model with texture, c) a Digital Terrain Model with a Digital Elevation Model.

Table 3 shows the $\Delta H_{i\text{-}j}$ heights differences of the f4, f5, f6, f11, f12, f13 and f14 photo points between their heights obtained in the SPC and the DEM, and measured by the use of the RTN method and checked in the polish geoportal (GEO).

Table 3. Heights differences of the photo points.

	SPC (i)			DEM (i)	
Photo point	RTN (j)	GEO	$\Delta H_{i\text{-}j}$ [m]	RTN (j)	GEO
f4	0.466	0.145		0.437	0.116
f5	0.354	0.108		0.224	-0.022
f6	0.475	0.196		0.391	0.112
f11	0.200	0.161		0.125	0.086
f12	0.469	0.243		0.413	0.187
f13	0.357	0.269		0.253	.165
f14	0.836	0.744		0.827	0.735
Mean	0.451	0.267		0.381	0.197

4.2 *Situational-altitude map*

A situational-altitude map of the terrain of the Nakło village was created based on the results of analysis of measurements taken directly in the research area using the RTN method and a previously created point cloud, within the Autodesk AutoCAD Civil 3D computer program.

A new working layer was created in the program, and a topographic surface was generated on it. The previously created point cloud was then imported. Based on the point cloud, a Triangulated Irregular Network (TIN) was built. To densify it, the program interpolated additional altitude points in the empty spaces between existing points. The coordinates of the photo points were also imported so that they could be used to check a fit of the orthophotomap at a later stage of the work. Then, the raster calibration was done. The file thus created in the program was integrated with the orthophotomap generated from the flight results. The program automatically fitted the raster in the correct location. Manual fitting was not required because the file obtained from the flight results had a saved, correct georeference. Based on previously imported coordinates of photo points, the correctness of raster matching was checked. The markers visible on the orthophotomap coincided with their corresponding matching points.

The obtained raster (Figure 5a) had a resolution that allowed for the proper differentiation of terrain details, which is necessary for the correct vectorisation of objects that are the content of a situational-altitude map. Objects previously measured using the RTN method were also vectorised on the raster. In this way, their characteristic points received the correct altitudes provided from the results of direct measurements. As a result of vectorising the previously calibrated raster with the orthophotomap, a situational-altitude map of Nakło village was obtained (Figure 5b).

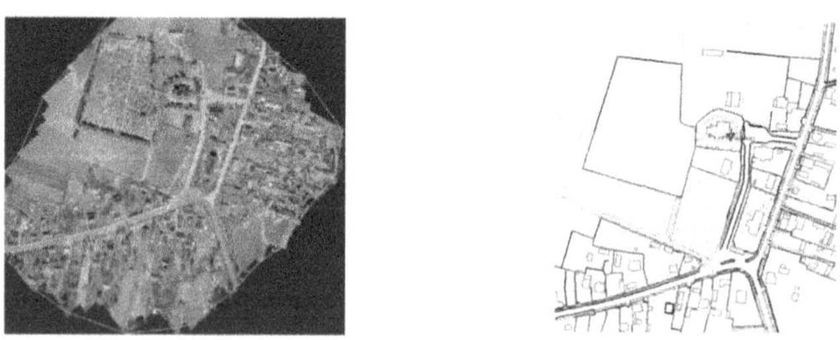

a) b)

Figure 5. Stages of creating a situational-altitude map of: a) a calibrated raster with the orthophotomap, b) a vectorised raster as a situational-altitude map.

4.3 *Discussion*

The accuracy of the altitude coordinate of the DEM was determined by comparing the heights of points directly measured in the research area using a GPS receiver with the RTN method with the heights of the same points on the DEM surface. 126 pickets were measured, representing the terrain details of the first and second groups of accuracy, among others: building corners, geodetic control points, boundary points of cadastral parcels (Dragomir *et al.* 2025), elements of the land infrastructure network, and so on. The average relative error in determining the point heights in the DEM was 0.88 m. This value is not satisfactory considering the elevation position of the Digital Terrain Model. However, this model can still be used to determine the situational position of selected terrain details, inventory objects (e.g. a building in ruins), land development (built or unbuilt area), or land use (agricultural land, wasteland, etc.). The following factors could have influenced the obtained average value of the relative error in determination of the altitude coordinate of the points constituting the DTM: too few points measured in the research area and analyzed in the DTM; number of flights, taken photos and photo points; the height of the UAV flight and the use of a camera with given parameters of lens; incorrectly assigned height to the measurement points in the DTM due to incorrect interpretation of taken photos (e.g. a point constituting the corner of a building could have been assigned the height of the roof overhang).

However, it is worth noting that the RMSE of the point height in the DEM is almost twice smaller than the average relative error of the point height and is 0.47 m. This value of the height determination error allows the use of DEM to assess the height of less important terrain details, e.g. those belonging to the third accuracy group.

Taking into account data presented in Table 3, it can be noticed that among the photo points the f14 photo point had the weakest match of the height in the DEM to the height measured (the RTN) and obtained from the geoportal (GEO) (successively 0.83 m and 0.74 m). This may be due to the distant location of the f14 photo point from the other photo points. Moreover, the average difference in the height of photo points in the DEM is approximately 21% lower than its value obtained in the SPC. The best fit of the height of photo points in the DEM was obtained for data derived from the geoportal and was on average about 0.20 m.

5 CONCLUSION

Nowadays, creating DTM is relatively quick and easy, provided that one has access to the right equipment and software and uses the proper measurement techniques and methods. Its creation allows for numerous analyses related to the way the land is developed and its use, as well as the technical condition of buildings, etc.

However, despite the highly developed technology of a low-altitude photogrammetry, the heights of points that consist of the DEM still do not achieve satisfactory accuracy (RMSE H_{DEM} = 0.47 m), sufficient for creation of studies in which the landform and the altitude coordinate is essential (e.g. the profiles and cross-sections through a mountain valley, a riverbed, an open pit excavation, etc.). The accuracy of determining the height of points in the DEM may also depend on the method used. Comparing point heights in the model with the results of direct geodetic measurements can yield low accuracy. However, comparing these heights with data from a geoportal can yield significantly higher accuracy.

Therefore, the altitude position of terrain and objects obtained from the low-altitude photogrammetry results should be trusted with limited confidence, and UAV flights should be planned meticulously. The DEM can provide general information about the landform; however, for some objects, their altitude coordinates should be determined using classical geodetic measurements, such as satellite or levelling techniques.

REFERENCES

Błaszczak-Bąk, W., Poniewiera, M., Sobieraj-Żłobińska, A., and Kowalik, M. (2019). Reduction of measurement data before Digital Terrain Model generation vs. DTM generalisation. *Survey Review* 51(368): 422–430. https://doi.org/10.1080/00396265.2018.1474685

Chen, Y., Tang, J., Jiang, C., Zhu, L., Lehtomäki, M., Kaartinen, H., Kaijaluoto, R., Wang, Y., Hyyppä, J., Hyyppä, H., Zhou, H., Pei, L., and Chen, R. (2018). The Accuracy Comparison of Three Simultaneous Localization and Mapping (SLAM)-Based Indoor Mapping Technologies. *Sensors* 18(10): 3228. https://doi.org/10.3390/s18103228

Chio, S.-H., and Hou, K.-W. (2021). Application of a Hand-Held LiDAR Scanner for the Urban Cadastral Detail Survey in Digitized Cadastral Area of Taiwan Urban City. *Remote Sensing* 13(24). https://doi.org/10.3390/rs13244981

Crommelinck, S., Bennett, R., Gerke, M., Nex, F., Yang, M.Y., and Vosselman, G. (2016). Review of Automatic Feature Extraction from High-Resolution Optical Sensor Data for UAV-Based Cadastral Mapping. *Remote Sensing* 8(8). https://doi.org/10.3390/rs8080689

Diara, F., and Roggero, M. (2022). Quality Assessment of DJI Zenmuse L1 and P1 LiDAR and Photogrammetric Systems: Metric and Statistics Analysis with the Integration of Trimble SX10 Data. *Geomatics* 2(3): 254–281. https://doi.org/10.3390/geomatics2030015

Dragomir, L.O., Popescu, C.A., Herbei, M.V., Popescu, G., Herbei, R.C., Salagean, T., Bruma, S., Sabou, C., and Sestras, P. (2025). Enhancing Conventional Land Surveying for Cadastral Documentation in Romania with UAV Photogrammetry and SLAM. *Remote Sensing* 17(13). https://doi.org/10.3390/rs17132113

Główny Urząd Statystyczny (2021). *NSP 2021—Wyniki ostateczne.* stat.gov.pl. https://stat.gov.pl/spisy-powszechne/nsp-2021/nsp-2021-wyniki-ostateczne/

Grzechca, D., Zięibiński, A., Paszek, K., Hanzel, K., Giel, A., Czerny, M., and Becker, A. (2020). How Accurate Can UWB and Dead Reckoning Positioning Systems Be? Comparison to SLAM Using the RPLidar System. *Sensors* 20(13): 3761. https://doi.org/10.3390/s20133761

Janusz, W., Niezabitowski, M., Orwat, J., and Sikora, P. (2016). The concept of combined measurements of the “Gliwice” radio tower deflection in relation to the speed and direction of the wind with use of quadrotor. *2016 17th International Carpathian Control Conference (ICCC)*, 280–285. https://doi.org/10.1109/CarpathianCC.2016.7501109

Jia, S., Liu, C., Wu, H., Zeng, D., and Ai, M. (2021). A cross-correction LiDAR SLAM method for high-accuracy 2D mapping of problematic scenario. *ISPRS Journal of Photogrammetry and Remote Sensing* 171: 367–384. https://doi.org/10.1016/j.isprsjprs.2020.11.004

Keitaanniemi, A., Virtanen, J.-P., Rönnholm, P., Kukko, A., Rantanen, T., and Vaaja, M. (2021). The Combined Use of SLAM Laser Scanning and TLS for the 3D Indoor Mapping. *Buildings* 11(9): 386. https://doi.org/10.3390/buildings11090386

Kim, P., Chen, J., and Cho, Y.K. (2018). SLAM-driven robotic mapping and registration of 3D point clouds. *Automation in Construction* 89: 38–48. https://doi.org/10.1016/j.autcon.2018.01.009

Liu, S., Sun, E., and Dong, X. (2024). SLAMB&MAI: A comprehensive methodology for SLAM benchmark and map accuracy improvement. *Robotica* 42(4): 1039–1054. https://doi.org/10.1017/S0263574724000079

Oniga, V.E., Chirilă, C., and Macovei, M. (2016). Low-cost unmanned aerial systems in cadastral applications. *International Multidisciplinary Scientific GeoConference: SGEM* 2: 947–954.

Paszek, J. (2024). The volume of stones stack calculated by different methods -a case study. *Journal of Physics: Conference Series* 2927(1): 012005. https://doi.org/10.1088/1742-6596/2927/1/012005

Poniewiera, M. (2018). Budowa numerycznego modelu terenu dla celów związanych z liczeniem objętości. *Systemy Wspomagania w Inżynierii Produkcji* 7(1): 470–483.

Poniewiera, M., Jelonek, I., and Błaszczak-Bąk, W. (2019). The optimal position of a pumping station based on the digital terrain model. *Journal of Water and Land Development* 40(I–III): 103–111. https://doi.org/10.2478/jwld-2019-0011

Popescu, G., Popescu, C.A., Dragomir, L.O., Herbei, M.V., Horablaga, A., Ţenche-Constantinescu, A.-M., Sălăgean, T., Bruma, S., Dinu-Roman (szabo), M., Colisar, A., Ceuca, V., Kader, S., and Sestras, P. (2024). Utilizing UAV technology and GIS analysis for ecological restoration: A case study on Robinia pseudoacacia L. in a mine waste dump landscape rehabilitation. *Notulae Botanicae Horti Agrobotanici Cluj-Napoca* 52(4): 13937–13937. https://doi.org/10.15835/nbha52413937

Puniach, E., Bieda, A., Świąkała, P., Kwartnik-Pruc, A., and Parzych, P. (2018). Use of Unmanned Aerial Vehicles (UAVs) for Updating Farmland Cadastral Data in Areas Subject to Landslides. *ISPRS International Journal of Geo-Information* 7(8). https://doi.org/10.3390/ijgi7080331

Ramadhani, S.A., Bennett, R.M., and Nex, F.C. (2018). Exploring UAV in Indonesian cadastral boundary data acquisition. *Earth Science Informatics* 11(1): 129–146. https://doi.org/10.1007/s12145-017-0314-6

Rozporządzenie Ministra Rozwoju z dnia 18 sierpnia 2020 r. W sprawie standardów technicznych wykonywania geodezyjnych pomiarów sytuacyjnych i wysokościowych oraz opracowywania i przekazywania wyników tych pomiarów do państwowego zasobu geodezyjnego i kartograficznego. (2020). https://isap.sejm.gov.pl/isap.nsf/DocDetails.xsp?id=WDU20200001429

Rozporządzenie Rady Ministrów z dnia 15 października 2012 r. W sprawie państwowego systemu odniesień przestrzennych. (2012). https://isap.sejm.gov.pl/isap.nsf/DocDetails.xsp?id=WDU20120001247

Šafář, V., Potůčková, M., Karas, J., Tlustý, J., Štefanová, E., Jančovič, M., and Cígler Žofková, D. (2021). The Use of UAV in Cadastral Mapping of the Czech Republic. *ISPRS International Journal of Geo-Information* 10(6). https://doi.org/10.3390/ijgi10060380

Sestras, P., Badea, G., Badea, A.C., Salagean, T., Oniga, V.-E., Roşca, S., Bilaşco, Ştefan, Bruma, S., Spalević, V., Kader, S., Billi, P., and Nedevschi, S. (2025a). A novel method for landslide deformation monitoring by fusing UAV photogrammetry and LiDAR data based on each sensor's mapping advantage in regards to terrain feature. *Engineering Geology* 346: 107890. https://doi.org/10.1016/j.enggeo.2024.107890

Sestras, P., Badea, G., Badea, A.C., Salagean, T., Roşca, S., Kader, S., and Remondino, F. (2025b). Land surveying with UAV photogrammetry and LiDAR for optimal building planning. *Automation in Construction* 173: 106092. https://doi.org/10.1016/j.autcon.2025.106092

Sestras, P., Roşca, S., Bilaşco, Ştefan, Şoimoşan, T.M., and Nedevschi, S. (2023). The use of budget UAV systems and GIS spatial analysis in cadastral and construction surveying for building planning. *Frontiers in Built Environment* 9. https://doi.org/10.3389/fbuil.2023.1206947

Stöcker, C., Nex, F., Koeva, M., and Gerke, M. (2020). High-Quality UAV-Based Orthophotos for Cadastral Mapping: Guidance for Optimal Flight Configurations. *Remote Sensing* 12(21). https://doi.org/10.3390/rs12213625

Štroner, M., Urban, R., Křemen, T., and Braun, J. (2023). UAV DTM acquisition in a forested area – comparison of low-cost photogrammetry (DJI Zenmuse P1) and LiDAR solutions (DJI Zenmuse L1). *European Journal of Remote Sensing* 56(1): 2179942. https://doi.org/10.1080/22797254.2023.2179942

Urban, R., Štroner, M., Braun, J., Suk, T., Kovanič, L'., and Blistan, P. (2024). Determination of Accuracy and Usability of a SLAM Scanner GeoSLAM Zeb Horizon: A Bridge Structure Case Study. *Applied Sciences* 14(12): 5258. https://doi.org/10.3390/app14125258

Yarovoi, A., and Cho, Y.K. (2024). Review of simultaneous localization and mapping (SLAM) for construction robotics applications. *Automation in Construction* 162: 105344. https://doi.org/10.1016/j.autcon.2024.105344

Mine Surveying and Mining Area Protection – Sokoła-Szewioła and Jarczyk (Eds)
© 2025 The Author(s), ISBN: 978-1-041-10531-2

Delineation of the potentially flooded area in the rock mass

M. Poniewiera, M. Pomykoł & P. Pierzyna
Silesian University of Technology, Gliwice, Poland

ABSTRACT: The correct determination of the mining space exposed to water infiltration is one of the key issues for ensuring the safe operation of a mining facility. Modern technologies allow for the creation of a spatial model of mining excavations in the context of the deposit model and the aquifer level. This, in turn, allows for the precise determination of spatial distances between the hazard and the location of the workforce. The article introduces the concept of a dynamic model of the mining space, which enables continuous monitoring of the relative position of hazards in space, taking into account the impact of exploitation on the rock mass. The paper briefly discusses issues related to creating a spatial model of mining excavations, generating a numerical deposit model, considering the impact of exploitation on the rock mass, and determining the boundary of the potentially flooded area within the rock mass. It presents an algorithm for calculating the intersection of two surfaces, such as the aquifer level and the coal seam. The technology for automatically determining the distance between the workings and hazards, such as planned drilling holes, is also discussed.

1 INTRODUCTION

The dynamic development of modern digital technologies, their widespread availability, as well as the applicable legal regulations, provide the opportunity to create a unified model that includes mining excavations, deposit elements, aquifer levels, and information on natural hazards. The model is expected to be dynamic, taking into account the impact of current and planned exploitation on the workings, rock mass, and the surface area. The dynamic model of the mining space is understood here as a database with software that enables various analyses and the generation of all technical documentation. This database stores XYZ coordinates and time, allowing for the generation of documentation for any moment in the past and forecasting the position of selected objects in the future.

The paper presents modern IT tools used in most mining companies (software: Geolisp, EDN-OPN, AutoCAD Civil, BricsCAD), with respect to their application for the automatic determination of the distance between mining excavations and hazards occurring in the rock mass. The algorithm presented in the article consists of several stages. The first stage outlines the process of building a spatial model of mining excavations. Next, the method of creating the deposit model is characterized, which includes the surface of the aquifer level, the coal seam, and the surface differences between them. The authors emphasize that the proper execution of the deposit model, including the consideration of faults and other discontinuities, has a significant impact on the results of predictive calculations (Poniewiera 2019). The next chapter is devoted to issues related to the impact of planned exploitation on the rock mass. The final part of the article presents the practical implementation of this problem in CAD systems.

DOI: 10.1201/9781003655558-17

2 CONSTRUCTION OF THE SPATIAL MODEL OF MINING WORKINGS

Spatial modeling of mining excavations forms the basis for modern information management in mines. In a spatial view, it is much easier to visualize hazards or newly designed elements than on a flat map (Kowalczuk *et al.* 2019, Zhang *et al.* 2020). A 3D model facilitates control and allows for quick identification of incorrectly entered objects. In the literature (Poniewiera 2007, Krawczyk 2023), examples of methods for creating spatial models of workings, data control based on which the model is created, and the practical use of such a model can be found. The development of modern technologies, including laser scanning, contributes to a significant increase in the accuracy of constructing spatial models of workings and, consequently, their usability (Janus 2016).

In mining practice, in Polish hard coal mines, Geolisp software is used, which enables the creation of a spatial model in several ways.

The first method involves the use of the excavation axis. This method is used for quick visualization of the entire mine's workings. The dimensions and shape of the solid are usually the same for the whole workings. Figure 1 shows the identification of hazard areas (discussed in Chapter 5) against the background of the excavation axis.

In this method, we draw the workings' lines roughly along their axis. The number of vertexes is selected based on the desired accuracy of depicting the workings. A larger number of points ensures more accurate representation, but displaying the model will place significant demands on computing resources. In all methods described below, it is advisable to draw the excavation axis, which is used for simplified visualization and various calculations, e.g., the distance between the workings and the aquifer level or changes in the position of the workings in the rock mass due to the exploitation of adjacent seams. Additional data attached to the axis contains attributes of the workings such as name, start and end dates, volume, mass, and the contamination level of the extracted material.

The axes should have common points at intersections. Topology control should be performed by running a water flow analysis. We can imagine the workings as water pipes. If there are no topology errors, connecting water at one point will flood all the workings.

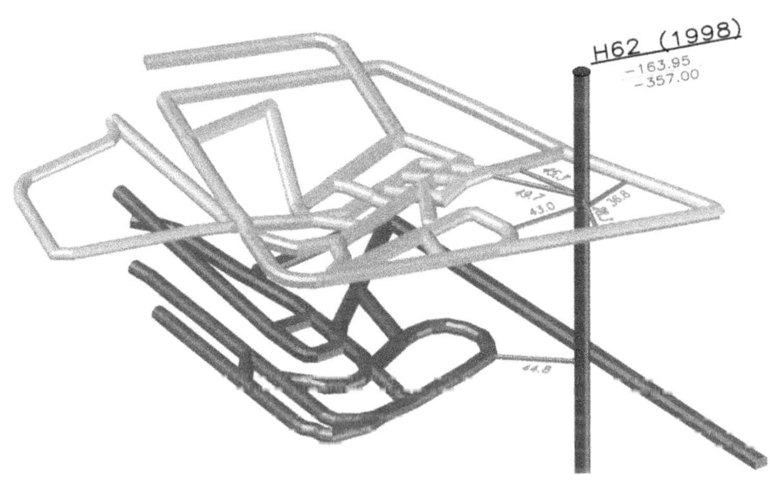

Figure 1. Identification of hazard areas in the spatial model of mining excavations.

The assignment of elevation coordinates to the points on the excavation axis can be done in two ways. The first method is to use elevation points found in the workings. In this case, it is necessary to check whether all workings have elevation points, if they have corrected values, and if each point is used exactly once (twice at intersections). The second method involves projecting the excavation axis onto the spatial model of the seam. The creation of such a model is described later in the article. This method more accurately interpolates the axis height at locations where no elevation points are present.

After the model is created, it should be checked, e.g., by verifying areas where the workings‘ slope is greater than expected. This is the most time-consuming stage of work. Proper control is the key to the high quality of the model created (Krawczyk 2018).

The final stage of model creation is transforming the created axes into solids, which can be done in several ways. First, CAD systems such as AutoCAD or BricsCAD allow for assigning width and height to lines. This is not a solid, so it hardly burdens the computer's memory, yet it allows for spatial visualization of the workings. Then, we can transform all the workings into solids created based on flat shapes corresponding to the used supports (circle, rectangle, LP support). There is an option to simultaneously create four solids: floor and sides, which allows for virtual tours inside the workings.

The second way to create the spatial model of the workings is by using its cross-sections (obtained from direct measurements or laser scanning). This is the most accurate method because the solid precisely reflects the shape of the workings. The modeling begins by creating 3D cross-sections where they were measured. The cross-sections can be vertical, horizontal, or oblique. When using a point cloud obtained from a laser scan, it is recommended to specify the cross-section width (distance between points of the same cross-section measured along the axis), the distance between cross-sections, and the degree of permissible generalization. Each of the cross-sections created can be dimensioned, yielding e.g., cross-sectional area, height, etc.

The third method is a combination of the two previously discussed methods. The program automatically creates identical cross-sections at every vertex of the excavation axis. Then, each of these cross-sections can be adjusted based on actual values obtained from measurements, such as height or width of the selected part of the workings. This allows for more accurate representation of the dimensions of selected sections, e.g., for visualizing equipment located in that part of the workings. On the other hand, most workings are the same size, so the workload for creating the model is relatively small.

A fourth method can also be identified: creating a model of the workings based on the sides. Similar to the above method, cross-sections are automatically inserted along the axis, but they are placed at the sides' vertexes, not the axis. The cross-section width equals the width of the workings, so it varies at each point. This method requires significantly more effort than the others because the map must be properly prepared. Each of the workings' sides should be created with a single polyline. These polylines should be approximately the same length and parallel to each other. At intersections, the workings‘ sides should intersect. If the workings connect to another working without intersecting it, the side lines should be extended approximately halfway to the width of the connected workings (Poniewiera *et al.* 2015). In this method, these solids are complex and place significant demands on computing resources and computer memory.

3 CONSTRUCTION OF THE NUMERICAL DEPOSIT MODEL

A numerical deposit model is understood here as a series of triangular meshes (TIN - Triangulated Irregular Network), whose vertices are points with specified X, Y, and Z coordinates. The X and Y coordinates indicate the measurement location, while the Z value corresponds to the selected attribute (seam floor height, thickness, ash content, etc.). The literature offers several sources describing the issues related to creating TIN surfaces (Jia *et al.* 2020, Jelonek *et al.* 2015, Krawczyk 2019, Sokoła & Poniewiera 2019, Wu *et al.* 2005). TIN surfaces can be created based on data retrieved from text files, point clouds obtained from laser scanning, objects found on maps (control points, elevation points, chemical analyses, geological boreholes, seam isohypses). For each seam, a separate triangular mesh is created. Additionally, each seam is defined by multiple meshes. representing various seam parameters such as sulfur content, ash content, etc. The deposit model can also be represented as a network of cuboids reflecting the entire rock mass (Białek *et al.* 2020).

When constructing a numerical model of a hard coal deposit, we usually encounter the problem of considering discontinuity lines: faults, seam junction lines, faulting, etc. In the articles (Sokoła-Szewioła & Poniewiera 2022a, b), proprietary methods for determining the spatial position of faults in the rock mass are presented. For these purposes, the following methods can be applied: a method based on using existing isohypses, a method assuming that the seam dip before and after the fault is constant, a method based on creating independent surfaces on both sides of the discontinuity line, and a method based on transferring the fault from an upper seam. The proposed methods have been implemented in most hard coal mines in Poland. An example deposit model with faults is shown in Figure 2.

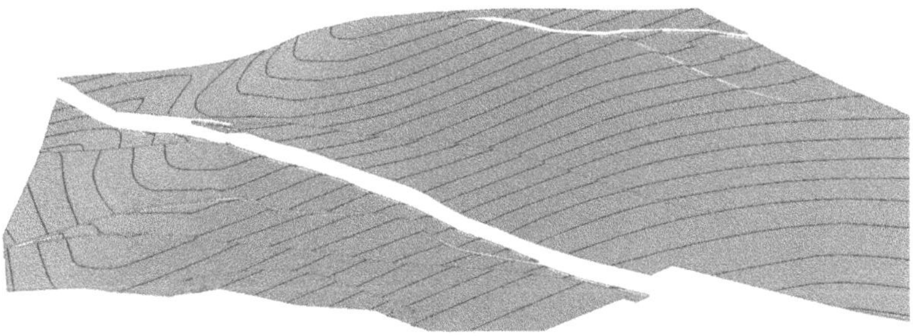

Figure 2. Deposit model with faults.

4 IMPACT OF EXPLOITATION ON THE ROCK MASS AND ITS WORKINGS

One of the consequences of underground mining exploitation is the emergence of continuous deformations of the mining area surface and the rock mass. The ability to predict such deformations allows for forecasting the adverse effects of mining exploitation and making corrections to exploitation plans in order to reduce these impacts.

In Poland, predictions of continuous deformations of the mining area are mainly made based on the integral geometric theory of influences. The essence of these theories is the assumption of the existence of an "influence function" $f(L, p_1...p_n)$, where L denotes the horizontal distance of the considered point P on the surface or in the rock mass from the element dS of the selected surface S, and the superposition (summation) principle of influences is applied. Integration in these theories is carried out over an area that is the projection of the exploited field onto a horizontal surface (Equation 1).

$$w_k(P, S(t)) = -\iint_{S(t)} w_{\max} f(L, p_1, ..p_n) dS \qquad (1)$$

where $w(P)$ – surface subsidence at point P; w_{max} maximum subsidence observed in a full subsidence basin, when the size of the exploitation field is sufficiently large compared to the depth of exploitation; L – horizontal distance of point P from the excavation surface dS; $p_1,...p_n$ – parameters of the influence theory depending on the properties of the rock mass, exploitation depth, seam dip, etc.

Further deformation indicators are generally obtained by differentiating the formula (1) with respect to the coordinates of point P. If the size and position of the exploitation field S change over time t, then the subsidence $w_k(P,S(t))$ is also a function of time t.

In the surface deformation forecasting programs (rock mass), EDN-OPN developed by Białek (2003), the S. Knothe formula is used in its general form, as shown in Equation 2.

$$w_k(x,y,z,t,r(z)\ldots) = -\iint\limits_{S(t)} \frac{ag}{r(z)^2}\exp\left(-\pi\frac{(\xi - x)^2 + (\eta - y)^2}{r(z)^2}\right)d\xi d\eta \tag{2}$$

where x, y – coordinates of the calculation point; $S(t)$ – surface of the exploited seam, which is generally a function of time t; ξ, η – coordinates of the surface element ds; $ag = w_{max}$ – maximum subsidence of the calculation point, which occurs when the selected surface S around this point is sufficiently large; $r = h(\xi, \eta, z, tg\beta)$ – radius of the dispersion of influences determined from measurements or for the surface, calculated using the formula $r = h/tg\beta$, where $tg\beta$ is a parameter of S. Knothe's theory.

The most important factor for forecasting inside the rock mass is knowing the actual distribution of the dispersion radius of influences $r(z)$. The literature offers several empirical formulas developed based on measurements carried out in underground mining excavations. One of the most frequently cited formulas is the relationship proposed by S. Knothe and modified by subsequent researchers, as presented in Equation 3.

$$r(z) = \frac{h}{tg\beta}\left(\frac{z - z_0}{h - z_0}\right)^n \tag{3}$$

where h – depth of the exploited element of the seam dS; $tg\beta$ – parameter of S. Knothe's theory; z – height of the calculation point above the surface element dS, z_0 – parameter proposed by Drzężla (1978) (in the programs, $z_0 = 5ag$ is assumed); n – exponent of formula (3) (in the programs, $n = 0.665$).

The values of the exponent n changed with the development of deformation research in the rock mass, as illustrated in Figure 3 and presented in Equations 4–8.

$$z_0 = 0;\ n = tg\beta\sqrt{2\pi}(\text{Budryk 1953}), \tag{4}$$

$$z_0 = 0;\ n = 1\ (\text{Knothe 1984}), \tag{5}$$

$$z_0 = 0;\ n = 0,5\ (\text{Litwiniszyn 1953}), \tag{6}$$

$$z_0 > 0;\ n = 0,665\ (\text{Drzężla 1978}), \tag{7}$$

$$z_0 = 2ag;\ n = 0,665\ (\text{Białek 2003}). \tag{8}$$

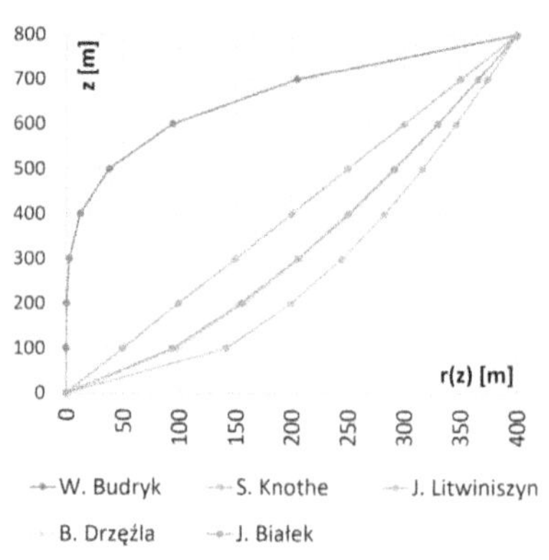

Figure 3. Distribution $r(z)$ according to formula (3).

Another approach to describing the distribution of deformations in the rock mass was proposed by (Chudek 1981), who developed a formula that takes into account the lithological characteristics of the rocks forming the rock mass (Equation 9).

$$r(z) = \sqrt{\frac{z \cdot R_{rs}}{\gamma_{śr}}} \tag{9}$$

where R_{rs} – weighted average tensile strength of rocks between the roof and the considered horizon; $g_{śr}$ – weighted average bulk density of rocks between the roof and the considered horizon; z – distance between the roof and the considered horizon.

5 ALGORITHM FOR DETERMINING THE INTERSECTION OF TWO TIN SURFACES

An essential element of the procedure for determining the mining space exposed to water infiltration is the algorithm for the intersection of the seam surface with the aquifer surface. Both of these surfaces should be defined as TIN meshes, the creation of which was discussed in Chapter 3. Typically, we are not only interested in determining the intersection trace but also in generating a new TIN (Triangulated Irregular Network) surface that is the difference between both meshes. For this purpose, we use the algorithm presented in Figure 4.

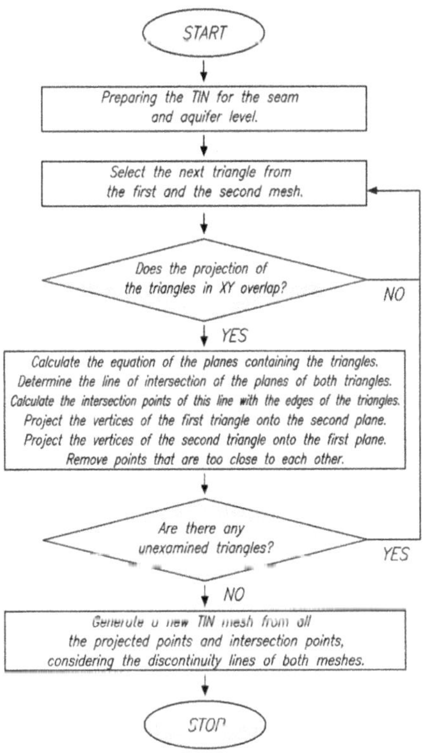

Figure 4. Algorithm for determining the intersection of two TIN (Triangulated Irregular Network) surfaces.

6 PRACTICAL IMPLEMENTATION OF DETERMINING HAZARD ZONES IN CAD PROGRAMS

In CAD systems, the surface of differences described in the previous section is called the difference surface. The difference surface has the same properties as any other surface. It can be visualized using: isohypses, spatial triangles or colored areas within a given value range. For example, coloring values from -X to +X will highlight the area where the seam is located at a distance less than X from the aquifer level, thus marking an area potentially at risk of water infiltration. If we generate an isohypse with the value "0.0", we will obtain the intersection trace of both surfaces, i.e., the point where the aquifer intersects the deposit. This line is the sought boundary of the potentially flooded area.

In AutoCAD Civil, the TIN (Triangulated Irregular Network) surface is created with the "AeccCreateSurface" command, and then we specify the objects that make up the surface. Surfaces can be created based on text files, point clouds, or objects in the drawing, such as points, contours, lines, etc. With the same command, we create the difference surface, except in this case, we select exactly two previously created TIN (Triangulated Irregular Network) surfaces. In BricsCAD Pro, the surface is created with the "TIN" command, and the difference surface is created with the "TinVolume" command. In the Geolisp system, the basic command for generating meshes is the "SIA" command.

Another method of obtaining the intersection line of surfaces in AutoCAD Civil is using the "MinimumDistBetweenSurfaces" command, which generates a closed 3D polyline without the need to create the difference surface.

Determining the boundary of the potentially flooded area in a given seam does not solve the problem of locating all the hazardous spots in the roadways. The workings could be located near the seam or the aquifer level but may not intersect it. Additional hazards may include boreholes passing through multiple aquifer layers. In the Geolisp program, the "ZnOs3d" command allows determining the workings located within a specified spatial distance from a selected object, such as a planned exploration borehole or aquifer level (Figure 1). This command requires the prior creation of a spatial model of the mining excavations, as presented in Chapter 2.

The position of the workings changes over time due to exploitation in adjacent seams. The distance between the workforce's location and the hazard may also change. The "TrP2P" command is used to display the predicted state of the workings on the map. The TrP2P program allows transforming the entire map (roadways, floor surface of the seam, etc.) based on calculations performed by the EDN-OPN program, which is discussed in Chapter 4.

7 CONCLUSIONS

Research is being conducted at the Department of Geoengineering and Raw Materials Extraction at the Silesian University of Technology in Gliwice aimed at developing technology for the automatic determination of the distance between mining excavations and hazards occurring in the rock mass. An essential element of this research is the development of algorithms that allow for the dynamic correction of databases, considering the impact of current and planned exploitation on the workings, rock mass, and surface area. The analysis of water-related hazards is an extremely important issue throughout the operational life of a mine. In Poland, it is regulated by mining law (Czaja *et al.* 2020, d'Obyrn *et al.* 2021). Fatal accidents related to water ingress into mining excavations occur sporadically.

In practical terms, determining the mining zones vulnerable to water infiltration involves the following key stages.

Based on a traditional map, it is very difficult to determine the distance to an existing water hazard in the rock mass. It is also not easy to automate the process of calculating the impact of rock mass deformation on the workings. Therefore, it is essential to create a spatial model of workings. The spatial visualization of the model allows for quick identification of incorrectly entered objects. This is a relatively simple process, described in Chapter 2.

The next stage involves creating the TIN (Triangulated Irregular Network) surface for the seam and aquifer surfaces (as discussed in Chapter 3).

Next, the impact of mining exploitation on the rock mass and surface must be determined. Chapter 4 presents the mathematical formulas and theoretical foundations for forecasting these deformations, which in Poland are based on the integral geometric theory of influences.

In the next step, a new TIN (Triangulated Irregular Network) surface, which is the difference between the seam surface and the aquifer surface, must be generated. Such an algorithm in pseudocode form is presented in Chapter 5.

The practical implementation of this issue in CAD systems is presented in Chapter 6. It discusses how to determine the boundary of the potentially flooded area in the rock mass. It also characterizes a program that allows the determination of the spatial distance between roadway and the hazard zone. The method for forecasting the state of hazards and workings for a selected period is also presented.

In summary, the algorithm proposed in this article allows for a quick and accurate determination of the boundary of the potentially flooded area in the rock mass. It also provides the ability to perform multi-variant analyses of the impact of a given phenomenon on the workings in the rock mass and conduct simulations of selected solutions.

REFERENCES

Białek, J. (2003). Algorytmy i programy komputerowe do prognozowania deformacji terenu górniczego. Gliwice: Wydawnictwo Politechniki Śląskiej.

Białek, J., Wesołowski, M., Mielimąka, R. and Sikora P. (2020). Deformations of mining terrain caused by the partial exploitation in the aspect of measurements and numerical modeling, *Sustainability*: vol. 12, no. 12, 5072, pp. 1–13. DOI:10.3390/su12125072.

Budryk, W. (1953). Wyznaczenie wielkości poziomych odkształceń terenu, *Archiwum Górnictwa i Hutnictwa* t. I, z. 1.

Chudek, M. (1981). *Mechanika górotworu*. Gliwice: Zeszyty Naukowe Politechniki Śląskiej.

Czaja, P., Kamiński, P., Dyczko, A. (2020). Polish experiences in handling water hazards during mine shaft sinking, *Mining techniques – past, present and future*. London, InTechOpen. DOI: 10.5772/intechopen.79241.

d'Obyrn, K., Kamiński, P., Motyka, J. (2021). Influence of Hydrogeological Investigation's Accuracy on Technology of Shaft Sinking and Design of Shaft Lining – Case Study from Southern Poland, *Energies*: Vol. 14(8), 2050. DOI:10.3390/en14082050.

Drzęźla, B. (1978). Rozwiązanie pewnego przestrzennego zadania liniowej teorii sprężystości w zastosowaniu do prognozowania deformacji górotworu pod wpływem eksploatacji górniczej wraz z oprogramowaniem, *Zeszyty Naukowe Politechniki Śląskiej*: s. Górnictwo, z. 91.

Janus, J. (2016). Zastosowanie skaningu laserowego do budowy modelu numerycznego wyrobiska górniczego, *Prace Instytutu Mechaniki Górotworu PAN*: t. 18, nr 3, pp. 27–34.

Jelonek, I. and Poniewiera, M. (2015). Qualitative Deposit Modelling: Example Kompania Węglowa S.A, *Schweizerbart Science Publishers, ICCP Program & Abstract Book*: t. 87, p. 91.

Jia, Q., Li, W. and Che, D. (2020). A Triangulated Irregular Network Constrained Ordinary Kriging Method for Three-Dimensional Modeling of Faulted Geological Surfaces. *IEEE Access*: 8, pp. 85179–85189. DOI: 10.1109/ACCESS.2020.2993050.

Knothe, S. (1984). *Prognozowanie wpływów eksploatacji górniczej*. Katowice: Wydawnictwo Śląsk.

Kowalczuk, J., Hadam, A. and Poniewiera, M. (2019). Tremors and rock bursts hazards characteristics using the digital model of the deposit in KWK ROW Site Marcel, *IOP Conf. Ser.: Earth Environ. Sci.*: 261 012020. DOI: 10.1088/1755-1315/261/1/012020.

Krawczyk, A. (2018). A concept for the modernization of underground mining master maps based on the enrichment of data definitions and spatial database technology, *E3S Web Conf.*: 26 (2018) 00010. DOI: 10.3390/ijgi12070278.

Krawczyk, A. (2019). Geomatics and mining geoinformation – their practical applications and development barriers, AGH University of Science and Technology Press, Kraków.
Krawczyk, A. (2023). Mining Geomatics, *ISPRS International Journal of Geo-Information*: 12, no.7: 278. DOI: 10.3390/ijgi12070278.
Litwiniszyn, J. (1953). Równanie różniczkowe przemieszczeń górotworu, *Archiwum Górnictwa i Hutnictwa* t. I.
Poniewiera, M. and Muras W. (2007). Budowa schematu przestrzennego wyrobisk górniczych, *Zeszyty Naukowe Politechniki Śląskiej*: nr 1752 Górnictwo; z. 278, pp. 313–321.
Poniewiera, M., (2010). Model numeryczny złoża węgla kamiennego i jego praktyczne zastosowania. *Wiadomości Górnicze*: R. 61 nr 7/8, pp. 458–465.
Poniewiera, M., Maciejowska-Figiel, E. and Wróbel M. (2015). Numeryczny model wyrobisk zabytkowej kopalni węgla kamiennego "Guido", *Czasopismo Inżynierii Lądowej, Środowiska i Architektury*: t. 32 z. 62 nr 3/II/2015, pp. 337–350.
Poniewiera, M. (2019). The impact of geometric data generalization on the accuracy of deformation predictions of the mining area - construction of a digital deposit model, *Systemy Wspomagania w Inżynierii Produkcji, Górnictwo - Perspektywy i zagrożenia*: vol. 8, Issue 1, pp. 136–147.
Sokoła-Szewioła, V. and Poniewiera, M. (2019). Application of a digital model of deposit in Polish hard coal mines on the example of Polish Mining Group Ltd. *In: Mining Goes Digital*. London: CRC Press/Balkema, pp. 344–354, DOI: 10.1201/9780429320774-40.
Sokoła-Szewioła, V. and Poniewiera, M. (2022a). Methods of introducing a discontinuity line into a numerical model of a hard coal deposit, *Scientific Journals of the Maritime University of Szczecin*: 71 (143).
Sokoła-Szewioła, V. and Poniewiera, M. (2022b). Verification of the deposit model and determining the course of the fault in the entire rock mass, *Scientific Journals of the Maritime University of Szczecin*: 72 (144).
Wu, Q., Xu, H. and Zou, X. (2005). An effective method for 3D geological modeling with multi-source data integration, *Computers & Geosciences*: 31, 1, pp. 35–43. DOI: 10.1016/j.cageo.2004.09.005.
Zhang, Y., Liu, Q., Wang, X. and Chen, J. (2020). 3D modeling of underground mine voids using integrated geospatial techniques, *Int J Min Sci Technol.*: 30(1):113–22. DOI: 10.1016/j.ijmst.2019.10.007.

Mine Surveying and Mining Area Protection – Sokoła-Szewioła and Jarczyk (Eds)
© 2025 The Author(s), ISBN: 978-1-041-10531-2

Correction of errors due to earth's curvature using selected photogrammetric software

M. Poniewiera
Silesian University of Technology, Gliwice, Poland

ABSTRACT: The main goal of this work is to prove that analytical aerotriangulation procedures should not apply corrections to photo coordinates due to the Earth's curvature. Instead, aerotriangulation should be adjusted in a three–dimensional geodetic topocentric coordinate system, disregarding the shape of the Earth. Tests conducted at the Silesian University of Technology compared the accuracy of obtained survey results (point clouds, control point coordinates, and orthophotomaps) produced by several popular photogrammetric programs, such as Agisoft Metashape Professional, Trimble Business Center, UAS-Master, Pix4Dmapper, Faro Scene, and Autodesk ReCap. The article presents an algorithm for the initial transformation of input data to a topocentric system, which should be performed before starting computations in the above–mentioned software. In the concluding section of the article, five arguments are presented that substantiate the advantages of performing aerotriangulation within the topocentric reference frame. The topocentric transformation approach is particularly relevant in the context of very large aerotriangulation blocks or project areas exhibiting substantial elevation differences.

1 INTRODUCTION

Photogrammetry is currently experiencing dynamic development. In geodesy, unmanned aerial vehicles are increasingly used; measurements cover ever larger areas, and the accuracy requirements for such studies are steadily increasing. The primary objective of this article is to present an algorithm that increases the accuracy of currently used aerotriangulation methods, as well as to assess the extent to which stereoscopic identification of photo points improves the precision of photogrammetric methods. This research expands upon technologies described in the doctoral thesis (Poniewiera 2001).

The conclusions from that dissertation have not yet been published. Existing numerical aerotriangulation methods (photogrammetric–geodetic network) are based on certain deformations of mathematical space. These deformations arise from applying corrections for Earth curvature to the image coordinates of each photograph. An alternative, recommended in this paper, is to adjust the aerotriangulation in a three–dimensional geodetic topocentric system. It should be emphasized that this approach does not require modifying the software being used; it only requires a correction to the input data.

At the Department of Geoengineering and Raw Material Exploitation of the Silesian University of Technology in Gliwice, research is being conducted to compare the results obtained (point clouds, control point coordinates, and orthophotomaps) in several popular photogrammetric software programs. As a side effect, brief user instructions for these programs were prepared. These will help in the development of an academic manual that will be published on the author's website. Although many comprehensive guides exist, there is a lack of short tutorials enabling quick generation of basic products (point clouds and orthophotomaps).

DOI: 10.1201/9781003655558-18

Fieldwork included numerous photo point measurements and drone images captured using DJI Mini 3 Pro, DJI Mavic 2 Enterprise, and DJI Phantom 4 drones. Results were processed using a stereoscopic photogrammetric station. The tested software includes Agisoft Metashape Professional, Trimble Business Center, UAS-Master, Pix4Dmapper, Faro Scene, and Autodesk ReCap

One part of the project was stereoscopic identification of control points. Faro Scene allows for importing a point cloud and performing measurements with virtual reality goggles. UAS-Master and Agisoft Metashape offer stereoscopic viewing with Nvidia 3D Vision glasses. This is especially helpful when identifying poorly visible elevated points, such as the top of a stake. In 2D view, it is easy to mistakenly mark the bottom of the stake instead of the top, potentially causing centimeter–scale errors. Several other programs can also be mentioned, such as ShapeCapture (Aguilar *et al.* 2005), Leica Photogrammetry Suite (Aguilar *et al.* 2013), ShapeNet (Niu *et al.* 2023).

Preliminary conclusions suggest that each of these programs can produce a correct orthophotomap (Figure 1). Accuracy can be improved by identifying photo points. The average coordinate deviation between points read from the orthophotomap and measured by total station was less than 2 cm (Figure 2a). Each program was also able to produce a correct point cloud. Coordinate deviation was below 2 cm as well (Figure 2b).

Figure 1. Fragment of the resulting orthophotomap.

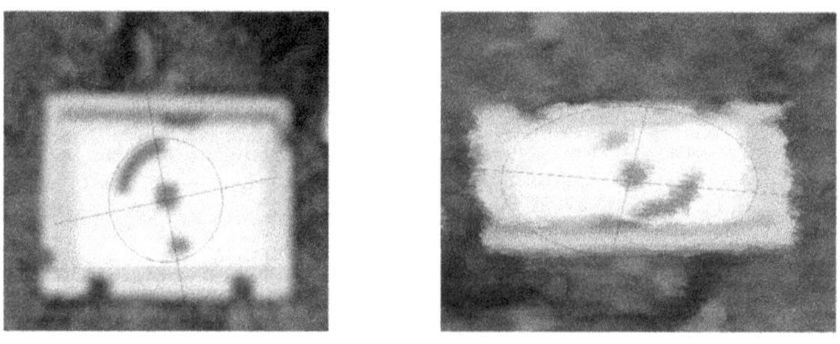

Figure 2. a. Enlargement of the orthophotomap. b. Enlargement of the point cloud.

Enlargement in the area of the control point. The difference between the raster readout (black dot) and the total station measurement (blue or red dot, respectively) is less than 2 cm.

2 ALGORITHMS ACCOUNTING FOR EARTH'S CURVATURE

The problem of Earth's curvature affecting aerotriangulation results is discussed in most photogrammetric textbooks (Schut 1978, Sitek 1991). Authors typically introduce curvature corrections in an approximate form (Equation 1). Kadaj (1990) proposed a more mathematically accurate correction (Equation 2).

$$\delta r = \frac{r^3 H}{2Rf^2} \tag{1}$$

$$\delta r = r \cdot \left[\frac{f}{r} \cdot \frac{R}{H} \cdot 2 \cdot \arctan\left(\frac{H}{R} \cdot \frac{1}{\frac{f}{r} + \sqrt{\frac{f^2}{r^2} - \frac{H}{R} \cdot \left(2 + \frac{H}{R}\right)}} \right) - 1 \right] \tag{2}$$

Where, in both equations, δr – correction for Earth's curvature; r – image distance measured from the principal point; H – flight height; f – camera focal length; R – Earth's radius.

These equations are derived from simplified geometric assumptions. They neglect the true shape of the Earth, cartographic projections, image rotation angles, unknown exact flight altitude, etc. Traditional methods can be improved iteratively (Poniewiera 2001), but they are still less accurate than the topocentric system method. Moreover, software source code modifications would be required, which is often not feasible.

Schut (1980) presented an algorithm that skips curvature corrections for image coordinates and instead introduces corrections to ground coordinates as shown in Equation 3.

$$\delta x = \frac{XZ}{R}; \delta y = \frac{YZ}{R}; \delta z = \frac{X^2 + Y^2}{2R} \tag{3}$$

Where δx, δy, δz – corrections to ground coordinates X, Y, Z; R – Earth's curvature.

This eliminates the influence of image tilt on aerotriangulation results, but these formulas still lack mathematical rigor (e.g., they omit cartographic projections). Hence, there is a need for an algorithm using an orthonormal reference frame to solve aerotriangulation, which is presented in the next section.

3 TRANSFORMATION TO TOPOCENTRIC COORDINATE SYSTEM

The geodetic topocentric coordinate system used here is an orthonormal system with its origin in the center of the study area, and the OX, OY axes lying in the plane tangent to the ellipsoid (Figure 3).

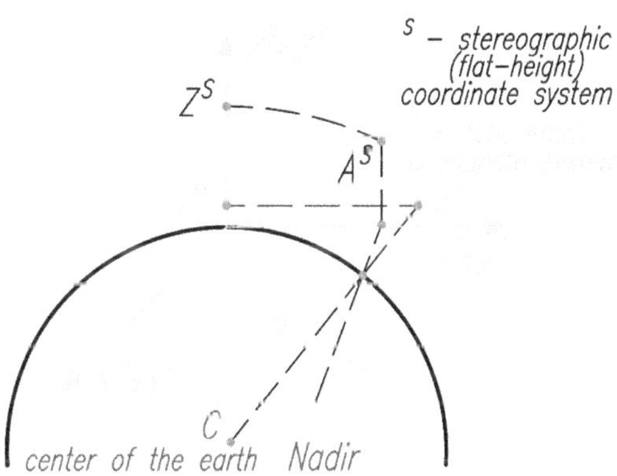

Figure 3. Fragment of the resulting orthophotomap.

Depending on available data, transformation problems can be grouped accordingly. Most often, we have reference point coordinates in the geocentric system and can transform them to the topocentric system using a 3D isometric transformation (Kadaj 2001, 2002; Sokoła-Szewioła *et al.* 2021).

A simplified algorithm was also developed (Poniewiera 2001), which assumes a spherical Earth and uses stereographic projection. It serves educational purposes and helps explain the essence of the topocentric method. This algorithm is insensitive to photo tilt angles or terrain elevation variations. It significantly reduces computation time while maintaining practical aerotriangulation accuracy. Code listing written in LISP transforming coordinates from the mapping to the topocentric system is shown in Figure 4.

```
(defun krzyw65 (lista / def1 R L)
 ; Transforms the XY coordinates of the plane to a tangent sphere
 ; Z coordinate calculated from the transformation of the plane into a sphere
 ; Pseudostereographic transformation
  (defun  def1 (pkt R dz / Xa Ya Zo Xo t Xk Zk Xt Zt)
    ; pkt - a point on the mapping plane
    ; R - radius of the Earth
    ; dz - distance of the mapping plane from the Earth
        (setq Xa (- (car pkt) (car globalny_srodek))
              Ya (- (cadr pkt) (cadr globalny_srodek)) Zo (caddr pkt))
        (setq Xo (distance '(0 0) (list Xa Ya))) ;guided ray
        (if (= 0 Xo)
          (setq Zt (- Zo dz))
          (setq t (/ (- (* 2 R) dz) Xo)
                Xk (/ (* 2 R t) (1+ (sqr t))) ; projection on the Earth
                Zk (- (/ (* Xk (- (* 2 R) dz)) Xo) (* 2 R))
                Xt (/ (* Xk (+ Zo R)) R) ; projection to the topocentric system
                Zt (- (/ (* Xt (+ Zk R)) Xk) R)
                Xa (/ (* Xa Xt) Xo)
                Ya (/ (* Ya Xt) Xo)
          )
        );i
        (setq Xa (+ Xa (car globalny_srodek)) Ya (+ Ya (cadr globalny_srodek)))
        (list Xa Ya Zt)
  );d
  (setq R       promien_Ziemi ; radius of the Earth
        dz      0.0 ;dz - distance of the mapping plane from the Earth
  )
  (mapcar '(lambda (l)
     (list
       (car l)  ; point number
       (cadr l) ; a point on the mapping plane
       (def1 (cadr l) R dz) ;a point in the topocentric system
     )
  ) lista)
);d
```

Figure 4. A program listing containing a procedure written in LISP that transforms coordinates from the mapping system to the topocentric system.

4 PROGRAM UAS-MASTER

Below is an excerpt from a tutorial demonstrating how to obtain an orthophotomap and point cloud using UAS-Master. The guide simplifies the steps to be followed. Similar instructions have been prepared for: Agisoft Metashape Professional, Trimble Business Center, UAS-Master, Pix4Dmapper, Faro Scene, and Autodesk ReCap. These are available at the author's website: geolisp.pl.

UAS-Master is an older program, and its interface somewhat resembles Windows 95. Nevertheless, it is regularly updated and remains competitive in terms of capabilities.

(1) Click the "New Project" button. A dialog window with options will appear.
(2) First, add images. Click the "Frame type" button and select "Import" (Figure 5).

Figure 5. Icon to add images to the project.

(3) Select a camera. Typically, the camera is embedded in the image metadata and does not need to be defined here. If camera internal orientation parameters need to be edited, click the "Cameras" button (Figure 6).

Figure 6. Icon to edit camera parameters.

(4) Add measurement points. Click the "Points" button and then choose "Import" (Figure 7).

Figure 7. Icon to add ground control points (GCP) to the project.

(5) Enter units and coordinate system (Figure 8).

Figure 8. Icon to select coordinate system.

(6) Group the images into strips. The software can automatically group images based on coordinates. Click the "Strips" button and select "Generate" (Figure 9).

Figure 9. Icon to group images into strips.

(7) Save the project. A new window opens. On the right, under the "UAS" tab, a checklist of tasks appears (Figure 10). Each task opens a dialog with additional options. Key questions include desired accuracy and point cloud resolution. Higher accuracy increases processing time.

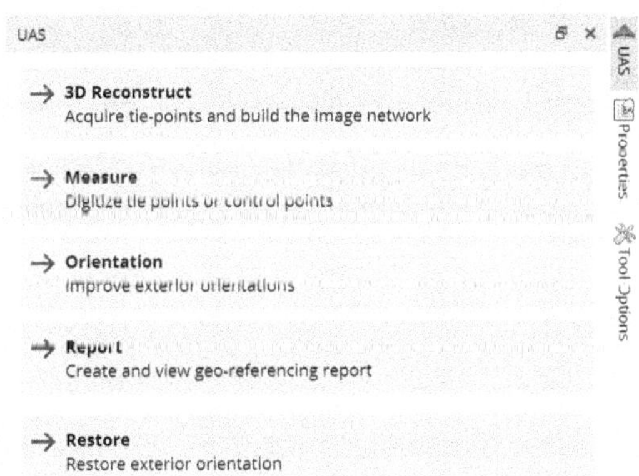

Figure 10. Task steps in UAS-Master.

(8) Aeroblock alignment control. Each processing stage generates a report that must be reviewed. Key checks include residuals on ground control points and the number of aligned images. The "Ground control point residuals" report shows these metrics (Figure 11). Large discrepancies suggest errors–most often incorrect photo point marking.

Ground control point residuals (given - adjusted)

ID	#	X [m]	Y [m]	Z [m]	Total [m]
1	7	0.0018	-0.0023	0.0010	0.0031
10	12	0.0026	0.0102	-0.0034	0.0110
2	10	-0.0007	-0.0044	0.0040	0.0060
3	12	0.0012	-0.0058	-0.0033	0.0068
4	10	0.0014	0.0030	0.0011	0.0035
5	3	-0.0040	-0.0046	-0.0012	0.0062
6	7	0.0022	0.0048	-0.0026	0.0059

Figure 11. Aeroblock alignment control (ground control point residuals).

5 CONCLUSIONS

This paper presents the main assumptions of research evaluating the influence of Earth's curvature on aerotriangulation results. Aerotriangulation is a method for densifying control networks using aerial images (Schut 1978, Zhang *et al.* 2025). Photogrammetry is used across many engineering fields, including mining (Krawczyk 2019). Studies on the impact of cartographic projections on aerotriangulation began in a doctoral dissertation (Poniewiera 2001) and are now being continued (Poniewiera 2022; Żogała *et al.* 2025). Planned research includes recalculation and comparison of real and simulated aerial blocks in different coordinate systems. Preliminary results confirm that such influence exists. For example, a block of 54 images at a 1:26,000 scale in the 1992 coordinate system showed up to 0.28 m deviation in aerotriangulation solutions when compared between topocentric and 1992 systems, depending on the block's position relative to the zone center.

Though this research is ongoing, early results suggest that calculations performed in an orthonormal system like the topocentric coordinate system can increase coordinate accuracy. The benefits of solving aerotriangulation in the topocentric reference frame include:

(1) Using raw, unprocessed image coordinates avoids hard–to–quantify errors from corrections for Earth's curvature, terrain height above a reference surface, and cartographic projection.
(2) It allows the use of unprocessed input data from field measurements, such as GPS–derived coordinate increments and distance measurements, which strengthens the network and improves result accuracy.
(3) The actual terrain surface is reconstructed without distortions introduced by projection corrections, important for differential measurements where the relative position of objects matters.
(4) Using the topocentric method significantly reduces the risk of network defects. Traditional methods may cause serious network disturbances from minor input errors under certain conditions.
(5) The topocentric method requires no modification to the photogrammetric software–only input data are adjusted.

Modern algorithms are highly flexible, and the impact of such deformations on calculation results may be small. However, with the increasing use of high–precision cameras and the growing expectation for millimeter–level accuracy, the precision of our algorithms becomes increasingly critical.

An additional goal of this study was to compare results (point clouds, control point coordinates, orthophotomaps) from several popular photogrammetric programs. Each program offers numerous options that cannot be fully covered in this short paper. What each user finds most convenient or important will vary. Therefore, it is premature to definitively judge which program delivers the best results.

Figure 12. Fragment of the resulting point cloud.

Nevertheless, all tested programs produced accurate orthophotomaps (Figure 1), allowed for increased accuracy by identifying photo points (Figure 2), and generated valid point clouds (Figure 12).

A particularly interesting experiment was computing aerotriangulation using images from the DJI Mini 3 Pro drone – a lightweight (249 g), low–cost, user–friendly UAV that requires no formal certification. Tests show that XYZ accuracy of the point cloud on marked points was within 2 cm, indicating that one of the major drawbacks of photogrammetry – high entry cost – has been effectively eliminated.

In summary, the implementation of the proposed topocentric transformation approach in photogrammetric data processing is especially advantageous for extensive aerotriangulation blocks or for project areas with pronounced elevation variation.

REFERENCES

Aguilar, M., Aguilar, F. and Agüera, F. (2005). Mapping small areas using a low-cost close range photogrammetric software package with aerial photography. *The Photogrammetric Record*. 20(112): 335–350. DOI: 10.1111/j.1477-9730.2005.00331.x.

Aguilar, M., Aguilar, F. and Fernandez, I. (2013). Accuracy assessment of commercial selfcalibrating bundle adjustment routines applied to archival aerial photography. *The Photogrammetric Record.* 28(141): 96–114. DOI: 10.1016/j.isprsjprs.2023.09.004.

Kadaj, R. (1990). Wyrównanie wielkiej sieci fotogrametryczno – geodezyjnej system PHOTONET C, Rzeszów: unpublished.

Kadaj R. (2001). Wytyczne techniczne G-1.10. *Poprawki odwzorowawcze państwowego układu współrzędnych.* Warszawa: COGiK.

Kadaj, R. (2002). Polskie układy współrzędnych. Formuły transformacyjne, algorytmy i programy. Rzeszów: Geoinwers.

Krawczyk, A. (2019). *Geomatics and mining geoinformation - their practical applications and development barriers.* Kraków: AGH University of Science and Technology Press.

Niu, W., Huang, X., Xiang, H., Wang, X., Ji, S. and Zhang, Fan. (2023). A symmetry-aware alignment method for photogrammetric 3D models. *ISPRS Journal of Photogrammetry and Remote Sensing.* 204: 184–208. DOI: 10.3390/app15115812.

Poniewiera, M. (2001). *Methodology of numerical elaboration of aerotriangulation in geodetical topocentric coordinate system.* AGH Kraków: unpublished doctoral dissertation.

Poniewiera, M. (2022). Developing a solution algorithm of aerotriangulation to minimize the influence of used map projection - project assumptions, *Energia, środowisko, eksploatacja kopalin, zarządzanie i zrównoważony rozwój*:113–121. Gliwice: STE Group.

Schut, G.H. (1978). *Adjustment of Bundles - Formulation and experiments*, Ottawa: National Research Council.

Schut, G.H. (1980). *Adjustment of models: formulation and program.* Ottawa: National Research Council.

Sitek, Z. and Mierzwa, W. (1972). Analityczna aerotriangulacja przestrzenna z niezależnych modeli. *Geodezja* 13.

Sitek, Z. (1991). *Fotogrametria ogólna i inżynieryjna*, Warszawa: PPWK.

Sokoła-Szewioła, V., Poniewiera, M. and Mierzejowska, A. (2021). Coordinate transformation using the author's software in GIS class system – case study. *Geoinformatica Polonica* 20.

Zhang, D., Chen, Y., Le, Y., Dong, Y., Dai, G., and Wang, L. (2022). Refraction and coordinate correction with the JONSWAP model for ICESat-2 bathymetry. *ISPRS Journal of Photogrammetry and Remote Sensing.* 186: 285–300. DOI: 10.1016/j.isprsjprs.2022.02.020.

Żogała, M., Sokoła-Szewioła V. and Poniewiera, M. (2025). Transformation of Coordinates of Boundary Points of Neighboring Mining Areas Using an Authorial Procedure Based on the Method of Independent Models - A Case Study. *Appl. Sci.* 15(11): 5812. DOI: 10.3390/app15115812.

Mine Surveying and Mining Area Protection – Sokoła-Szewioła and Jarczyk (Eds)
© 2025 The Author(s), ISBN: 978-1-041-10531-2

Sustainable post-mining transition in Europe: Integrating environmental engineering and urban regeneration

R. Rodríguez-Galindo Díez & M. Menéndez Fernández
Universidad de León, León, Spain

L.A. Menéndez-García
Universidad de Oviedo, Oviedo, Spain

V. Sokoła-Szewioła
Silesian University of Technology, Gliwice, Poland

ABSTRACT: The decommissioning of coal mines across Europe poses a dual challenge: mitigating residual pollution and regenerating post-industrial landscapes. We present a conceptual methodological framework that integrates empirical data preprocessing, dynamic emission simulations, nature-based solutions (NbS), quantified health impact assessment, and multi-criteria optimization, all aligned with the EU's just transition framework. Building on Menéndez-García (2021) and validated with Menéndez-García *et al.* (2022), the approach offers a replicable roadDogan map to design, evaluate, and adapt remedial measures in post-mining territori

1 INTRODUCTION

Coal—once an economic and social engine for many European regions—saw its share fall from 0.071% of GDP (Gross Domestic Product) in 1985 to 0.016% in 2021, while direct employment declined from 450,000 to 230,000 workers (World Bank 2025; Eurostat 2025) – Figure 1. These trends reflect EU-level policies: ECSC Decision 3501/85 (Commission 1985) constrained state aid to the sector; Directive 2001/80/EC set SO_2 and NO_x limits that raised the cost of large combustion plants; and Regulation (EU) 2018/1999 requires closure schedules and reconversion plans under the Just Transition Mechanism (JTM, €17.5 bn). Consistent with Regulation (EU) 2021/1056 establishing the Just Transition Fund (JTF), our study adopts a policy-driven design: eligibility via Territorial Just Transition Plans, exclusions of fossil-fuel investments, and priorities for economic diversification, reskilling, and site regeneration are encoded as constraints and objectives within the NSGA-II optimization. This alignment enhances regulatory traceability and practical deployability in post-mining regions such as Upper Silesia (European Union 2021).

Upper Silesia is among Central Europe's most industrialized and polluted regions, where 4.5 million inhabitants live alongside coal mines, steelworks, thermal power plants, and heavy urban traffic. The complex intermixing of residential, commercial, and industrial areas—without clear boundaries—makes strict application of Directive 2008/50/EC (limit values for $PM_{2.5}$, NO_2, and VOCs) (European Union 2008) particularly challenging. In this context, conventional dispersion modeling demands highly detailed emission inventories and is not well-suited to rapidly evaluating mitigation scenarios.

At the same time, the situation represents an opportunity to improve air quality and well-being through NbS and advanced engineering tools. We propose a five-block conceptual framework—preprocessing, dynamic simulation, NbS, health assessment, and optimization—to guide the transition of Europe's post-mining territories.

DOI: 10.1201/9781003655558-19

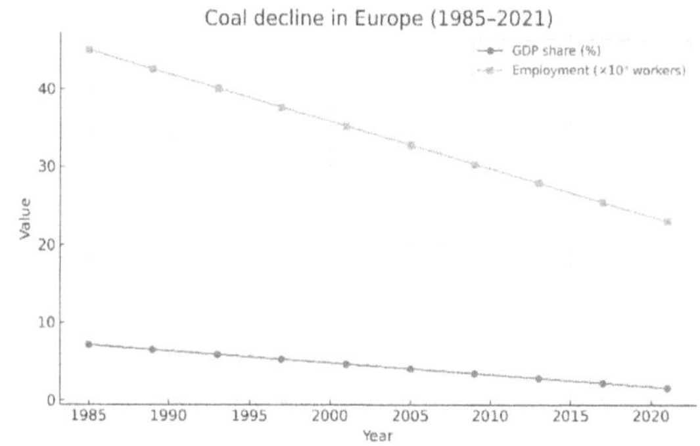

Figure 1. Coal decline in Europe (1985-2021).

2 METHODOLOGY

The chosen methodology integrates data, modeling, and decision tools to address heterogeneous sources, spatiotemporal dynamics, and the need for multi-criteria, adaptive policies (Figure 2).

2.1 *Empirical data preprocessing*

Following Menéndez-García *et al.* (2022), we applied pruning + random replacing (20% artificial gaps, 5 rounds) to $PM_{2.5}$, NO_2, SO_2, and benzene time series in Silesia. Six imputation methods were compared; Kalman smoothing yielded the lowest RMSE ($\approx 2.1\ \mu g/m^3$) and the greatest stability, ensuring clean series for modeling.

2.2 *Dynamic emissions and mining phase-out*

We conceptualize an hourly inventory combining:

(1) Urban traffic simulated with MATSim (Horni *et al.* 2016), modulating flows according to the mine closure trajectory in the NECP (European Union. 2018. Regulation (EU) 2018/1999).
(2) A hybrid CTM+ML model corrected with meteorological and urban variables to identify emission hotspots at each phase.
(3) Temporary interventions (mobile filtration units, road dust cleaning) synchronized with the closure schedule.

Integration with EMEP/EEA guidance: Emission categories, default emission factors, and QA/QC procedures follow the EMEP/EEA Air Pollutant Emission Inventory Guidebook 2023 (European Environment Agency 2023). Tiered methodologies (Tier 1–3) are adopted by sector according to data availability, with temporal consistency checks and recommended spatial downscaling proxies. This alignment ensures CLRTAP/NEC comparability and robust time-series coherence across the phase-out trajectory.

2.3 *Nature-based solutions: Green corridors*

Grounded in the post-industrial NbS literature (Krzysztofik *et al.* 2022; Dogan *et al.* 2023), we define:

- Mapping of spoil/slag heaps and industrial platforms.
- Corridor typologies (width, native species, connectivity).
- Integration into Urban Development Plans as dispersion barriers and recreational spaces.

2.4 *Health impact assessment*

We use WHO dose–response functions to estimate DALYs averted from pollutant reductions and produce vulnerability maps that overlay population density with proximity to mines (World Health 2021). This health criterion prioritizes interventions where benefits are maximized.

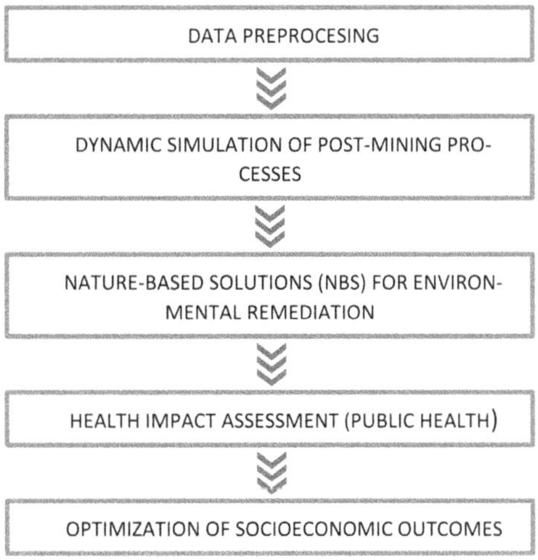

Figure 2. Methodology. Sequential flow diagram.

2.5 *Multi-objective optimization & continuous monitoring*

We propose a conceptual NSGA-II (Deb *et al.* 2002) approach to balance emission reduction, costs, job creation, and urban accessibility, yielding a Pareto front of staged policy packages. In parallel, an IoT sensor + GPS feedback loop supports real-time strategy adjustment (Figure 2).

3 EXPECTED OUTCOMES & CONCEPTUAL VALIDATION

This section presents a conceptual heat map of Upper Silesia (southern Poland) highlighting emission hotspots (Figure 3). The region is used as an emblematic case study as one of Europe's main historical coal-mining hubs: the industry accounted for around 6.9% of regional GVA (2017) and still employed ∼73,000 people (references: official EU energy sources). The infographic uses color gradients (dark blue → yellow) to show areas of higher residual methane and CO_2 emissions associated with former mining operations, with elevated concentrations in historical mining districts. Key cities (e.g., Katowice) are labeled, and three main hotspots are marked. This visualization quickly identifies priority areas for remediation and monitoring, aligning with real-world evidence reported for the Silesian basin.

Conceptual validation and expected metrics:

- High-quality time series. Replication of the imputation protocol confirms the applicability of Kalman smoothing in Silesia.
- Dynamic emission inventories. Evolving hotspot maps to guide temporary interventions according to phase-out stages and operating conditions.

- Effectiveness of green corridors. Local $PM_{2.5}$ reductions up to 12%, consistent with the literature on NbS in post-industrial settings.
- Quantified health benefits. ≈0.5 DALYs averted per 10,000 inhabitants with combined intervention packages.
- Pareto frontier of solutions. Clear trade-offs among impact, cost, and employment, useful for multi-criteria decision-making.
- Regulatory alignment. Consistency with ECSC 3501/85, Directive 2001/80/EC, Regulation 2018/1999, and the EU's Just Transition Strategy.

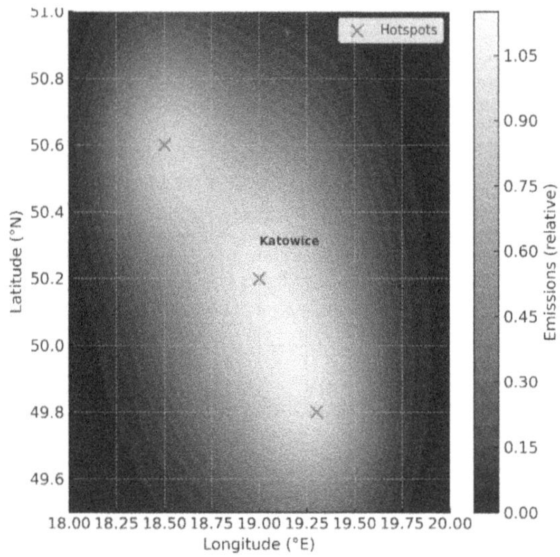

Figure 3. Upper Silesia- Emission Hotpots (conceptual).

Relative intensity of residual emissions (methane/CO_2) in historical mining districts. Three main hotspots (cyan X) and the location of Katowice are indicated. Source: authors' elaboration using representative synthetic data and official EU institutional references.

4 DISCUSSION

The Conclusions image in Figure 4, synthesizes the conceptual framework and its alignment with European Union regulation in an infographic. At the center, the node "Post-Mining Transition – Conceptual Framework" connects to four strategic domains: (1) Environmental Engineering (soil and water remediation, and nature-based solutions); (2) Urban Regeneration (rehabilitation of cities and former mining land for new uses); (3) Public Health & Community (assessment of health and social impacts); and (4) Socioeconomic Optimization (balancing costs, financing, and re-employment). Each component is represented by soft-colored ovals (green, blue, coral, and yellow) linked to the central node to convey their interdependence. A fifth node, in white with a black border, captures "Alignment with EU Regulation"—including the European Green Deal and the Just Transition Mechanism, which mobilizes around €55 billion (2021–2027) to support regions most affected by coal phase-out—and a dashed arrow underlines the plan's coherence with EU objectives (climate neutrality by 2050 and the mitigation of negative socioeconomic impacts).

This framework integrates regulatory requirements, engineering rigor, and urban-planning criteria, and its validity is supported both by alignment with EU policies and by empirical validation of its components. It nevertheless acknowledges residual uncertainty in data imputation and modeling, which motivates the incorporation of Bayesian methods to better capture variability and strengthen inference. A pilot implementation in Silesia would enable an operational assessment of practical effectiveness—from intervention prioritization and health-impact assessment to cost–impact–jobs trade-offs—and would provide solid evidence for scaling. Ultimately, scalability to other European contexts will depend on the quality of local data and close inter-institutional collaboration (authorities, universities, operators, and communities), a necessary condition to sustain a just, effective, and measurable transition.

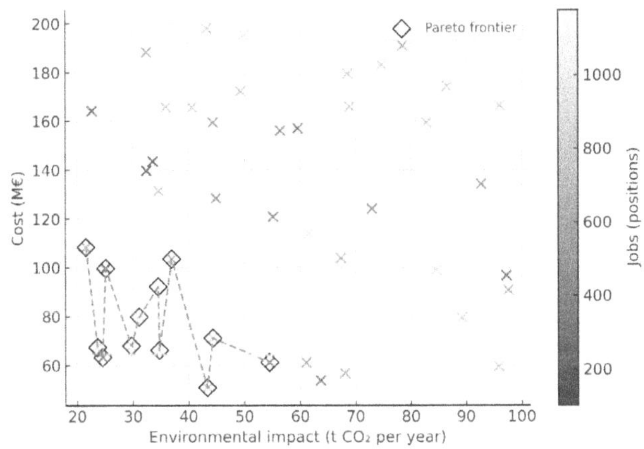

Figure 4. Trade-offs and Pareto Frontier (conceptual).

5 CONCLUSIONS

- Integrated, empirically grounded framework. A modular pipeline—empirical preprocessing (Kalman), hourly emissions inventory (MATsim + CTM–ML), NbS corridors, who dose-response HIA, and NSGA-II—delivers end-to-end decision support, from data curation to policy optimization.
- Policy-driven and audit-ready. Design choices are aligned with EU regulation and the just-transition mandate, with transparent assumptions, reproducible workflows, and traceable inputs/outputs to facilitate regulatory scrutiny and uptake.
- Quantified benefits and explicit trade-offs. Conceptual validation for upper Silesia indicates local $PM_{2.5}$ reductions up to ~12% and ≈0.5 DALYS averted per 10,000 inhabitants under combined measures, while pareto fronts map impact–cost–jobs–accessibility compromises for evidence-based prioritization.
- Replicable roadmap and scalability conditions. The approach specifies data requirements, templates, and an IoT+GPS feedback loop for adaptive control, transferability hinges on local data quality, institutional capacity, and governance arrangements
- Limitations and next steps. Residual uncertainty (imputation, model structure) motivates Bayesian updating, sensitivity/ablation analyses, and a staged pilot in Silesia—including quasi-experimental evaluation and equity metrics—to strengthen external validity and operational effectiveness.

6 SUMMARY

This work presents a modular framework, aligned with public policy and regulation, for a sustainable post-mining transition in Europe. Empirical preprocessing (Kalman) yields clean multi-pollutant time series; an hourly inventory couples MATSim traffic with CTM–ML to localize evolving hotspots; NbS corridors act as dispersion barriers with co-benefits; WHO dose–response functions convert concentration changes into DALYs; and NSGA-II reveals policy-relevant Pareto sets. A lightweight IoT+GPS loop enables adaptive management. Conceptually validated in Upper Silesia, the approach anticipates local $PM_{2.5}$ reductions up to ~12% and ≈0.5 DALYs averted per 10,000 inhabitants under combined measures. Transferability depends on data quality and inter-institutional coordination; remaining uncertainty (imputation, model structure) motivates Bayesian updating and a pilot in Silesia.

REFERENCES

Commission of the European Communities. (1985). Commission Decision No 3501/85/ECSC of 11 December 1985 amending Decision No 528/76/ECSC regarding the Community system of measures taken by the Member States to assist the coal-mining industry. *Official Journal of the European Communities.* https://eur-lex.europa.eu/eli/dec/1985/3501/oj/eng

Deb, K., Pratap, A., Agarwal, S., and Meyarivan, T. (2002). A fast and elitist multiobjective genetic algorithm: NSGA-II. *IEEE Transactions on Evolutionary Computation*, 6(2), 182–197. https://doi.org/10.1109/4235.996017

Dogan, E., Cuomo, F., and Battisti, L. (2023). Reviving Urban Greening in Post-Industrial Landscapes: The Case of Turin. *Sustainability*, *15*(17), 12760. https://doi.org/10.3390/su151712760

European Environment Agency. (2023). EMEP/EEA air pollutant emission inventory guidebook 2023. https://www.eea.europa.eu/en/analysis/publications/emep-eea-guidebook-2023

European Union. (2008). Directive 2008/50/EC of the European Parliament and of the Council of 21 May 2008 on ambient air quality and cleaner air for Europe. Official Journal of the European Union. https://eur-lex.europa.eu/eli/dir/2008/50/oj/eng

European Union. (2018). Regulation (EU) 2018/1999 on the Governance of the Energy Union and Climate Action. Official Journal of the European Union. https://eur-lex.europa.eu/legal-content/EN/ALL/?uri=celex%3A32018R1999

European Union. (2021). Regulation (EU) 2021/1056 establishing the Just Transition Fund. Official Journal of the European Union. https://eur-lex.europa.eu/legal-content/EN/TXT/HTML/?uri=CELEX%3A32021R1056

Eurostat. (2025). Employment by sex, age and detailed economic activity (from 2008 onwards, NACE Rev. 2) [LFSQ_EGAN22D] – B05 Mining of coal and lignite (EU-27). https://ec.europa.eu/eurostat/databrowser/view/LFSQ_EGAN22D__custom_1040052/default/table?lang=en

Horni, A., Nagel, K., and Axhausen, K.W. (Eds.). (2016). The multi-agent transport simulation MATSim. *Ubiquity Press.* https://doi.org/10.5334/baw

Krzysztofik, R., Rahmonov, O., Kantor-Pietraga, I., and Dragan, W. (2022). The Perception of Urban Forests in Post-Mining Areas: A Case Study of Sosnowiec-Poland. *International Journal of Environmental Research and Public Health*, *19*(7), 3852. https://doi.org/10.3390/ijerph19073852

Menéndez García, L.A. (2021). Predicción de valores del contaminante atmosférico benceno en Madrid con metodologías de Machine Learning, y análisis mediante georreferenciación de datos [Thesis doctoral, Universidad de León]. BULERIA. https://doi.org/10.18002/10612/13353

Menéndez García, L.A., Menéndez Fernández, M., Sokoła-Szewioła, V., Álvarez de Prado, L., Ortiz Marqués, A., Fernández López, D., and Bernardo Sánchez, A. (2022). A method of pruning and random replacing of known values for comparing missing data imputation models for incomplete air quality time series. *Applied Sciences*, 12(13), 6465. https://doi.org/10.3390/app12136465

World Bank. (2025). Coal rents (% of GDP) [NY.GDP.COAL.RT.ZS] – European Union. https://data.worldbank.org/indicator/NY.GDP.COAL.RT.ZS?locations=EU

World Health Organization. (2021). WHO global air quality guidelines: Particulate matter ($PM_{2.5}$ and PM_{10}), ozone, nitrogen dioxide, sulfur dioxide and carbon monoxide. https://apps.who.int/iris/bitstream/handle/10665/345329/9789240034228-eng.pdf

Mine Surveying and Mining Area Protection – Sokoła-Szewioła and Jarczyk (Eds)

Remote sensing and low-altitude photogrammetry in monitoring the condition of mining waste disposal sites

Z. Różański, A.P. Niewiadomski, G. Pach & P. Wrona
Silesian University of Technology, Gliwice, Poland

L. Balcarczyk, J. Pawliński & A. Wala
3D FORMAT Sp. z o.o., Gliwice, Poland

ABSTRACT: The ability to observe and record imagery from a low height and across multiple electromagnetic spectrum bands enables both in-field inspection and detailed off-site analysis of terrain conditions in various aspects. This paper provides a brief overview of remote sensing methods used for the identification and assessment of hazards on mining waste disposal sites. Based on the authors' own experience, the study presents data acquisition techniques employing remote sensing (including thermal imaging and LiDAR) and low-altitude photogrammetry, used either as standalone tools or as complementary approaches for early threat detection and technical condition assessment of post-mining sites. Selected case studies from the Upper Silesian Coal Basin are presented, including results of thermal condition assessments, terrain morphology measurements, slope stability analyses, and assessment of slope degradation due to water erosion.

1 INTRODUCTION

Numerous spoil heaps and mining waste management facilities located in coal basins continue to pose significant challenges for both environmental protection and public safety. Their presence is associated with a range of adverse impacts on the surrounding environment, including degradation of air, soil, and water quality, disruption of ecosystems, alteration of the landscape, and potential threats to human health and safety. These issues result from emissions of gases and dust into the atmosphere, leaching of pollutants into soils and water bodies, as well as the occurrence of subsidence and landslides (Fabiańska *et al.* 2019; Nguyen 2025; Różański 2018; Smoliński *et al.* 2018; Suponik *et al.* 2021).

These phenomena are further intensified by the risk of fire, which is common in waste sites containing sufficient amounts of combustible materials with a high tendency toward self-ignition.

The likelihood of such adverse occurrences is significantly influenced by factors such as the size, condition, and surface morphology of the waste disposal and management sites, the effectiveness of drainage systems, and the condition of the biological cover (Nguyen 2022; Querol *et al.* 2011; Różański *et al.* 2022;). These characteristics determine the manner and intensity of the impact of external factors primarily meteorological conditions such as wind, precipitation, atmospheric pressure, and solar radiation—on the given site.

Therefore, periodic surface condition monitoring is a crucial preventive measure, enabling early detection of irregularities and anomalies. This, in turn, allows for timely implementation of corrective and preventive actions to restore safe and stable conditions. A comprehensive understanding of potential hazards is essential for selecting appropriate preventive strategies, as well as for designing technologies and safe methods for conducting technical and biological land reclamation works. Modern geodetic methods based on remote sensing and photogrammetry are rapidly advancing, enabling efficient observation and recording of

DOI: 10.1201/9781003655558-20

various terrain features from elevated platforms. These technologies significantly reduce the time required for inspections and offer extensive possibilities for analysing surface conditions (Colomina and Molina 2014; Thiruchittampalam *et al.* 2023). Remote sensing systems used for imaging land areas—including post-mining sites—are deployed on satellites (Nádudvari *et al.* 2021; Sobrino *et al.* 2004, 2021; Xiao and Tian 2024), manned aircraft (Allison *et al.* 2016; Chung *et al.* 2014), unmanned aerial vehicles (UAVs) (Kovanič *et al.* 2023; Liu *et al.* 2025; Khumalo *et al.* 2025) or terrestrial stations (Lewińska and Dyczko 2016).

The observation altitude influences both the size of the area being surveyed and the achievable image resolution. While satellite imagery is advantageous for covering large areas and conducting long-term change detection, UAV-based photogrammetry and remote sensing are generally more suitable for detailed measurements and analyses over smaller areas (ranging from a few to several hundred hectares), due to their significantly higher spatial resolution.

The aim of this article is to present remote sensing and photogrammetric methods utilizing unmanned aerial vehicles (UAVs), demonstrated through specific case studies related to the investigation of post-mining areas, primarily those associated with the management of mining waste.

2 REVIEW OF PHOTOGRAMMETRIC AND REMOTE SENSING METHODS USING UAV GETTING STARTED

2.1 *Photogrammetry in the visible light spectrum*

UAV-based photogrammetry in the visible light range involves capturing a series of overlapping RGB images using a camera mounted on a drone, followed by automatic reconstruction of scene geometry (3D point positions) from these images using techniques such as Structure-from-Motion (SfM) and Multi-View Stereo (MVS) (Kovanič *et al.* 2023). The outputs include point clouds, digital terrain and surface models (DTM/DSM), orthophotos, and 3D object models.

To obtain these products, a typical photogrammetric workflow must be followed, consisting of the following steps:

(1) Flight Planning – Selection of flight altitude, speed, and flight path layout to ensure the required ground sampling distance (GSD) and images overlap are achieved.
(2) High-Resolution Image Acquisition – Collection of RGB (visible spectrum) images with stabilized cameras and controlled exposure settings; optionally including oblique imagery.
(3) Camera Calibration and Image Orientation – Detection of tie points and bundle adjustment to estimate internal and external camera orientation parameters.
(4) Dense Point Cloud Reconstruction and Model Generation – Creation of dense point clouds, mesh surfaces, DSM, and DTM.
(5) Georeferencing – Application of ground control points (GCPs) or integration of Real-Time Kinematic (RTK) and Post-Processed Kinematic (PPK) GNSS data to improve absolute spatial accuracy.
(6) Orthophoto and Thematic Products Generation – Production of orthomosaics and derivative products such as maps, volume calculations, and change detection analyses.

2.2 *Thermal infrared photogrammetry*

The photogrammetric process in the infrared spectrum is essentially similar to the one described above. However, image acquisition is performed using a thermal infrared camera. The images are recorded in the thermal infrared range (8–14 μm). These cameras typically have significantly lower spatial resolution (i.e. 640×512 pixels up to a maximum of 1280×1024 pixels). As a result, the ground sampling distance (GSD) is approximately 8–10 cm/pixel, leading to considerably lower detail in thermal maps. This limitation can be partially compensated by lowering the flight altitude and significantly increasing the number of images acquired. Additionally, the field georeferencing approach differs from that used in visible-spectrum photogrammetry. Ground control points (GCPs) are usually implemented using high-emissivity targets that stand out clearly in thermal imagery—such as aluminium or mirrored plates—which allow for the accurate identification of GCPs in thermograms (Kelly *et al.* 2019).

2.3 *Multispectral photogrammetry*

Multispectral photogrammetry involves capturing images across multiple narrow spectral bands, including not only visible light (RGB) but also near-infrared (NIR), red-edge, and other selected spectral bands (Adão *et al.* 2017; Wang 2021). Multispectral cameras mounted on drones enable the simultaneous acquisition of several synchronized spectral channels. Similar to RGB photogrammetry, Structure-from-Motion (SfM) and Multi-View Stereo (MVS) techniques are used to generate orthophotos and 3D models; however, these products contain extended spectral information for each pixel. The resulting multispectral orthomosaics can be analyzed to calculate vegetation indices such as NDVI, NDRE, SAVI, among others, as well as other biophysical parameters (Adão *et al.* 2017; Bendig *et al.* 2013). A distinguishing feature of multispectral photogrammetry compared to standard RGB photogrammetry is the requirement for radiometric calibration. This is typically performed using calibration panels and/or light sensors (sun sensors) to compensate for variable lighting conditions (Wang 2021).

2.4 *LiDAR*

LiDAR (Light Detection and Ranging) is an active remote sensing technology that involves emitting laser pulses and measuring the time it takes for the reflected beam to return from the surface of an object. Based on this, the distance to the object is calculated. When integrated with GNSS (RTK/PPK) and an IMU (Inertial Measurement Unit), the system precisely determines the position and orientation of each laser pulse. In drones, linear or multi-beam laser scanners mounted on UAV platforms are commonly used (Ko *et al.* 2025; Kovanič *et al.* 2023). Unlike photogrammetry, which relies on passive imaging (visible or multispectral light), LiDAR directly captures the three-dimensional geometry of surfaces, even in areas with low radiometric contrast or dense vegetation cover. The result is a georeferenced, high-density 3D point cloud.

Table 1 provides a summary of imaging methods used in post-mining areas along with examples of their applications.

Table 1. Imaging methods and examples of applications in mining area monitoring.

Technology / Method	Applications in Post-Mining Area Monitoring
UAV + digital camera UAV + LiDAR	Terrain topography mapping, volume calculations, morphology. Monitoring of deformations, subsidence, and slope stability. Earthworks supervision.
UAV + IR camera	Thermal anomalies, localization of thermal activity zones.
UAV + multispectral camera	Vegetation status analysis (NDVI, NDRE, SAVI, etc.), moisture assessment, thermal anomalies.
Method Integration and Field Verification	The most complete and accurate assessment of on-site conditions

3 EXAMPLES OF APPLICATIONS

Below, selected examples of the application of remote sensing and photogrammetry in the assessment of areas used for the disposal or management of mining waste within the Upper Silesian Coal Basin are described based on the authors' own experience.

3.1 *Terrain topography studies using photogrammetry and LiDAR*

For comprehensive inventorying of the geometry of constructed waste dumps, low-altitude photogrammetry is currently most commonly used—airborne images captured by a drone equipped with a non-metric camera. Such cameras represent a relatively inexpensive and widely accessible solution, and with proper calibration and georeferencing using ground control points (GCPs) with known geodetic coordinates, orthophotomaps and digital terrain models with accurate positioning and scale can be obtained.

Currently, significant support in photogrammetry comes from the use of the RTK (Real-Time Kinematic) system, which, by receiving corrections from the national ASG-EUPOS reference network, allows the assignment of geographic coordinates to images with centimeter-level accuracy, thus improving the accuracy of the resulting model. Low-altitude photogrammetry was employed to inventory a post-mining waste dump in Trachy. Using an unmanned aerial vehicle equipped with a 20 MP RGB camera and an RTK system, images were autonomously acquired and subsequently processed into an orthophotomap (Figure 1b) and a point cloud using Pix4D Mapper photogrammetric software. Due to the dynamically progressing vegetation succession on the waste dump, capturing the surface geometry solely through photogrammetry would be challenging. In such cases, an ideal complement is airborne LiDAR (Light Detection and Ranging) scanning. Using a multi-return pulsed laser scanner mounted on an unmanned aerial vehicle (Figure 1a), the area of the dump subject to inventory was scanned. Over areas covered with trees or dense grass, the scanner recorded multiple laser beam returns, acquiring information on the terrain surface beneath the vegetation. The LiDAR data were processed using DJI Terra software. Thanks to the RTK system functioning during measurements and the inclusion of ground control points distributed across the area, a metric point cloud was obtained. Integration with photogrammetric measurements also enabled the assignment of RGB colors to the point cloud, facilitating subsequent identification of terrain details. These prepared products enabled, among other things, calculation of the capacity reserve relative to the designed volume of the waste dump (Figure 1c), identification of reworked areas, and the creation of terrain cross-sections.

Figure 1. Measurement and products obtained from the integration of photogrammetry with laser scanning technology.

3.2 *Material density studies*

At the central waste dump in Przezchlebie (Upper Silesian Coal Basin), photogrammetry was used to determine the density of accumulated coal waste. The applied method involved excavations (Figure 2a) within the dump, weighing the excavated material, and precisely determining the volume of the selected space. This approach allowed for testing a significantly larger sample, with the mass limited only by the load capacity of the vehicle transporting the material to the industrial scale for weighing (Figure 2b and 2c). This positively influenced the accuracy of density determination. The precise volume measurement was conducted using photogrammetric methods, which ensured high accuracy. To provide calculation verification, two alternative photogrammetric techniques were employed: ground-based close-range photogrammetry and low-altitude photogrammetry using unmanned aerial vehicles (UAVs). Photographic equipment with sensors of 36 MP and 14 MP resolution was used. The photographic documentation was processed using specialized software: Pix4D Mapper (image stitching, georeferencing, point cloud densification – Figure 2d), AutoCAD Civil 3D (importing point clouds, creating triangular meshes – Figure 2e, and volume calculation – Figure 2f). Based on the results obtained in this manner regarding mass and volume, the bulk density of three coal waste samples under actual dump conditions was determined, with an average value of 1.706 Mg/m^3. In situ studies, unlike laboratory tests, allowed accounting for the real compaction of the highly heterogeneous waste material located on the dump. The obtained results formed the basis for financial settlements between the municipality owning the land and the company conducting re-exploitation of the dump.

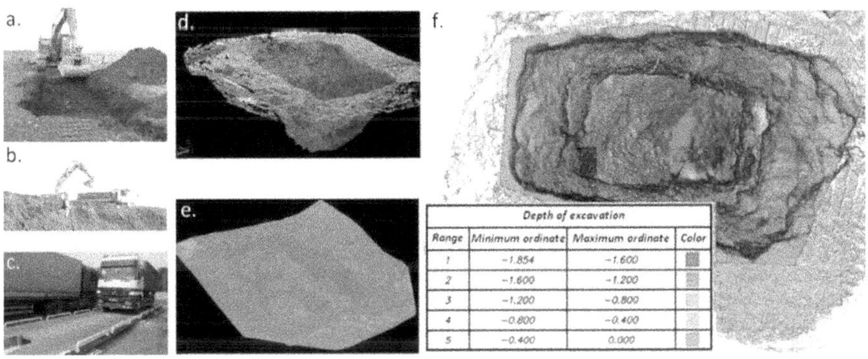

Figure 2. Selected stages of activities related to determining the volume of excavated coal waste (Różański *et al.* 2019).

3.3 *Thermal condition studies using infrared photogrammetry*

In areas designated for coal waste management, one of the most frequent and troublesome hazards is the risk of fire. The use of infrared photogrammetry provides a convenient method for thermal inventory of such sites. Conducting aerial surveys under appropriate conditions (no precipitation, negligible wind, overcast skies, or flights before sunrise) enables the localization of thermal anomalies on the surface of the studied area, as well as observation of changes and development of thermal activity. The resulting thermal maps serve as a basis for detailed ground-based thermal assessments (temperature measurements and gas concentration monitoring within the substrate), which are subsequently used in developing fire suppression technologies and designing site reclamation. An example of periodically generated thermal maps for a site located in Bytom is shown in Figure 3. The maps illustrate changes in the thermal state within the studied area.

Figure 3. Changes in the thermal state of the coal waste dump in Bytom (Upper Silesian Coal Basin).

3.4 *Thermal condition studies based on vegetation health*

In cases where an infrared imaging camera is unavailable or the survey area is covered by dense vegetation that prevents direct measurement of ground temperature, multispectral imaging using a drone can be employed. A multispectral survey in the R (red), RE (red edge), NIR (near infrared), and G (green) bands was conducted over a fire-affected waste dump near the former KWK Pokój mine in Ruda Śląska. The multispectral survey (Figure 4a) was performed concurrently with thermal imaging (Figure 4b). From the collected data, orthomosaics were generated for the recorded spectral bands, and vegetation indices NDVI and NDRE were calculated. Comparison of the NDVI index with the thermal images revealed a correlation between poor vegetation condition and elevated ground temperatures due to ongoing smoldering combustion. The results are presented in Figure 4, where correlated areas are marked with blue ellipses. Additionally, gas concentration measurements were carried out using an MRU-NOVA+ analyzer, which detected increased carbon dioxide (CO_2) levels in the soil within these areas.

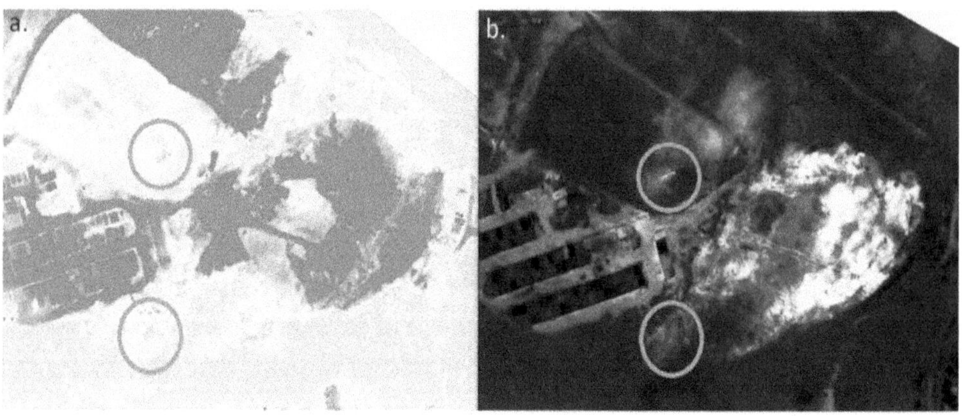

Figure 4. Comparison of the NDVI index (a) with the thermogram (b) in the fire-affected area of the site.

3.5 *Stability and slope erosion studies*

Photogrammetric products or LiDAR-derived results provide valuable data that can be utilized in the assessment of slope conditions of coal waste dumps. Digital terrain models supply essential information for slope stability calculations by providing detailed geometry and profile of the slope. This, in turn, enables the determination of the scale of landslide and erosion phenomena. Such data were employed in an expert analysis aimed at identifying the causes of a landslide that occurred spontaneously in 2023 on a fire-affected slope of a coal waste dump in Radlin (Upper Silesian Coal Basin), which resulted in significant dust pollution over a large part of the city (Figure 5). By comparing archival digital terrain models available from the Geoportal (Figure 5a) with the current model generated using photogrammetric techniques (Figure 5b) and applying differential volume calculations, the volumes of colluvial material displaced by the landslide were determined. Additionally, slope geometry data were obtained to perform stability calculations (Figure 5c). These data were complemented by thermographic survey results (Figure 5) and geotechnical probing to establish the material parameters of the analysed embankment. Based on these inputs, an analysis of the landslide causes was conducted.

Figure 5. Landslide at the KWK ROW Ruch Marcel spoil heap in Radlin (source of the left photo: Ireneusz Wilgucki - Aeroklub ROW Rybnik).

Figure 6. Digital terrain models: a. archival model obtained from Geoportal (status as of 22.04.2019), b. current photogrammetrically derived model (status as of 13.10.2023), c. stability analysis model.

Another example is the use of photogrammetry-derived data to analyze the development of water erosion on the slopes of a mining waste disposal site in Libiąż (Upper Silesian Coal Basin). Within the TEXMIN project, the impact of erosion on fire hazard was investigated (Różański *et al.* 2022). The progression of an erosion gully (Figure 7a,b) and its influence on thermal activity in the vicinity (Figure 7c) were studied. It was demonstrated that degradation of the upper slope layer due to precipitation increases air access to the interior of the site, thereby promoting the development of endogenous fires.

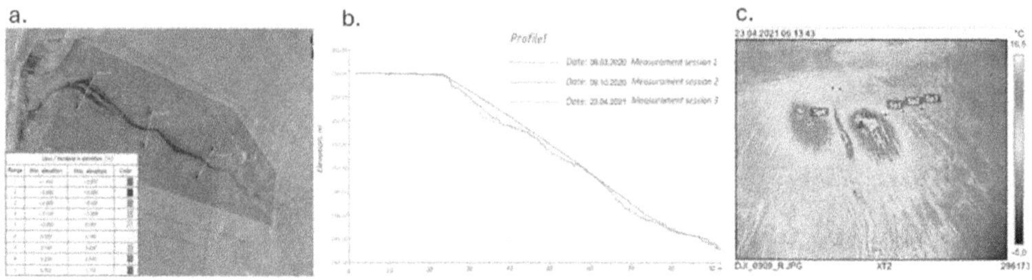

Figure 7. Analysis of Water Erosion Development and Thermal Activity on the Slope of a Coal Waste Disposal Site in Libiąż (Różański *et al.* 2022).

4 ADVANTAGES AND LIMITATIONS

The choice of method depends on the type of terrain to be mapped as well as the desired results. Accuracy plays a crucial role. LiDAR provides highly precise data, delineating tree canopies and penetrating dense vegetation to deliver information about the ground surface, whereas photogrammetry offers accurate and realistic visual perspectives. Regarding accuracy in topographic mapping, photogrammetry using geotagged images and LiDAR achieve similar horizontal accuracy, which can reach 1–2 cm. Vertical accuracy in both methods is lower; photogrammetry can achieve 3–5 cm in optimal conditions, while LiDAR typically attains slightly better vertical accuracy of 2–3 cm (Jiang *et al.* 2021; Niu *et al.* 2024; Nguyen *et al.* 2024; Wallace *et al.* 2020; Torres-Sánchez *et al.* 2023). It is important to note, however, that accuracy depends on multiple factors such as the scanner class, absolute trajectory alignment precision, reference station accuracy, and the precision of photo control points. When selecting a technique, factors like vegetation presence, texture homogeneity, shadows, lighting variability, and ground visibility must also be considered. In homogeneous areas (e.g., water bodies, snow, dense tree canopies), photogrammetry may produce gaps or erroneous surfaces. Shadows and variable lighting conditions during flights reduce feature matching consistency between images. Studies show that UAV photogrammetry is less effective in challenging lighting environments, while LiDAR remains largely unaffected. For complex surfaces with limited ground visibility (dense tree canopies, water), combining photogrammetry with LiDAR or incorporating oblique image captures is advisable. Infrared detection has become a common method for assessing thermal conditions. However, for drone-based surveys, flight altitude has a direct and significant effect on results. Beyond spatial resolution, temperature measurement accuracy decreases with increasing altitude due to greater infrared scattering and absorption in the atmosphere. Furthermore, UAV-based thermal measurements alone are insufficient for comprehensive assessment of the extent and intensity of thermal phenomena. Infrared photogrammetric surveys provide data for thermal mapping of the entire site and preliminary identification of thermally active zones. In densely vegetated areas, multispectral photogrammetry can additionally support this task. Nonetheless, a full evaluation of thermal phenomena intensity must be supplemented by ground-based temperature measurements and gas concentration analyses within the surveyed objects. This approach allows for the assessment of the development stage of thermal processes (self-heating, fire, or extinction).

5 CONCLUDING REMARKS

UAV-based low-altitude observations offer a cost-effective and high-resolution tool for monitoring post-mining sites, including spoil heaps and waste management facilities. The acquired data support reliable assessments of environmental hazards, geotechnical stability, and safety. While each method has its limitations, their complementary use, combined with advanced processing software, enhances the accuracy and practicality of monitoring. These results underline the potential of UAV techniques as an effective support tool for environmental and safety management on mining waste disposal sites.

REFERENCES

Adão, T., Hruška, J., Pádua, L., Bessa, J., Peres, E., Morais, R., and Sousa, J.J. (2017). Hyperspectral Imaging: A Review on UAV-Based Sensors, Data Processing and Applications for Agriculture and Forestry. *Remote Sensing*, 9(11), 1110. https://doi.org/10.3390/rs9111110

Allison, R.S., Johnston, J.M., Craig, G., and Jennings, S. (2016). Airborne Optical and Thermal Remote Sensing for Wildfire Detection and Monitoring. *Sensors*, 16(8), 1310. https://doi.org/10.3390/s16081310

Bendig, J., Bolten, A., and Bareth, G. (2013). UAV-based imaging for multi-temporal, very high resolution crop surface models to monitor crop growth variability. *Unmanned aerial vehicles (UAVs) for multi-temporal crop surface modelling*, 44.

Chung, C.F., *et al.* (2014). Using airborne hyperspectral data to characterize surface pH and mineralogy of pyritic mine tailings. *International Journal of Applied Earth Observation and Geoinformation*, 28, 17–29.

Colomina, I. and Molina, P. (2014). Unmanned aerial systems for photogrammetry and remote sensing: A review, ISPRS. *Journal of Photogrammetry and Remote Sensing*, 92, 79–97, ISSN 0924-2716, https://doi.org/10.1016/j.isprsjprs.2014.02.013

Fabiańska, M., Ciesielczuk, J., Nádudvari, Á., Misz-Kennan, M., Kowalski, A., and Kruszewski, Ł. (2019). Environmental influence of gaseous emissions from self-heating coal waste dumps in Silesia, Poland. *Environmental Geochemistry and Health*, 41(2), 575–601.

Jiang, S., Jaakkola, A., Hyyppä, J., and Wang, Y. (2021). Accuracy assessment of UAV photogrammetry with direct georeferencing using onboard RTK/PPK GNSS. *Remote Sensing*, 13(19), 3812. https://doi.org/10.3390/rs13193812

Kelly, J., Kljun, N., Olsson, P.-O., Mihai, L., Liljeblad, B., Weslien, P., Klemedtsson, L., and Eklundh, L. (2019). Challenges and Best Practices for Deriving Temperature Data from an Uncalibrated UAV Thermal Infrared Camera. *Remote Sensing*, 11(5), 567. https://doi.org/10.3390/rs11050567

Khumalo, S., *et al.* (2025). Monitoring slope stability: A comprehensive review of UAV applications in open-pit mining. *Land*, *14*(6), 1193.

Ko, S.J., Hong, S., Kwak, T.Y., Gu, K.Y., and Kim, S.R. (2025). UAV LiDAR–based 3D point cloud data processing for monitoring consolidation settlement during construction: a case study in Busan, South Korea. *Environmental Earth Sciences*, 84(13), 362.

Kovanič, ⊠., Topitzer, B., Pe⊠ovský, P., Bliš⊠an, P., Gerge⊠ová, M.B., and Bliš⊠anová, M. (2023). Review of Photogrammetric and Lidar Applications of UAV. *Applied Sciences*, 13(11), 6732. https://doi.org/10.3390/app13116732

Lewińska, P. and Dyczko, A. (2016). Thermal digital terrain model of a coal spoil tip – a way of improving monitoring and early diagnostics of potential spontaneous combustion areas. *Journal of Ecological Engineering*, 17(4), 170–179. https://doi.org/10.12911/22998993/64605

Liu, Y., *et al.* (2025). A novel method for landslide deformation monitoring by fusing UAV photogrammetry and LiDAR. *Engineering Geology*, *330*, 107439.

Nádudvari, Á., Abramowicz, A., Fabianska, M., Misz-Kennan, M. and Ciesielczuk, J. (2021). Classification of fires in coal waste dumps basedon Landsat, Aster thermal bands and thermal camera in Polish and Ukrainian mining regions. *International Journal Coal Science Technology* 8,441–456.

Nguyen, H.T., Le, T.H., and Pham, M.T. (2024). Accuracy assessment of UAV photogrammetry system with RTK measurements for direct georeferencing. *International Journal of Remote Sensing*. Advance online publication. https://doi.org/10.1080/01431161.2024.2356789

Nguyen, P.M.V., Wrana, A., Rajwa, S., Różański, Z., and Frączek, R. (2022). Slope stability numerical analysis and landslide prevention of coal mine waste dump under the impact of rainfall—a case study of Janina Mine, Poland. *Energies*, 15(21), 8311.

Nguyen, P.M.V. (2025). A Review of the Impact of Spontaneous Combustion on Slope Stability in Coal Mine Waste Dumps. *Applied Sciences* 15.13, 7138.

Niu, Z., Xia, H., Tao, P., and Ke, T. (2024). Accuracy Assessment of UAV Photogrammetry System with RTK Measurements for Direct Georeferencing, ISPRS Ann. Photogramm. *Remote Sens. Spatial Inf. Sci.*, X-1-2024, 169–176, https://doi.org/10.5194/isprs-annals-X-1-2024-169-2024, 2024.

Querol, X., Zhuang, X., Font, O., Izquierdo, M., Alastuey, A., Castro, I., van Drooge, B.L., Moreno, T., Grimalt, J.O., Elvira, J. *et al.* (2011). Influence of soil cover on reducing the environmental impact of spontaneous coal combustion in coal waste gobs: A review andnew experimental data. *International Journal Coal Geology*, 85, 2–22.

Różański, Z. (2018). Fire hazard in coal waste dumps–selected aspects of the environmental impact. In *IOP Conference Series: Earth and Environmental Science*, 174, 1, 12013). IOP Publishing.

Różański, Z., Konior, J., and Balcarczyk, L. (2019). Testing the in situ Bulk Density of Mining Waste Stored in Dumping Grounds. *Polish Journal of Environmental Studies*, 28(3). DOI:10.15244/pjoes/84863

Różański, Z., Wrona, P., Pach, G., Niewiadomski, A.P., Markowska, M., Wrana, A., Frączek, R., Balcarczyk, L., Vaquero Quintana, G. and de Paz Ruiz, D. (2022). Influence of water erosion on fire hazards in a coal waste dump—A case study. *Science Total Environ.* 834, 155350.

Sciortino, A., Antonielli, B., Scancella, S., Bozzano, F., Mazzanti, P., and Moretto, S. (2021). Satellite InSAR for assessing deformation processes at the Zelazny Most Tailings Dam (Poland). *In 7th International Conference on Tailings Management.*

Smoliński, A., Howaniec, N. and Kuna-Gwoździewicz, P. (2018). *Chemometric Exploration of the Data Concerning Gases Emitted from Burning Mine Waste Dump.* In: Nižetić, S., Papadopoulos, A. (eds) *The Role of Exergy in Energy and the Environment. Green Energy and Technology*. Springer, Cham. https://doi.org/10.1007/978-3-319-89845-2_33

Sobrino, J.A., Jiménez-Muñoz, J.C. and Paolini, L. (2004). Land surface temperature retrieval from LANDSAT TM 5. *Remote Sensing of Environment*, 90(4), 434–440.

Suponik, T., Franke, D., Frączek, R., Nowińska, K., Pierzyna, P., Różański, Z., and Wrona, P. (2021). The impact of precipitation on the groundwater of coal waste dump. *Journal of Sustainable Mining*, 20(1), 13–19.

Thiruchittampalam, S., Singh, S.K., Banerjee, B.P. *et al.* (2023). Spoil characterisation using UAV-based optical remote sensing in coal mine dumps. *International Journal of Coal Science and Technology* 10, 65. https://doi.org/10.1007/s40789-023-00622-4

Torres-Sánchez, J., Peña, J.M., and López-Granados, F. (2023). Performance of UAV-based photogrammetry in precision agriculture: Accuracy analysis of different flight configurations. *Applied Sciences*, 13(11), 6732. https://doi.org/10.3390/app13116732

Wallace, L., Lucieer, A., Watson, C., and Turner, D. (2020). Development of UAV LiDAR for environmental applications. *ISPRS Journal of Photogrammetry and Remote Sensing*, 168, 156–170. https://doi.org/10.1016/j.isprsjprs.2020.08.010

Wang, C. (2021). At-Sensor Radiometric Correction of a Multispectral Camera (RedEdge) for sUAS Vegetation Mapping. *Sensors*, 21(24), 8224. https://doi.org/10.3390/s21248224

Xiao W and Tian W. (2024). Remote sensing identification and hazard assessment methods for spoil sites. *Science Report.* 27, 14(1), 22121. doi:10.1038/s41598-024-72942-7.

Mine Surveying and Mining Area Protection – Sokoła-Szewioła and Jarczyk (Eds)
 ISBN: 978-1-041-10531-2

Numerical assessment of building safety under mining-induced ground curvature: Verification of ITB instruction 364/2007

L. Słowik & L. Chomacki
Building Research Institute, Warsaw, Poland

J. Rusek
AGH UST of Krakow, Cracow, Poland

ABSTRACT: Underground mining activity induces ground deformations that may threaten the safety and serviceability of buildings. Current Polish guidelines, such as 364/2007, ITB Instruction specify threshold segment lengths of rigid buildings for which additional calculations of mining curvature effects are not required. In this study, more than 100 numerical FEM models were created to evaluate the relevance of these thresholds and to determine the structural response under different soil stiffness and curvature conditions. The results showed that in several cases, the tensile forces in floor ring beams exceeded the load-bearing capacity of the minimum rate of reinforcement prescribed in the ITB Instruction. Further analyses indicated that introducing the recommended 4 Ø 12 mm reinforcement or applying Eurocode 2 provisions significantly changes the permissible segment lengths, in some cases shortening them considerably. The research results suggest that the current ITB recommendations require updating and that numerical modeling provides a more reliable basis for assessing building safety in mining-affected areas.

1 INTRODUCTION

Underground mining activities inevitably affect the ground surface, posing risks to existing buildings structures and infrastructure (Kawulok 2015). These impacts can be either static or dynamic in nature and depend on various factors, including the geological structure of the deposit, the depth and type of exploited seams, the condition of previously mined areas, groundwater levels, and—most importantly—the applied mining methods. In Poland, the most severe consequences are observed in the USCB: Upper Silesian Coal Basin (Strzałkowski and Strzałkowska 2023) and the LGCD: Legnica–Głogów Copper District. At the surface, mining activity typically manifests as subsidence troughs, which may be accompanied by flooding, discontinuous deformations, or mining-induced tremors (Jiang *et al.* 2020, Tajduś *et al.* 2018, 2021).

The assumption that well-controlled mining operations produce only continuous deformations that can be safely resisted by properly designed buildings structures has often proven overly simplistic. In reality, mining-induced ground movements may cause excessive stresses in structural elements, leading to characteristic damage patterns (Chomacki *et al.* 2022) such as: significant cracks and scratches of load-bearing elements, differential vertical displacements, horizontal deformations, or excessive tilting (Gromysz 2019), particularly in linear structures (Kalisz and Zięba 2014). These effects become especially complex in areas subjected to repeated mining, where soil–structure interactions are influenced by both rock mass behavior and the properties of the soil foundation.

The assessment of mining impacts on buildings requires reliable methods to determine the forces and displacements at the soil–structure interface. The accuracy of such evaluations

DOI: 10.1201/9781003655558-21

directly affects both the assessment of structural safety and the design of safety structural solutions (Drobiec *et al.* 2019; Szojda and Kapusta 2023). Depending on the adopted approach, the analysis may be simplified—using analytical structural mechanics and beam models with parametric soil representation—or advanced, employing finite element methods (FEM) where both the building and the soil mass are modeled explicitly (Fedorowicz *et al.* 2021). Increasingly, the use of numerical methods is becoming the standard, as traditional simplified approaches are being replaced by models capable of representing geometry, material properties and soil–structure interaction with much greater realism (Szojda and Strzałkowski 2025).

European standards, such as Eurocode 7 (PN-EN 1997, 2008), explicitly recognize the application of numerical methods in geotechnical design whenever soil–structure interaction needs to be considered. In the context of mining engineering, this means that FEM-based analyses must account for realistic boundary conditions, including ground stiffness degradation or changes in stress state caused by ongoing mining operations (Wysokiński *et al.* 2011).

The present study builds on these foundations by employing numerical analyses to identify cases in which simplified methods are sufficient and when more detailed modeling is necessary. The aim is to provide guidelines for distinguishing situations where additional structural assessments are required, thereby improving both the efficiency and reliability of engineering practice in mining-affected areas.

2 MATERIALS AND METHODS

2.1 *Bending structure on mining ground, assumptions used in modeling the Building (B) - Mining ground (Mg) system*

The essence of contact analyses between buildings and mining-affected subsoil lies in two fundamental objectives (Fedorowicz 2008):

- a reliable assessment of structural stresses, ensuring adequate load-bearing capacity, and
- an evaluation of deformations, verifying serviceability conditions.

According to established 416/2006 ITB guidelines (Kawulok 2006) and 364/2007 (Kawulok *et al.* 2007):

- approximate internal forces in building structures can be estimated using grate or beam static schemes, while
- for spatial structural systems, computer-based methods are recommended, enabling explicit modeling of both the soil and the superstructure.

The concept of a computational model for the structural subsystem (*B*) may therefore refer either to a simplified classical beam or grate model, or to a more advanced numerical representation of reality. Since finite element models of the (*B*)-(*Mg*) system (building–mining ground) can be highly complex, it is neither feasible nor practical to prescribe detailed modeling rules. Nevertheless, the credibility of numerical results must be verified against appropriate criteria.

In basic analyses, the least controversial approach is the implementation of predicted vertical ground displacements (and the corresponding movements of the cooperating soil layer) in a parametric soil model. The coefficient of vertical subgrade compliance, C_0, is typically determined using the Boussinesq model, with Budzianowski's correction for strip footings. This engineering method, developed by Budzianowski, allows estimation of foundation displacements, generalized internal forces (M, Q, N) and deflection adjustments of a rigid building resting on deformable mining subsoil. A key assumption of this approach is that the coefficient C_0 remains constant beneath the entire foundation, which requires a preliminary correction of the foundation footprint geometry to ensure uniform values under each strip footing.

It should also be noted that in building–soil systems, the parametric soil model relies on the assumption of equal settlements (Kwiatek *et al.* 1997) under all footings of a grate foundation placed on a homogeneous subsoil of sufficient thickness.

2.2 *Model geometry*

As part of the numerical studies, static calculations were performed using the Autodesk Robot Structural Analysis Professional 2024 computer program. To verify the criteria specified in 364/2007 ITB Instruction (Kawulok *et al.* 2007), it was necessary to prepare numerous calculation models with variable geometry: 10.0, 12.0, 15.0, 18.0, 24.0, 27.0, 30.0, and 36.0 m. The model width remained constant and was 10 m for all variants, and all buildings were two-bay. The building consisted of a basement and three above-ground stories. The floors were designed as reinforced concrete structures with reinforced concrete ring beams. The foundations were designed as reinforced concrete strip footings, and the walls were designed as masonry structures. A simplified model was used without window openings, where stress disturbances would arise as a result of the notch effect. Computational analysis was carried out for linear-elastic work material. The wall parameters (the elastic model with the following parameters: Young's modulus E=10000 MPa; Poisson's ratio v=0.3; Kirchhoff's coefficient G=9000 MPa and specific weight g = 18,00 kN/m^3) were based on Eurocode 6 standard (PN-EN 1996 2010), and the reinforced concrete element parameters (the elastic model with the following parameters: Young's modulus E=30000 MPa; Poisson's ratio v=0.2; Kirchhoff's coefficient G=12500 MPa and specific weight g = 24.53 kN/m^3) were based on Eurocode 2 standard (PN-EN 1992 2008).

An example three-dimensional model of a 24.0 m long building is shown in Figure 1a, and a view of the longitudinal wall of the building, for axial symmetry assumed at half its length, is shown in Figure 1b.

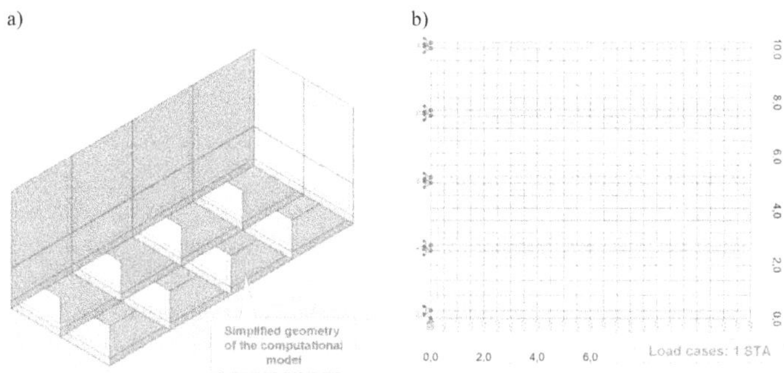

Figure 1. View of an example model of a 24 m long building: a) 3D view of the entire model from below, b) view of a fragment of the longitudinal internal wall for axial symmetry at half of its length.

2.3 *Boundary conditions and curvature loads of the mining terrain*

The model was supported along its lower horizontal edge (at the foundation base level) using flexible supports. The horizontal direction was blocked at the mid-span (Figure 1b). Vertical flexibility was calculated according to 416/2006 ITB instruction (Kawulok 2006), namely:

$$C_V = 2\omega \frac{E_0}{b_i} \tag{1}$$

where E_0 – primary soil deformation modulus; ω – coefficient depending on the ratio of foundation dimensions, adopted as for type D in accordance with Table 6 of the manual (Kawulok 2006); b_i –foundation width.

In order to analyze all the values included in the instruction, vertical compliance was determined for all analyzed cases, the results of which are summarized in Table 1.

Table 1. Determination of the vertical susceptibility value C_V.

L [m]	L/B [-]	ω [-]	Vertical susceptibility value C_V [MN/m]			
			E_0 = 10 MN/m^2	E_0 = 20 MN/m^2	E_0 = 50 MN/m^2	E_0 = 80 MN/m^2
10	20	0.29	11.6	23.2	58.0	92.8
12	24	0.29	11.6	23.2	58.0	92.8
15	30	0.27	10.8	21.6	54.0	86.4
18	36	0.26	10.4	20.8	52.0	83.2
24	48	0.24	9.6	19.2	48.0	76.8
27	54	0.23	9.2	18.4	46.0	73.6
30	60	0.23	9.2	18.4	46.0	73.6
36	72	0.23	9.2	18.4	46.0	73.6

The model was loaded with mining terrain curvature by displacing individual nodes at its base by values resulting from the radius of curvature. Displacement values were determined for threshold values for each mining area category (Table 2).

Table 2. Mining area categories (Kwiatek *et al.* 1997).

Category	Limit values of terrain deformation indicators		
	tilt T [‰]	radius of curvature R [km]	horizontal deformation ε [‰]
0	$T \leq 0.5$	$\lvert R\rvert \geq 40$	$\lvert\varepsilon\rvert \leq 0.3$
I	$0.5<T \leq 2.5$	$40>\lvert R\rvert \geq 20$	$0.3<\lvert\varepsilon\rvert \leq 1.5$
II	$2.5<T \leq 5$	$20>\lvert R\rvert \geq 12$	$1.5<\lvert\varepsilon\rvert \leq 3$
III	$5<T \leq 10$	$12>\lvert R\rvert \geq 6$	$3<\lvert\varepsilon\rvert \leq 6$
IV	$10<T \leq 15$	$6>\lvert R\rvert \geq 4$	$6<\lvert\varepsilon\rvert \leq 9$
V	$T>15$	$\lvert R\rvert<4$	$\lvert\varepsilon\rvert>9$

Table 3 summarizes the vertical displacement values of individual nodes along the section from the axis of symmetry to 4 m from the axis, at 0.5 m distances. The displacement values were determined for each radius of curvature up to a distance of 18.0 m from the center of symmetry (equivalent to a building length of 36.0 m). An example of applying loads due to mining terrain curvature for the selected model (Figure 5b) is shown in Figure 2.

Table 3. Values of vertical displacements of nodes depending on the distance of the node from the axis of symmetry and the radius of curvature.

Parameter	Value								
Radius of curvature R_d [km]	20								
Distance of the node from the axis x [m]	0	0.5	1	1.5	2	2.5	3	3.5	4
The value of the vertical displacement of the node y [mm]	0	0.01	0.04	0.10	0.17	0.27	0.38	0.52	0.68
Radius of curvature R_d [km]	12								
Distance of the node from the axis x [m]	0	0.5	1	1.5	2	2.5	3	3.5	4
The value of the vertical displacement of the node y [mm]	0	0.02	0.07	0.16	0.28	0.44	0.64	0.87	1.13
Radius of curvature R_d [km]	6								
Distance of the node from the axis x [m]	0	0.5	1	1.5	2	2.5	3	3.5	4
The value of the vertical displacement of the node y [mm]	0	0.04	0.14	0.32	0.57	0.89	1.28	1.74	2.27

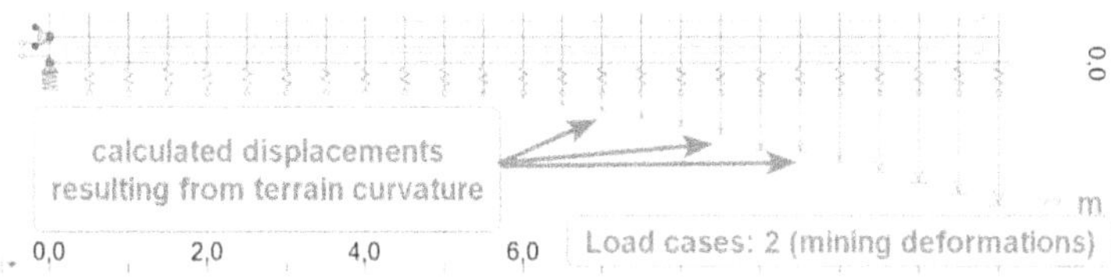

Figure 2. View of an example model of a 24.0 m long building with applied loads due to the curvature of the mining area

3 RESULTS

3.1 *Results for the values given in the ITB instructions*

The number of analyses performed was dictated by the need to reproduce the allowable segment lengths, as specified in 364/2007 ITB Instruction (Kawulok *et al.* 2007). Over 100 numerical models were analyzed to reproduce all variants (Table 4). The threshold lengths, which will have their equivalents in the subsequent analysis, are marked in yellow. Selected stress and membrane forces maps of the selected numerical model are presented in Figure 3.

Table 4. Permissible lengths (L) of rigid object segments, at which calculations for terrain curvature effects are not required (Kawulok *et al.* 2007).

Modulus of primary soil deformation E_0 [MPa]	Permissible length of segments L in m, at the forecasted radius of terrain curvature $\|R\|$ [km]			
	$\|R\| \geq 20$	$\|R\| > 12$	$\|R\| \geq 6$	$\|R\| \geq 4$
$\leq$ 10	36.0	36.0	30.0	27.0
$\leq$ 20	36.0	30.0	24.0	18.0
$\leq$ 50	27.0	24.0	15.0	12.0
$\leq$ 80	24.0	18.0	12.0	10.0

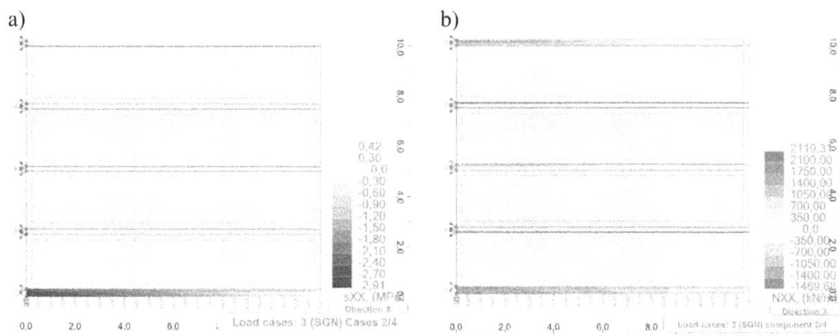

Figure 3. Results in the form of maps, for the selected numerical model, of the inner longitudinal wall: a) horizontal stresses σ_{xx}, b) membrane forces N_{xx},

It was determined that the most reliable comparison would be made of horizontal forces occurring at the level of reinforced concrete ceilings or foundations. For this purpose, global extremes were analyzed for each model, and the extreme values of tensile forces were recorded. These results apply to the ultimate limit state load combinations, taking into account the building's own weight and mining impacts. The values of these forces are summarized in Tables 5 and 6, where the force values corresponding to the lengths from the ITB Instruction (Kawulok *et al.* 2007) are highlighted in yellow (see Table 4).

Table 5. Determined extreme values of horizontal tensile forces that can occur in the model for given input parameters. Values for radius of terrain curvature of 20 and 12 km (mining terrain categories I and II).

Length L [m]	Module E_0 [MPa]	Radius R_d [km]	Horizontal force F_x [kN]	Length L [m]	Module E_0 [MPa]	Radius R_d [km]	Horizontal force F_x [kN]
36	10	20	84.72	36	10	12	141.01
36	20	20	176.57	36	20	12	211.76
36	50	20	215.78	36	50	12	301.94
27	10	20	55.58	30	10	12	75.67
27	20	20	78.85	30	20	12	119.59
27	50	20	105.24	30	50	12	183.56
27	80	20	114.7	30	80	12	211.7
24	20	20	55.83	24	20	12	57.32
24	50	20	76.62	24	50	12	94.76
24	80	20	84.42	24	80	12	113.32
18	50	20	57.65	18	50	12	58.09
18	80	20	61.59	18	80	12	70.73

Table 6. Determined extreme values of horizontal tensile forces that can occur in the model for given input parameters. Values for radius of terrain curvature of 6 and 4 km (mining terrain categories III and IV).

Length L [m]	Module E_0 [MPa]	Radius R_d [km]	Horizontal force F_x [kN]	Length L [m]	Module E_0 [MPa]	Radius R_d [km]	Horizontal force F_x [kN]
30	10	6	150.34	27	10	4	154.19
30	20	6	237.93	27	20	4	250.86
24	10	6	68.45	18	10	4	57.34
24	20	6	113.46	18	20	4	96.25
24	50	6	188.09	18	50	4	167.07
24	80	6	225.08	18	80	4	205.02
15	10	6	32.29	12	10	4	34.08
15	20	6	54.9	12	20	4	59.9
15	50	6	99.64	12	50	4	114.93
15	80	6	125.96	12	80	4	150.21
12	50	6	77.54	10	50	4	82.76
12	80	6	101.1	10	80	4	111.12

Based on the above results, the obtained tensile forces F_x were analyzed, where the value:

- maximum was 176.57 kN,
- average was 113.63 kN,
- minimum was 70.73 kN.

These values locally exceed the load-bearing capacity of 4 Ø 12 mm of minimum peripheral floor ties reinforcement recommended in the ITB Instructions (Kawulok 2006). This conclusion was analyzed in further stages of the research.

3.2 *Results for recommended minimum reinforcement 4 Ø 12 mm*

In the next step of the analyses, the minimum reinforcement of the peripheral floor ties recommended in the ITB instructions, which is 4 Ø 12, was taken into account. Such reinforcement, assuming the use of the most popular B500SP steel, allows for the transfer of a horizontal tensile force of approximately 115 kN. For this limiting force, the allowable segment lengths, which do not require calculations for curvature, have changed. The results are summarized in Table 7. The values marked in yellow have not changed, but the values marked in green have.

Table 7. Permissible lengths (L) of rigid object segments, with 4 Ø 12 mm of minimum peripheral floor ties reinforcement recommended in the ITB Instructions.

Modulus of primary soil deformation E_0 [MPa]	Permissible length of segments L in m, at the forecasted radius of terrain curvature $\|R\|$ [km]			
	$\|R\| \geq 20$	$\|R\| \geq 12$	$\|R\| \geq 6$	$\|R\| \geq 4$
$\leq$ 10	36	30	27	24
$\leq$ 20	30	27	24	18
$\leq$ 50	27	24	18	12
$\leq$ 80	24	24	12	10

Five of the previous segment lengths were shortened by 3 m or 6 m, and two of the previous lengths were lengthened also by 3 m or 6 m. These values should ensure the safety of the structure and users, they were determined on the basis of numerical analyses and should be taken into account when updating the ITB instructions. The next step was to assess the natural resistance of the building, designed in accordance with Eurocode 2, to the influence of mining curvature.

3.3 *Results for minimum horizontal force according to Eurocode 2*

According to Eurocode 2 (PN-EN 1992 2008), commonly used for the design of RC and masonry buildings in Europe, the minimum horizontal force transferred by a peripheral ties is 70 kN. For this horizontal force value, a separate analysis of the maximum segment length was performed, which did not require calculations for the influence of mining curvature (Table 8).

Table 8. Permissible lengths (L) of rigid object segments, with minimum horizontal force transferred by a peripheral ties is 70 kN (according to Eurocode 2).

Modulus of primary soil deformation E_0 [MPa]	Permissible length of segments L in m, at the forecasted radius of terrain curvature $\|R\|$ [km]			
	$\|R\| \geq 20$	$\|R\| \geq 12$	$\|R\| \geq 6$	$\|R\| \geq 4$
$\leq$ 10	27	27	24	18
$\leq$ 20	24	24	18	12
$\leq$ 50	18	18	10	-
$\leq$ 80	18	15	10	-

All lengths have been shortened by from 2 m to 12 m, and for two variants the lengths have not been determined because they must be shorter than 10 m.

4 DISCUSSION AND CONCLUSIONS

The numerical analyses demonstrated that the threshold segment lengths specified in ITB Instruction 364/2007 are not always conservative. In particular, for less stiffer soils, the calculated tensile forces in peripheral floor ties locally exceeded the capacity of the minimum reinforcement, which questions the universal applicability of the existing guideline values.

When the recommended by ITB Instructions 4 Ø 12 mm reinforcement was introduced, the maximum tensile forces decreased to values within the safe range, but this adjustment also resulted in modified permissible segment lengths. Some lengths were shortened, while others were extended, indicating that reinforcement requirements and segment thresholds should be treated jointly rather than independently. This highlights the importance of not relying solely on guideline values, but instead assessing whether specific building and ground conditions justify additional calculations.

A comparison with Eurocode 2 revealed significantly shorter permissible segment lengths, in some cases below 10 m, which effectively eliminates the possibility of avoiding detailed calculations. This discrepancy highlights the differences between national and European standards and underlines the importance of updating ITB guidelines based on numerical modeling results and soil–structure interaction considerations. The study demonstrates that numerical modeling provides a more realistic picture of structural behavior than simplified analytical schemes, and therefore should be used as a benchmark for revising national guidelines.

Future work should focus on validating the numerical predictions against field observations and damage surveys of buildings located in mining areas. Such validation would provide the missing empirical evidence needed to confirm whether the updated segment lengths and reinforcement recommendations can be safely generalized in practice.

REFERENCES

Chomacki, L., Rusek, J. and Słowik, L. (2022) Machine Learning Methods in Damage Prediction of Masonry Development Exposed to the Industrial Environment of Mines, *Energies*, 15(11), pp. 1–23. Available at: https://doi.org/10.3390/en15113958.

Drobiec, Ł. *et al.* (2019) The method of strengthening the church building in terms of the planned mining exploitation, *MATEC Web of Conferences*, 284, p. 05003. Available at: https://doi.org/10.1051/matecconf/201928405003.

Fedorowicz, J. (2008) Criteria for evaluation of masonry-structure behaviour in mining areas, *Architecture Civil Engineering Environment*, 2, pp. 131–135.

Fedorowicz, J., Fedorowicz, L. and Kadela, M. (2021) Description of limit states in the subsurface layer of loosened subsoil in view of critical state soil mechanics, *Materials*, 14(23). Available at: https://doi.org/10.3390/ma14237288.

Gromysz, K. (2019) Analysis of removal method of a 19 th church's deflection, *MATEC Web of Conferences*, 284, p. 05005. Available at: https://doi.org/https://doi.org/10.1051/matecconf /20192840500 5.

Jiang, Yue *et al.* (2020) A new prediction model of surface subsidence with Cauchy distribution in the coal mine of thick topsoil condition, *Archives of Mining Sciences*, 65(1), pp. 147–158. Available at: https://doi.org/10.24425/ams.2020.132712.

Kalisz, P. and Zięba, M. (2014) Impact of mining exploitation on pipelines, *Acta Montanistica Slovaca*, 19(3), pp. 111–117.

Kawulok, M. (2006) Projektowanie budynków na terenach górniczych, *Instrukcja ITB nr 416/2006*. Warszawa: Instytut Techniki Budowlanej.

Kawulok, M. *et al.* (2007) Wymagania techniczne dla obiektów budowlanych wznoszonych na terenach górniczych, *Instrukcja ITB nr 364/2007*. Warszawa: Instytut Techniki Budowlanej.

Kawulok, M. (2015) Szkody górnicze w budownictwie (Mining damages in buildings). *Monografia*. Warszawa: Instytut Techniki Budowlanej.

Kwiatek, J. *et al.* (1997) Ochrona obiektów budowlanych na terenach górniczych. *Praca pod kierunkiem J. Kwiatka.* Katowice: Główny Instytut Górnictwa.

PN-EN 1992 (2008) PN-EN 1992 Eurokod 2: Projektowanie konstrukcji z betonu (Eurocode 2: Design of concrete structures).

PN-EN 1996 (2010) PN-EN 1996 Eurokod 6: Projektowanie konstrukcji murowych (Eurocode 6 - Design of masonry structures).

PN-EN 1997 (2008) Eurokod 7: Projektowanie geotechniczne (Eurocode 7: Geotechnical design).

Strzałkowski, P. and Strzałkowska, E. (2023) An assessment of the impact of the degree of the filling of shallow voids on the possibility of sinkhole formation on the surface, *Gospodarka Surowcami Mineralnymi / Mineral Resources Management*, 39(1), pp. 173–191. Available at: https://doi.org/10.24425/gsm.2023.144627.

Szojda, L. and Kapusta, L. (2023) Numerical Analysis of Buildings Located on the Edge of the Post-Mining Basin, *Archives of Mining Sciences*, 68(1), pp. 125–140. Available at: https://doi.org/10.24425/ams.2023.144321.

Szojda, L. and Strzałkowski, P. (2025) Numerical analysis of building damage caused by the exposure of a terrain threshold as a result of underground mining-induced subsidence, *Scientific Reports*, 15(1), pp. 1–17. Available at: https://doi.org/10.1038/s41598-025-96648-6.

Tajduś, K. *et al.* (2021) Analysis of Mining-Induced Delayed Surface Subsidence, *Minerals*, 11(11). Available at: https://doi.org/10.3390/min11111187.

Tajduś, K., Misa, R. and Sroka, A. (2018) Analysis of the surface horizontal displacement changes due to longwall panel advance, *International Journal of Rock Mechanics and Mining Sciences*, 104(February), pp. 119–125. Available at: https://doi.org/10.1016/j.ijrmms.2018.02.005.

Wysokiński, L., Kotlicki, W. and Godlewski, T. (2011) Projektowanie geotechniczne według Eurokodu 7 (Geotechnical design according to Eurocode 7). *Instytut Techniki Budowlanej.*

Mine Surveying and Mining Area Protection – Sokoła-Szewioła and Jarczyk (Eds)
© 2025 The Author(s), ISBN: 978-1-041-10531-2

Forecasting changes in seismicity induced by mining operations based on variations in mining depth

V. Sokoła-Szewioła & P. Jarczyk
Silesian University of Technology, Gliwice, Poland

ABSTRACT: Induced seismicity resulting from underground mining operations constitutes a significant threat both to the safety of mine workers-due to the potential occurrence of rock bursts, and to public safety, due to the effects of vibrations caused in the surface environment by seismic phenomena. These vibrations can pose a threat to infrastructure, including damage to buildings, roads and transmission networks. Additionally, they can negatively affect the health and well-being of residents in mining areas, causing stress and a sense of danger. Seismic hazard forecasting enables appropriate measures to be taken to reduce the probability of a seismic event. As previous studies have shown one of the key factors influencing the magnitude and characteristics of seismic events is the depth at which mining operations are carried out. The paper presents the results of a study on the possibility of prediction of changes in induced seismic activity based on changes in the depth of mining operations. The research utilized archival data from selected mines located in Poland, in the Upper Silesian Coal Basin, taking into account mining history and recorded seismic events. Prognostic models were developed and applied to forecast seismic activity in the studied areas. The research showed that the proposed forecasting method can be a valuable tool for supporting seismic risk management in mining planning.

1 INTRODUCTION

Seismicity induced by mining activity is the subject of numerous research studies due to the fact that it affects many mining areas and that seismic hazards—along with the associated risk of rock bursts—constitute one of the most dangerous threats in underground mining (Białek 1999; Descour 1979, Kijko *et al.* 1982 1985; Lasocki 1996, Marcak 2010, Sokoła-Szewioła 1996, 2008).

An important research direction is the forecasting of seismic hazards, which involves determining the location, energy, and time of seismic events. Such forecasting makes it possible to implement appropriate measures to reduce the likelihood of a mining seismic event occurring, in particular above a certain energy threshold, and thus helps to avoid negative consequences. Energy forecasting refers to estimating the maximum energy of anticipated events. Typically, data collected from mining operations conducted under similar geological and mining conditions is used, with the application of statistical or analytical methods (Biliński 1981, Jaworski 2000, Kłeczek and Zorychta 1991, Kijko *et al.* 1982, Lasocki 1992, 1996; Marcak 2010, Makówka and Kabiesz 2005).

The significance of this issue is associated with the fact that mining is currently being carried out at increasing depths, under conditions of heightened stress within the rock mass (e.g., Wu *et al.* 2024, Zhou *et al.* 2024). This leads to an increased risk of high-energy seismic events (seismic energy $E \geq 10^5$ J), which may be accompanied by rock bursts and can have negative effects on the surface, including damage to buildings, roads, and transmission

DOI: 10.1201/9781003655558-22

networks. Moreover, these events can negatively affect the health and well-being of residents in mining areas, causing stress and a sense of danger.

In coal mines located in Poland's Upper Silesian Coal Basin, a total of 51,849 high-energy seismic events occurred between 1981 and 2018, with a cumulative measured seismic energy release of 162.94 GJ. On average, 11 high-energy seismic events occurred per one million tons of coal extracted. The average annual number of such events was 1,364 (Drzewiecki 2017; Gibowicz 1963, 2009; Kabiesz 2017; Kłeczek 2017; WUG 2017, 2018, 2019).

Despite a reduction in coal output and the closure of many underground coal mines in Poland in accordance with EU policy, between 2019 and 2024 a total of 10,215 high-energy mining seismic events were recorded, with a cumulative measured energy output of 19.33 GJ (WUG 2023, 2024). During this period, there were 32 high-energy seismic events per one million tons of coal extracted, and the average annual number of events increased to 1,702.

The causes of this trend were studied, among others, at the Silesian University of Technology (Sokoła-Szewioła 2019). The research covered selected mines operating in the Upper Silesian Coal Basin. It was found that the increase in seismic hazard was related to the growing depth of mining operations, the geological structure of the rock mass (which includes thick sandstone layers and complex tectonics), and the challenging and confined mining conditions (e.g., goafs and remnant areas). As a result of the research, statistically significant relationships were identified between the increasing cumulative seismic energy, the number of seismic events, and the increasing depth of mining. Figure 1 presents the distribution of cumulative seismic activity in relation to mining depth for one of the mines belonging to Jastrzębska Spółka Węglowa (JSW S.A.) in the years 2010–2017, developed by V. Sokoła-Szewioła (2019).

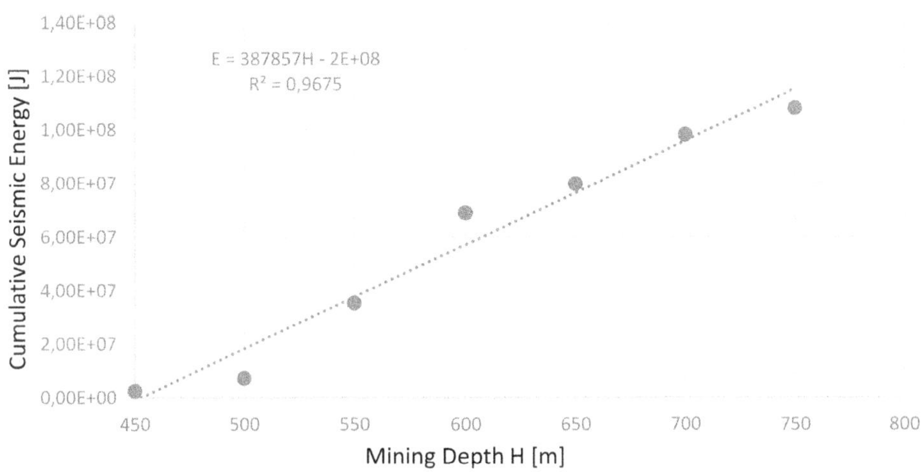

Figure 1. Seismic activity distribution as a function of mining depth in one of the JSW S.A. mines.

The results of the analysis of mining depth in relation to recorded seismic activity in Polish hard coal mines were also presented in the study by Mirek *et al.* (2022), in which the authors point out an increase in seismic hazard despite the reduction in coal output, linking this trend to the descent of mining operation to greater depth. The justification for pursuing research aimed at developing new methods of forecasting induced seismicity is underscored by the number of accidents resulting from rock bursts. Between 1981 and 2018, a total of 293 rock bursts occurred in Polish hard coal mines, resulting in injuries to 878 people, including 179 fatal accidents. In the years 2019–2024, there were 14 rock bursts, causing injuries to 134 people, including 25 fatalities.

Due to the significance of the issue, studies on the relationship between seismic activity and mining depth are being conducted in various countries (Ge 2022, Király-Proag *et al.* 2016, Lenhardt 1989, 2009, 2020; Lenhardt and Hagan 1990, Lenhardt and Pascher 1996, Li *et al.* 2024, Martinsson and Törnman 2020). In his work, W. Lenhardt (1992) presented the results of research on the correlation between seismic activity and mining depth in gold mines in South Africa, which is described by a linear functional relationship, with a coefficient of determination of 0.9499.

The results of the aforementioned studies motivated the authors to undertake research whose main objective was to determine whether it is possible to forecast induced seismic activity based on observed changes in mining depth in a given region, and whether such forecasts are acceptable in terms of accuracy. The article presents the methodology for developing predictive models of changes in induced seismic activity based on variations in mining depth, as well as its application in selected coal mines located in the Upper Silesian Coal Basin and operated by JSW S.A.

2 METHODOLOGY

The study was based on regression analysis methods (Draper and Smith 1973) to forecast changes in induced seismicity depending on mining depth The following seismic activity indicators were used: - N – the number of seismic events within a given depth interval, - E – the total energy of seismic events within that interval.

The research process included: selection of study areas, collection of seismic and mining data, development of a model relating seismicity to depth, model verification, and performing forecasts and ex-post analysis. Calculations were performed using Statistica 13.3 software.

Mining depth data was obtained from mine surveying and geological departments, while seismic data came from mine geophysical stations. Due to the highly nonlinear nature of cumulative energy, only high-energy seismic events were analyzed, excluding regional or tectonic events (with energies in the range of 10^8 J, 10^9 J). The research was conducted in three JSW S.A. coal mines: "Borynia-Zofiówka", "Knurów-Szczygłowice", and "Pniówek", in areas with low geological variability.

2.1 *Regression model*

A linear regression model was applied to describe the relationship between cumulative seismic energy (E_i) or number of seismic events (N_i) and mining depth (H_i). The model has the form:

$$E_i = \alpha + \beta H_i + \varepsilon_i \quad i = 1, 2 \ldots n \tag{1}$$

where: α – intercept, β – regression coefficient, ε_i is a random component, n - number of observation (i = 1,2,...n).

Parameter estimation was carried out using the least squares method. The estimated values of the parameters α and β in the linear model were denoted by a and b (2, 3):

$$b = \frac{\sum_{i=1}^{n} H_i E_i - \left[\left(\sum_{i=1}^{n} H_i\right)\left(\sum_{i=1}^{n} E_i\right)\right]/n}{\sum_{i=1}^{n} H_i^2 - \left(\sum_{i=1}^{n} H_i\right)^2/n} \quad i = 1, 2 \ldots n \tag{2}$$

$$a = \bar{E} - b\bar{H} \tag{3}$$

where:
$\bar{E} = \frac{\left(\sum_{i=1}^{n} E_i\right)}{n}$, $\bar{H} = \left(\sum_{i=1}^{n} H_i\right)/n$, n- number of observations.

After accounting for the notation, the regression function based on the sample was written as (4):

$$\hat{E}_i = a + bH_i \quad i = 1, 2 \ldots n \tag{4}$$

The values of this function $\hat{E}_i$, calculated for the values of H_i from the sample, are called theoretical values. The values of the differences are referred to as residuals e_i (5):

$$e_i = E_i - \hat{E}_i \quad i = 1, 2 \ldots n \tag{5}$$

where: E_i – value of the sample regression function, $\hat{E}_i$ – estimated value.

Standard errors of parameters evaluations (S_a, S_b) and residual standard deviation (Se) were also estimated.

2.2 *Model verification*

The verification of the model was based on an examination of: the quality of the structural parameter evaluations, the degree of conformity of the model with empirical data, and the distribution of random deviations (Nowak 2006).

The following analyses were conducted:

- Assessment of the significance of structural parameters α, β (Student's t-test).
- Analysis of variance – determining the influence of the independent variable *Hi* on the dependent variable E_i (*F*-Snedecor test).
- Assessment of the degree of fit of the model is based on the coefficient of determination R^2, which determines the proportion of variation in the dependent variable explained by the model to the total variation observed in the sample. The square root of the coefficient is the Pearson's correlation coefficient *R*, which indicates the degree of linear interdependence of the two variables.
- Assessment of the value of the random variability coefficient W_e, which indicates what percentage of the arithmetic mean of the variable H_i is represented by the standard deviation of the residuals.

The study of the properties of random deviations was carried out using statistical tests:

- Symmetry test – checking if the number of positive and negative residuals is equal to 0.5 (*t*-test).
- Shapiro-Wilk normality test – verifying the normality of the distribution of residuals.
- Homoskedasticity test – comparing variance of residuals from initial and final observations (*F*-test).
- Runs test – assessing randomness of residuals.
- Durbin-Watson test – examining first-order autocorrelation in residuals

All tests were performed at a significance level of α = 0.05. The same procedures were applied for the number of seismic events (N_i), replacing E_i with N_i in formulas and analyses.

2.3 *Forecast of induced seismic energy based on the established regression model*

If the model was found to be accurate, forecasts were generated. Forecasts of total induced seismic energy were calculated using the forecasting principles of linear econometric models (Dittmann 2004).

- If *T* is the forecast step, and E_{iT} is the value of the independent variable at the forecast step *T* then the forecast of E_{iT} is of the form (6):

$$E^*_{iT} = a + bH_{iT}^* \tag{6}$$

where: $T=n+1,..., n+h, h>0$ (h-horizon of the forecast), H_{iT}*– the depth H at the end of the prediction step „T"[m], E_{iT}*– total cumulative seismic energy at the end of a step „T" [J].

- Calculation of the relative error of ex ante forecasts according to the formula (7):

$$V(E_T^*) = \frac{D(E_{iT}^*)}{E_{iT}^*} 100\% \; T = n+1, ..., \quad n+h \tag{7}$$

where: $D(E_{iT}*)$ means absolute ex ante forecast calculated according to the formula (Dorosiewicz *et al.* 1996) (8):

$$D(E_{iT}^*) = S_e\sqrt{1 + c'(H_i'H_i)^{-1}c} \tag{8}$$

where: $c' = (1, H_{iT}^*)$, S_e – the standard deviation of the residuals, the symbol ′ (prim) is a transposition mark.

After the prediction time, an ex-post analysis was performed - comparing the total cumulative seismic energy predicted in step T with the actual E_T values. For this purpose, it was determined:

- average absolute percentage error of ex post forecasts calculated for the forecast horizon h (9):

$$\psi = \frac{1}{h}\sum_{T=n+1}^{n+h} \left|\frac{E_{iT} - E_{iT}^*}{E_{iT}}\right| 100\% \tag{9}$$

where: E_T – total actual seismic energy in the forecast period, E_T^*– total seismic energy predicted.

The forecast for the total number of seismic events was made in a similar manner. In the above formulas, *Ei* was replaced by *Ni.*

3 RESULTS

Verification of the correctness of the adopted model and the forecast was performed for the previously mentioned three coal mines belonging to JSW S.A. The number of recorded high-energy events (with energies ranging from $1\text{x}10^5$ J to $1\text{x}10^7$ J) that induced mining operations to the depth of 1,200 m was analyzed.

3.1 *"Knurów–Szczygłowice" coal mine*

In the area of the "Knurów -Szczygłowice" coal mine, data covering the years 2011-2019 was analyzed. In this period, 119 high-energy phenomena were recorded with energies ranging from $1\text{x}10^5$ J to $1\text{x}10^7$ J, which induced ongoing mining operations up to a depth of 1,200 m.

Table 1. Results of estimation of the models‘ parameters in the area of the "Knurów -Szczygłowice" coal mine- Szczygłowice Section.

Coal mine, Section, Seam	Parameter	Estimated parameter values	Standard errors of the estimated parameters	Significance level	R	R^2
Szczygłowice Section E_i, H_i	Intercept *a*	−2.5284E8	3.5260E7	0.05	0.9426	0.8884
	Regression coefficient *b*	2.964E5	4.288E4	0.05		
Szczygłowice Section N_i, H_i	Intercept *a*	−341.393	53.096	0.05	0.9500	0.9025
	Regression coefficient *b*	0.403	0.05409	0.05		

The regression models were determined for the Szczygłowice section, assuming a class interval width for depth of 50 meters. The results of the parameter estimations for the models (E_i, N_i) are presented in Table 1.

Figure 2 shows the distribution of *Ei* in respect to *Hi* with a regression line and confidence curves in the Szczyłowice Section. Accordingly, Figure 3 shows the distribution of *Ni* in respect to *Hi* with a regression line and confidence curves in the Szczyłowice Section. The relevant distributions were also prepared for the remaining research areas.

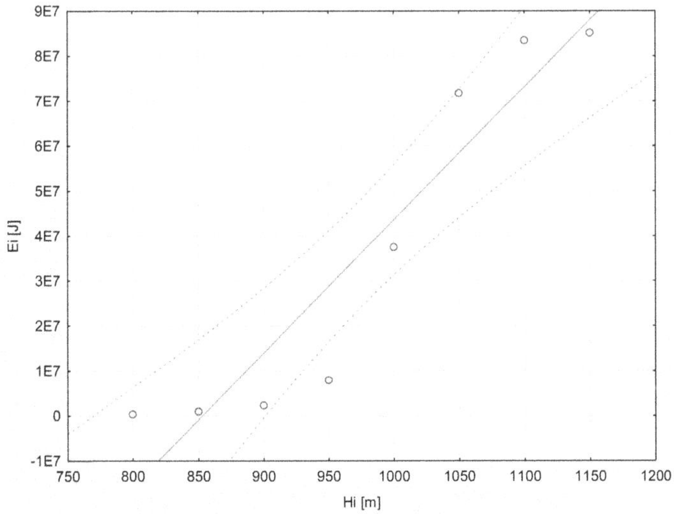

Figure 2. Cumulative seismic energy *Ei* induced in 2011–2019 as a function of *Hi* – regression line, confidence curves – Section Szczygłowice.

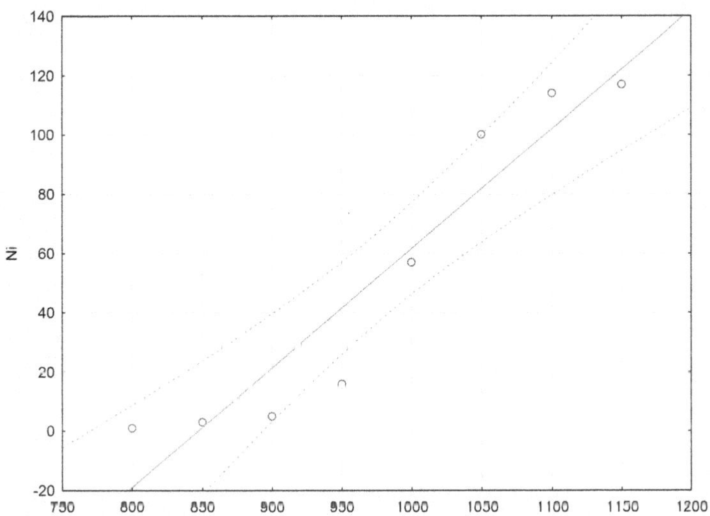

Figure 3. Cumulative number of seismic events N*i* induced in 2011–2019 as a function of *Hi* – regression line, confidence curves – Section Szczygłowice.

For the established regression relationships, it was found that the correlation coefficient, the regression coefficient, and the intercept are significantly different from zero at the assumed significance level.

- The determined regression model of E_i with respect to H_i explains over 88% (R^2 = 0.8884) of the variability in the cumulative seismic energy. The value of the random variability coefficient is 0.04, which indicates that the model is well-fitted to the empirical data. An analysis of the residuals confirmed the model's applicability for forecasting. The ex-ante forecast error did not exceed 10%. The forecast was made for the next step, i.e., 50 meters. The ex-post forecast error was 15%.
- The determined regression model of N_i with respect to H_i explains over 90% (R^2 = 0.9025) of the variability in the cumulative number of seismic events. However, the value of the random variability coefficient was 1.16, indicating that the model was not well-fitted to the empirical data. Therefore, no further analysis was conducted.

3.2 *"Borynia–Zofiówka" coal mine*

Research in the area of the "Borynia-Zofiówka" mine was conducted taking into account seismic activity recorded between 2010 and 2019. The analysis covered 173 high-energy seismic events with energies ranging from 1×10^5 J to 1×10^9 J, induced during this period by mining operations conducted at depths of up to 1,020 meters in the Zofiówka section. In accordance with the study assumptions (see the Section 2), seismic events with energies on the order of 10^8 J and 10^9 J (2 events) were excluded.

Regression models were determined assuming a class interval width of 20 meters for depth. Due to significant variability in mining and geological conditions (including significantly different degrees of exploitation of the deposit, significantly different depths of exploitation, and the division of the deposit into sections by faults), the analysis was conducted separately for seam 505/1 and seam 409/3. The results of the parameter estimations for the models (E_i, N_i) are presented in Table 2.

Table 2. Results of estimation of the models' parameters in the area of the "Borynia – Zofiówka" coal mine - the seam 505/1, 409/3.

Coal mine, Section, Seam	Parameter	Estimated parameter values	Standard errors of the estimated parameters	Significance level	R	R^2
Zofiówka Section: the seam 505/1 E_i, H_i	Intercept a	−1.5434E8	3.0185E7	0.05	0.9068	0.8222
	Regression coefficient b	1.8244E5	3.2064E4	0.05		
Zofiówka Section : the seam 505/1 N_i, H_i	Intercept a	−187.200	31.8036	0.05	0.9223	0.8507
	Regression coefficient b	0.2130	0.03378	0.05		
Zofiówka Section : the seam 409/3 E_i, H_i	Intercept a	−1.2609E8	3.229E7	0.05	0.8963	0.8033
	Regression coefficient b	1.6953E5	3.7515E4	0.05		
Zofiówka Section : the seam 409/3 N_i, H_i	Intercept a	−173.643	16.2118	0.05	0.9818	0.9639
	Regression coefficient b	0.218	0.0188	0.05		

For the established regression relationships, it was found that the correlation coefficient, the regression coefficient, and the intercept are significantly different from zero at the assumed level of significance.

For the seam 505/1:

- The determined regression model of E_i with respect to H_i explains over 82% (R^2 = 0.8222) of the variability in the cumulative seismic energy. The value of the random variability coefficient is 0.14, which indicates that the model is well-fitted to the empirical data. An analysis of the residual properties confirmed the applicability of the model for forecasting. The ex-ante forecast error did not exceed 10%. The forecast was made for the next prediction step, i.e., 20 meters. The ex-post forecast error was 11%.
- The determined regression model of N_i with respect to H_i explains over 85% (R^2 = 0.8507) of the variability in the cumulative number of seismic events. However, the value of the random variability coefficient was 1.34, indicating that the model was not well-fitted to the empirical data. Therefore, no further analysis was carried out.

For the seam 409/3:

- The determined regression model of E_i with respect to H_i explains over 80% (R^2 = 0.8033) of the variability in the cumulative seismic energy. The value of the random variability coefficient is 0.20, indicating that the model is well-fitted to the empirical data. An analysis of the residual properties confirmed the possibility of applying the model for forecasting. The ex-ante forecast error did not exceed 14%. The forecast was made for the next prediction step, i.e., 20 meters. The ex-post forecast error was 20%.
- The determined regression model of N_i with respect to H_i explains over 96% (R^2 = 0.9639) of the variability in the cumulative number of seismic events. However, the value of the random variability coefficient was 1.14, indicating that the model was not well-fitted to the empirical data. Therefore, no further analysis was conducted.

3.3 *"Pniówek" coal mine*

Between 2012 and 2019, 35 high-energy seismic events with energies ranging from 1×10^5 J to 1×10^7 J were recorded at the "Pniówek" mine, induced by mining operations conducted at depths of up to 1,050 meters. Regression models were determined assuming a class interval width for depth of 50 meters. The results of the parameter estimations for the models (*Ei, Ni*) are presented in Table 3.

Table 3. Results of estimation of the models‘ parameters in the area of the "Pniówek" coal mine.

Coal mine, Section, Seam	Parameter	Estimated parameter values	Standard errors of the estimated parameters	Significance level	*R*	R^2
Pniówek coal mine *Ei, Hi*	Intercept *a*	−2.8298E7	3.0326E7	0.05	0.9872	0.9746
	Regression coefficient *b*	4.0446E4	3.265E4	0.05		
Pniówek coal mine *Ni, Hi*	Intercept *a*	−76.20	8.5427	0.05	0.9847	0.9696
	Regression coefficient *b*	0.1040	0.0092	0.05		

For the established regression relationships, it was found that the correlation coefficient, the regression coefficient, and the intercept are significantly different from zero at the assumed level of significance.

- The determined regression model of E_i with respect to H_i explains over 97% (R^2 – 0.9746) of the variability in the cumulative seismic energy. The value of the random variability coefficient is 0.07, indicating that the model is well-fitted to the empirical data.

An analysis of the residual properties confirmed the applicability of the model for forecasting. The ex-ante forecast error did not exceed 7%. The forecast was made for the next prediction step, i.e., 50 meters. The ex-post forecast error was 10%.

- The determined regression model of N_i with respect to H_i explains over 96% (R^2 = 0.9696) of the variability in the cumulative number of seismic events. The value of the random variability coefficient is 0.09, which also indicates a good fit to the empirical data. The ex-ante forecast error did not exceed 11%. The forecast was made for the next prediction step, i.e., 50 meters. The ex-post forecast error was 7%.

4 CONCLUSIONS

The research results presented in this study on the possibility of forecasting changes in the seismic activity of the rock mass induced by ongoing mining operations in underground coal mines in relation to changes in mining depth demonstrated the following findings in the investigated areas:

- It is possible to apply a linear regression model with a single independent variable to determine acceptable forecasts of the cumulative seismic energy induced by mining operations in a given area based on changes in mining depth. The functional relationships determined for each area between E_i and H_i explained between 80% and 97% of the variability in the cumulative seismic energy. It was found that the correlation and regression coefficients were significantly different from zero at the assumed significance level. The proportion of variation in the variable E_i explained by the model, based on the random variability coefficient, ranged from 0.04 to 0.20, which indicates that, in general, the models were well-fitted to the empirical data. The analysis of the residuals showed no irregularities. The average ex-ante relative forecast error during the tested forecast periods was approximately 10%, and the average ex-post absolute error was 14%. Considering the above, it can be concluded that the determined forecasts are acceptable.
- The use of a linear regression model with one independent variable to forecast the cumulative number of seismic events (N_i) based on changes in mining depth (H_i) proved significantly less effective. The validation of the regression models for the number of seismic events in relation to mining depth showed that the functional relationships between N_i and H_i explained between 83% and 98% of the variability in the cumulative number of seismic events. The correlation and regression coefficients were significantly different from zero at the assumed level. However, the value of the random variability coefficient exceeded 1 in three out of the four models, which in principle eliminated the possibility of using the model to determine acceptable forecasts. Only in one of the examined areas ("Pniówek" coal mine) was the value of the random variability coefficient satisfactory, with ex-ante forecast errors not exceeding 11% and ex-post errors at 7%. The significantly less satisfactory results of analyses concerning the predicted number of seismic events may be related to the strong non-linearity of high-energy mode phenomena.

Based on the above, it can be concluded that it is possible to use a linear regression model with one independent variable to predict cumulative induced seismic energy based on mining depth. While such forecasts should be considered supplementary to existing forecasting methods, they may provide valuable insight, especially when planning deeper mining in the same area.

The results presented in this study support the rationale for further research on the relationship between mining depth and induced seismic activity. The obtained results may be applied during the implementation of measures aimed at reducing seismic hazards, including rock-bursts, and can support the improvement of rock-burst risk assessment methods.

REFERNCES

Białek J. (1999). Ocena stanu zagrożenia tąpaniami i wstrząsami górniczymi w oparciu o obliczone ekstremalne w czasie naprężenia górotworu. *Zeszyty Naukowe Politechniki Śląskiej*, s. Górnictwo z. 239, Gliwice, s. 19–34.

Biliński A. (1981). Metoda określenia stanu zagrożenia tąpaniami stropowymi na podstawie rozeznania warunków naturalnych i górniczych. *Przegląd Górniczy* nr 7–8, s. 415–423.

Descour J. (1979). Aktywność sejsmiczna i osiadanie powierzchni a mechanizm tąpań. *Cuprum* nr 2, s. 3–7.

Dittmann P. (2004). *Prognozowanie w przedsiębiorstwie*. Kraków. Oficyna Ekonomiczna..

Dorosiewicz S., Gruszczyński M., Kołatkowski D., Kruszewski T., Podgórska M. Syczewska E. (1996). *Ekonometria*. Warszawa, SGH.

Draper N.R., Smith H. (1973). *Analiza regresji stosowana*. PWN, Warszawa.

Drzewiecki, J. (2017). Prognozowanie zagrożeń sejsmicznych na podstawie analizy energetycznej aktywności sejsmicznej. *Prace Naukowe GIG*. Górnictwo i Środowisko, 2, 7–18.

Ge, S. (2022). Mechanisms and mitigation of induced seismicity. *Annual Review of Earth and Planetary Sciences*, 50, 31–56. https://doi.org/10.1146/annurev-earth-032320-083004

Gibowicz, S.J. (2009). Seismicity Induced by Mining: Recent Research. In H. Gupta (Ed.), *Advances in Geophysics* (Vol. 51, pp. 1–53). Elsevier. https://doi.org/10.1016/S0065-2687(09)05101-3

Gibowicz, S.J. (1963). Klasyfikacja energetyczna wstrząsów podziemnych na Górnym Śląsku i częstotliwość ich występowania w zależności od wielkości energii. *Archiwum Górnictwa*. t. VIII, z.1, Warszawa, 17–41.

Jaworski A. (2000). Poszukiwanie związków poziomu sejsmiczności z analitycznie obliczanym przyrostem energii odkształcenia sprężystego. *Acta Montana* IRSM AS CR Series A, No.16 (118), Praga, s. 85–98.

Kijko A., Dessokey M., Głowacka E. (1982). Wstępna ocena powtarzalności silnych wstrząsów górniczych w kopalniach LGOM. *Publ. Inst. Geophys. Pol. Acad. Sc.* M-5 (155), Warszawa.

Kijko A., Drzęźla B., Mendecki A. (1985) Dlaczego rozkłady ekstremalnych zjawisk w kopalniach mają charakter bimodalny? *Acta Motana*, nr 71.

Király-Proag, E., Wiemer, S., and Giardini, D. (2016). Induced Seismicity Test Bench: A Tool to Validate Induced Seismicity Models. *Journal of Geophysical Research: Solid Earth*, 121, 9035–9047. https://doi.org/10.1002/2016JB013236

Kłeczek Z. (2017). Tąpania w kopalniach węgla kamiennego oraz w kopalniach rud miedzi w Polsce. *Monografia* nr 19, Komag, Gliwice.

Kłeczek Z., Zorychta A. (1991). Geomechaniczne warunki powstawania wstrząsów górniczych. Materiały III Konferencji Naukowo-Technicznej. *Zastosowanie metod geofizycznych w górnictwie kopalin stałych*, Kraków, s. 175–191.

Lasocki S. (1992). Statystyczna ocena prawdopodobieństwa silnych wstrząsów górniczych w rejonie ścianowego frontu eksploatacyjnego. *Publ. Inst. Geophys. Pol. Acad. Sc.* M-5(155), Warszawa.

Lasocki S. (1996). Ocena i prognoza lokalnego ryzyka sejsmicznego poprzez analizę danych sejsmologicznych. *Biblioteka Szkoły Eksploatacji Podziemnej*, s. wykłady 12, s. 49–66.

Lenhardt, W.A. (1989). Seismic event characteristics in a deep-level mining environment. In: Maury, V., and Fourmaintraux, D. (Eds.), *Rock at Great Depth: Proceedings of the ISRM Symposium* (pp. 727–733). Rotterdam: A.A. Balkema.

Lenhardt, W.A. (1992). Seismicity associated with deep level mining. *Acta Montana*, Series A, No.2 (88), 179–192.

Lenhardt, W.A. (1992). Seismicity associated with deep-level mining at Western Deep Levels Limited. *Journal of the South African Institute of Mining and Metallurgy*, 92(S), 113–120.

Lenhardt, W.A. (2009). The impact of earthquakes on mining operations. *Bergund Hüttenmännische Monatshefte*, 154(6), 249–254. https://doi.org/10.1007/s00501-009-0001-2

Lenhardt, W.A. (2020). Seismic events associated with large mining-induced fractures. In: Hudson, J.A. (Ed.), *Mechanics of Jointed and Faulted Rock* (pp. 687–694). London: Taylor & Francis. https://doi.org/10.1201/9781003078975-100

Li, G., Zhu, J., Zhang, Z., Liu, W., Liu, G., and Wu, H. (2024). Study on dynamic hazard risk in deep mining based on microseismic monitoring and numerical simulation. *Journal of Low Frequency Noise, Vibration and Active Control*, 43(1), 322–335. https://doi.org/10.1177/01445987241230593

Makówka J., Kabiesz J. (2005). Prognoza miejsc i energii wstrząsów sejsmicznych- metoda i wyniki. *Materiały XII Międzynarodowej Konferencji Naukowo-Technicznej*, Górnicze Zagrożenia Naturalne, GIG, Katowice 2005, s. 166–171.

Marcak H. (2010). Geomechaniczna interpretacja struktury zjawisk geofizycznych, w szczególności sejsmologicznych, gromadzonych dla oceny hazardu sejsmicznego. *Przegląd Górniczy* 6, s. 1–8.

Martinsson, J., and Törnman, W. (2020). Modelling the Dynamic Relationship Between Mining Induced Seismic Activity and Production Rates, Depth and Size of Mining Areas. *Pure and Applied Geophysics*, 177, 3883–3905. https://doi.org/10.1007/s00024-019-02378-y.

Mirek A. i in.(2022): Głębokość eksploatacji a poziom generowanej wydobyciem aktywności sejsmicznej w kopalniach węgla kamiennego w latach 2005–2020. *Przegląd Górniczy T.* 78, nr 1, strony 26–36

Nowak E. (2006). *Zarys metod ekonometrii.* Warszawa, PWN.

Praca zbiorowa pod red. Kabiesza J. (2017). Raport roczny 2017 o stanie zagrożeń naturalnych i technicznych w górnictwie węgla kamiennego. GIG, Katowice.

Sokoła-Szewioła V. (1996). Rozkłady energii i ilości wstrząsów w wybranych ścianach KWK „Katowice". *Zeszyty Naukowe Politechniki Śląskiej*, s. Górnictwo, z. 232, Gliwice, s. 253–264.

Sokoła-Szewioła V. (2008). Models of dependences between induced seismic activity and observed deformations of surface in the area of the conducted longwall exploitation. *Gospodarka Surowcami Mineralnymi*, t. 24, z. 31, Kraków, s. 159–172.

Sokoła-Szewioła V. (2019). BK nr 06/050/BK_19/0072. Analiza aktywności sejsmicznej indukowanej eksploatacją złóż węgla kamiennego w aspekcie prowadzenia wydobycia na dużych głębokościach.

Wu, T., Zhang, C., Li, Y., *et al.* (2024). Real-time intelligent location of induced seismic events using deep learning with microseismic networks in coal mines. *Tunnelling and Underground Space Technology*, 143, 105367. https://doi.org/10.1016/j.tust.2023.105367.

WUG (2017). Ocena stanu bezpieczeństwa pracy, ratownictwa górniczego oraz bezpieczeństwa powszechnego w związku z działalnością górniczo-geologiczną w roku 2016. Katowice, Wyższy Urząd Górniczy. https://www.wug.gov.pl/bhp/stan_bhp_w_gornictwie.

WUG (2018). Ocena stanu bezpieczeństwa pracy, ratownictwa górniczego oraz bezpieczeństwa powszechnego w związku z działalnością górniczo-geologiczną w roku 2017. Katowice, Wyższy Urząd Górniczy. https://www.wug.gov.pl/bhp/stan_bhp_w_gornictwie.

WUG (2019) Ocena stanu bezpieczeństwa pracy, ratownictwa górniczego oraz bezpieczeństwa powszechnego w związku z działalnością górniczo-geologiczną w roku 2018. Katowice, Wyższy Urząd Górniczy. https://www.wug.gov.pl/bhp/stan_bhp_w_gornictwie.

WUG (2019). http://www.wug.gov.pl/bhp/zdarzenia. Accessed on July 15, 2025.

WUG (2023). Ocena stanu bezpieczeństwa pracy, ratownictwa górniczego oraz bezpieczeństwa powszechnego w związku z działalnością górniczo-geologiczną w 2023 roku. https://www.wug.gov.pl/bhp/stan_bhp_w_gornictwie. Accessed on July 15, 2025.

WUG (2024). Ocena stanu bezpieczeństwa pracy, ratownictwa górniczego oraz bezpieczeństwa powszechnego w związku z działalnością górniczo-geologiczną w 2024 roku, https://www.wug.gov.pl/bhp/stan_bhp_w_gornictwie. Accessed on July 15, 2025.

Zhou, K., Zhao, Y., Deng, Q., He, M., and Song, D. (2024). Determination of the Advanced Mining Influence Range Using Microseismic Monitoring in Coal Mines. *Applied Sciences*, 14(17), 7737. https://doi.org/10.3390/app14177737

Mine Surveying and Mining Area Protection – Sokoła-Szewioła and Jarczyk (Eds)
© 2025 The Author(s), ISBN: 978-1-041-10531-2

Towards the use of low-cost sensors in automated systems for monitoring ground surface vibrations

L. Stolecki & K. Fuławka
KGHM Cuprum Research and Development Centre, Wroclaw, Poland

ABSTRACT: This article presents results from the grant-funded projects ASMOW and IPMOS conducted at KGHM Cuprum. The objective of ASMOW was to develop an information platform that reports local industrial impacts on the environment. As part of the research, we built an integrated measurement system comprising geodetic stations and a seismic network, enabling the estimation and visualization of ground-vibration parameters generated by high-energy seismic events (magnitude $M > 3.0$). In the IPMOS project, research focused on the feasibility of employing low-cost sensors to monitor mining-induced effects at the ground surface. Such low-cost receivers constitute a novel approach to surface monitoring and, prior to wide deployment, require thorough assessment of their operational performance and measurement accuracy. The validation campaign covered both time-domain and frequency-domain analyses. The results confirm the suitability of the proposed solution for seismic monitoring of the ground surface.

1 INTRODUCTION

Monitoring the impacts of underground mining on the land surface has long been a research priority due to risks to infrastructure and public safety (Cieślik *et al.* 2017, Fuławka *et al.* 2020). Damage can arise from dynamic loads induced by mining-related seismicity (Tatara 2012, Zembaty *et al.* 2015), typically assessed using surface ground-vibration measurements (Mutke and Siata 2004, Stolecki 2011). Growing environmental awareness reinforces the need for transparent, credible communication to local communities to mitigate conflict.

This need is reflected in research efforts, including large-scale infrastructures such as the European Plate Observing System. At the project scale, ASMOW—Automatic System for Monitoring the Impacts of High-Energy Seismic Tremors on the Land Surface Using GNSS/PSInSAR Satellite Observations and Seismic Measurements—delivered an automated system for visualizing surface effects of high-energy mining tremors (Paziewski *et al.* 2021, Stolecki *et al.* 2021, Włodarczyk *et al.* 2021). Building on these results, IPMOS—Innovative Precise Monitoring System Based on the Integration of Low-Cost GNSS and MEMS IMU Sensors—developed a deformation- and dynamics-monitoring solution employing low-cost GNSS (Global Navigation Satellite System) and IMU (Inertial Measurement Unit) sensors (Paziewski *et al.* 2023). Low-cost instrumentation enables denser monitoring networks at reduced cost, improving accuracy and reliability of results (Stolecki *et al.* 2024).

In this paper, we present solutions for automating acquisition, estimation, and visualization of ground-acceleration amplitudes and evaluate the feasibility of low-cost sensors for seismic monitoring of the land surface.

DOI: 10.1201/9781003655558-23

2 AUTOMATING THE ANALYSIS AND VISUALIZATION OF GROUND VIBRATIONS

The ASMOW project developed an integrated system to visualize surface impacts of high-energy mining tremors by integrating geodetic platforms, seismic instrumentation, and IT (Information Technology) tools. The solution automates event detection and impact quantification, with outputs delivered in a GIS (Geographic Information System) to provide a comprehensive picture of mining-induced surface effects. The architecture comprises three cooperating layers: (1) field measurement devices with peripherals, (2) database and computation, and (3) a Web-GIS presentation portal.

Field investigations addressed impacts from underground copper-ore extraction in south-western Poland (LGOM, Polkowice area). A surface seismic network of ten stations (Figure 1) was deployed above or near mining areas to capture vibrations and deformation.

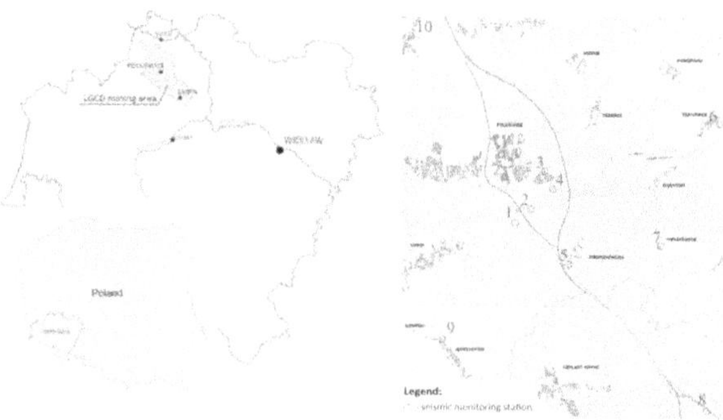

Figure 1. Seismic monitoring station locations (Stolecki *et al.* 2021).

Stations used short-period accelerometers (REFTEK 130-SMHR) with a ±4 g measurement range and 138 dB dynamic range. Remote access to monitoring units was provided via an industrial LTE (Long Term Evolution) link, while data acquisition from accelerographs was handled by an RTPD server over UDP/IP (Figure 2), enabling near-real-time streaming and processing. Dynamic impact levels were characterized using the horizontal peak ground acceleration (PGA_{H10}), determined in accordance with the GSI-2004/18 scale (Stolecki and Jaśkiewicz-Proć 2021).

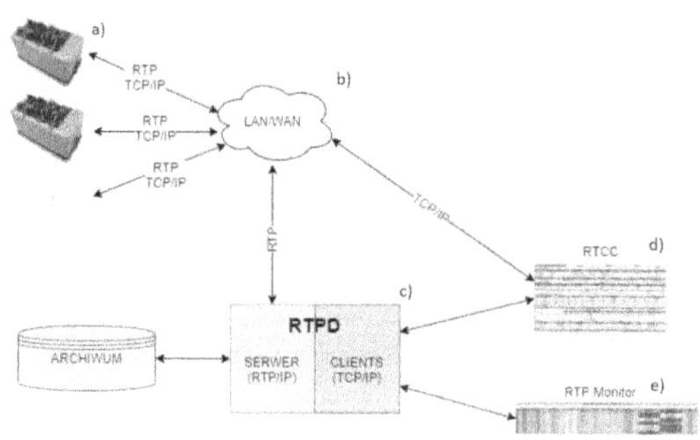

Figure 2. Network operation diagram (a- accelerometer, b – internet network, c – data exchange server, d - command and control program, e - network monitor program (Stolecki *et al.* 2021).

Surface interpolation of ground-motion parameters commonly relies on ground-motion prediction equations (GMPEs), which estimate amplitudes from source size (magnitude, seismic energy), distance, and near-surface geology. To ensure generality, we adopt a region-agnostic approach. GMPEs are typically calibrated for specific areas with local geology (Lasocki 2013, Chodacki 2016, Kozłowska *et al.* 2016) or for broader regions (Adams 2013, Campbell and Bozorgnia 2017, Goulet *et al.* 2021), and thus involve substantial averaging.

In contrast, the ASMOW model uses a two-dimensional concentric Gaussian. Using PGA_{H10} from 10 seismic stations, we interpolate acceleration isopleths across the observation area. To first order, the accelerations are described by a two-dimensional concentric Gaussian:

$$f(r) = a[\exp(-r/b)(1 + r/b)] \tag{1}$$

where r is the distance from epicentre (x_0, y_0); (x,y) location of the seismic station and a, b the Gaussian model parameters:

$$r = r(x, y) = \left[(x - x_0)^2 + (y - y_0)^2\right]^{1/2} \tag{2}$$

The Gaussian model parameters a, b, x_0, y_0 are estimated via least squares by fitting to the PGA_{H10} values recorded at the ten seismic stations. An example of the fitted Gaussian surface is shown in Figure 3-left.

The fitted Gaussian model $f(x,y)$ is treated as a first-order approximation to the target surface distribution of ground-motion amplitudes PGA_{H10}:

$$PGA_{H10}(x, y) = f(x, y) + \varepsilon(x, y) \tag{3}$$

where the residual ε(x, y) is represented by the deviations $\varepsilon_i = PGA_{H10i} - f_i$ at the measurement points i = 1,2, ...10

The residual field ε(x, y) can be modelled using standard surface-interpolation methods such as kriging, minimum-curvature surfaces, or inverse distance weighting (IDW) (Osada 2014). We employ IDW, which exactly honors the observations at the measurement points and yields a near-flat distribution between stations.

$$\varepsilon(x, y) = \sum_{i=1}^{10} (\varepsilon_i / d_i^n) / (1 / d_i^n) \tag{4}$$

where ε_i are the residuals at the measurement points and d_i is the distance from the interpolated location (x, y) to the i-th measurement point.

$$d_i = \left[(x - x_i)^2 + (y - y_i)^2\right]^{1/2} \tag{5}$$

Based on the analytical model PGA_{H10}(x, y), values of PGA_{H10} are computed at the nodes of a regular 100 m grid covering the monitored area encompassing the ten seismic stations. Example acceleration isopleth (contour) maps are shown in Figure 3-right.

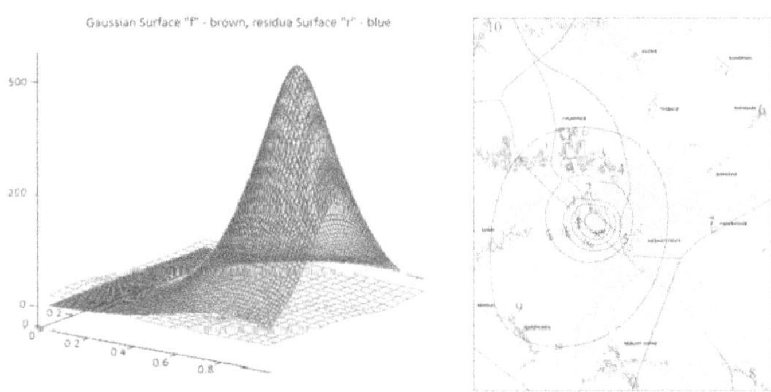

Figure 3. Gaussian surface fitted to the measured PGA_{H10} data and the contour (isopleth) distribution computed for the tremor of 22 July 2020 with energy 1.1×10^7 (M = 3.0) (Stolecki *et al.* 2021).

The principal advantage of the ASMOW system is its continuous, automated operation and web-based, near-real-time visualization of surface changes induced by seismic tremors. The workflow automates the analysis and interpolation of recorded ground-acceleration amplitudes. For compact monitoring domains (several km^2), the approach provides high accuracy and a faithful representation of dynamic impacts. Scaling to mine-wide areas, however, requires densifying the station network. Because conventional seismometers/accelerometers are costly, network resolution is constrained; a practical remedy is the use of next-generation Micro-Electro-Mechanical Systems (MEMS) accelerometers (Li 2021). The feasibility of such low-cost instrumentation for seismic monitoring was examined in IPMOS — *Innovative Precise Monitoring System based on the integration of low-cost GNSS and MEMS IMU sensors.*

3 LOW-COST SENSOR SOLUTIONS FOR VIBRATION MEASUREMENTS

The IPMOS project aimed to develop an innovative, nationally scalable system for precise monitoring of surface deformation and dynamic effects using low-cost GNSS and IMU sensors. To assess feasibility for a prototype monitoring station, we tested GNSS chipsets (Septentrio Mosaic-X5, u-blox ZED-F9P, SkyTrack PX1122), GNSS antennas (u-blox ANN-MB1-00, Tallysman TW7972, ELT0149), and inertial systems (ADIS 16470, ADXL 354BZ, TDK-DK 2352). Instrument characteristics were benchmarked against reference devices: a Trimble Alloy receiver with a Trimble choke-ring antenna and a Syscom Instruments MR3003C accelerograph.

Based on the evaluation, the Septentrio Mosaic-X5 receiver and ADIS 16470 IMU were selected for further work. Using these sensors, we built a prototype integrated low-cost station (Figure 4) and software modules for high-rate GNSS/IMU data fusion and continuous long-term processing. Nineteen prototype units were manufactured and deployed in a designated test area (Figure 5), with a measurement and control network installed to validate system performance under real field conditions (Figure 6).

Figure 4. Integrated low-cost monitoring station incorporating GNSS (Septentrio Mosaic-X5) - left and IMU (ADIS 16470) sensors - right.

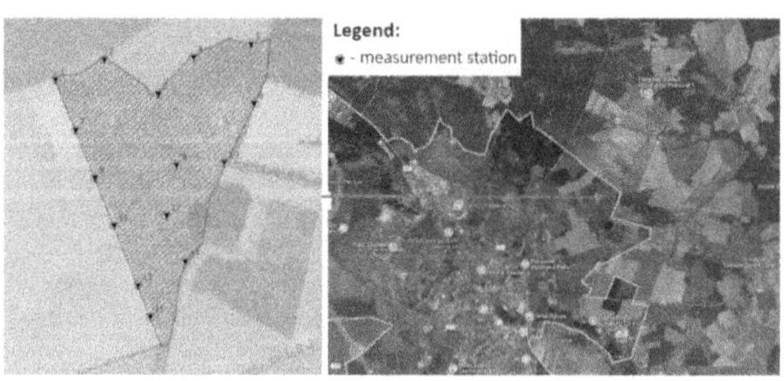

Figure 5. Location of the study area with the installed measurement and control network.

Figure 6. Photograph of the measurement station with the installed prototype monitoring system.

4 FEASIBILITY OF LOW-COST SENSORS FOR MONITORING MINING-INDUCED SEISMIC IMPACTS AT THE GROUND SURFACE

To evaluate the suitability of the low-cost MEMS-IMU ADIS 16470 for measuring vibrations generated by mining-induced seismicity, we verified its performance under real field conditions. Validation consisted of co-locating a reference accelerograph (SYSCOM MR3003C) and the prototype monitoring station. Analyses compared time-history records from both sensors and the peak particle acceleration (*PPA*). Additionally, frequency-domain comparisons were performed by transforming the time signals to the frequency domain via Fourier analysis, computing amplitude spectra for each component, and identifying dominant frequencies and corresponding amplitudes. The campaign spanned approximately three months (September–November 2023) and captured eight seismic events, summarized in Table 1.

Table 1. Summary of analyzed seismic events and computed correlation coefficients between signals recorded by the MEMS–IMU accelerometer and the classical reference accelerometer.

				Correlation coefficient for the full time history r_{xy}		
No.	Date	Time	M_L	X	Y	Z
1	19/09/2023	10:10:01	2.3	0.842800	0.735258	0.894078
2	22/09/2023	19:34:31	2.0	0.730907	0.633919	0.859396
3	28/09/2023	03:45:28	2.1	0.691446	0.624487	0.693733
4	12/10/2023	03:01:16	2.1	0.834049	0.679523	0.935465
5	13/10/2023	14:44:40	2.5	0.929384	0.804275	0.923567
6	17/10/2023	12:04:51	2.1	0.708869	0.684197	0.870140
7	17/11/2023	20:38:57	2.1	0.837859	0.919151	0.824089
8	18/11/2023	03:04:12	2.1	0.795800	0.933120	0.757327

4.1 *Time-history comparison*

Reliable characterization of the amplitude of surface-recorded vibrations is essential for assessing risks to surface infrastructure from mining-induced seismicity. We applied a correlation-matrix analysis to quantify the agreement between time-domain records from the MEMS–IMU accelerometer and the reference classical accelerometer. For each seismic event and all three orthogonal components, Pearson correlation coefficients were computed; the results are in Table 1.

The computed correlation coefficients range from 0.6245 to 0.9355 across events and components. While interpretation is discipline-specific, Hinkle *et al.* (2003) classify 0.60–0.79 as strong and 0.80–1.00 as very strong, which supports the utility of the low-cost MEMS-IMU (ADIS16470) for field measurements of mining-induced seismicity. Figure 7 presents a representative comparison of the seismic time histories recorded by both sensors for the 13/10/2023 event with energy 1.4×10^6 J (M = 2.5).

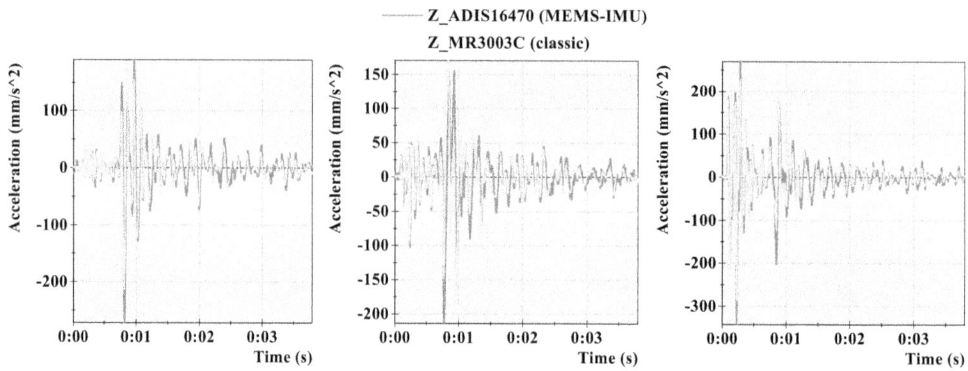

Figure 7. Comparison of seismic time histories recorded by the MEMS–IMU accelerometer (red) and the classical accelerometer (green) for the X (left), Y (middle), and Z (right) components.

4.2 *Frequency-domain comparison*

Frequency-domain analysis of acceleration records from mining-induced seismicity is critical for protecting surface engineering structures. Structural classes are particularly sensitive to excitation near their resonant frequencies; hence identifying dominant frequencies in the recorded signals informs impact assessment. Our analyses show close agreement in the dominant frequencies derived from both sensors. The maximum discrepancy was 18.3% (equivalent to 0.9 Hz). Figure 8 presents representative Fourier spectra for the event on 13/10/2023.

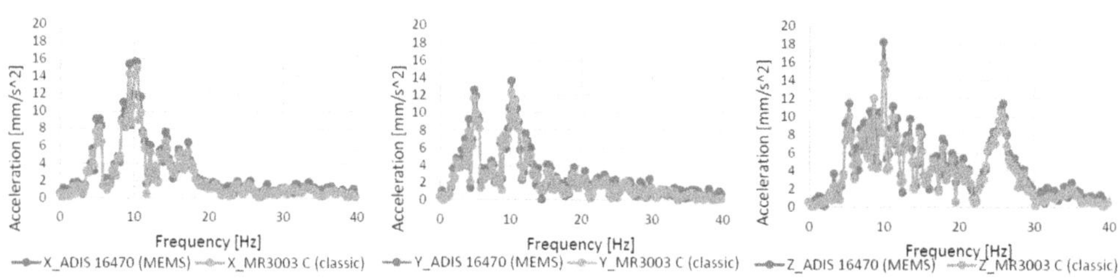

Figure 8. Comparison of frequency spectra obtained with the MEMS–IMU accelerometer (blue) and the classical accelerometer (orange) for the X (left), Y (middle), and Z (right) components.

4.3 *Noise-level comparison between acceleration recordings from the prototype monitoring station and the classical accelerometer*

Processing, analysis, and comparison of the seismic data were performed in NI DIAdem. NI DIAdem is an application for data processing, visualisation, and analysis that enables full task automation via Visual Basic and Python scripting (NI, 2024). To characterise the noise spectrum, the power spectral density (PSD) was computed, providing the distribution of signal power as a function of frequency.

Calculations followed the widely used Fourier-transform approach to seismic-noise quantification, which is now standard practice (Fuławka *et al.* 2022, Jana *et al.* 2017, McNamara *et al.* 2009). Based on the recorded datasets and comparison of the seismic-noise characteristics from both measurement systems, the MEMS-IMU ADIS16470 accelerometer exhibited substantially higher noise levels than the classical accelerometer, for both horizontal and vertical components (Figure 9). The noise also spanned a markedly broader frequency band. This indicates lower effectiveness of the ADIS16470 for microseismic observations relative to classical accelerometry. Elevated noise further complicates epicentral and hypocentral location due to reduced precision in picking P- and S-wave onsets at the station; however, these limitations primarily affect low- to moderate-energy events with distant sources.

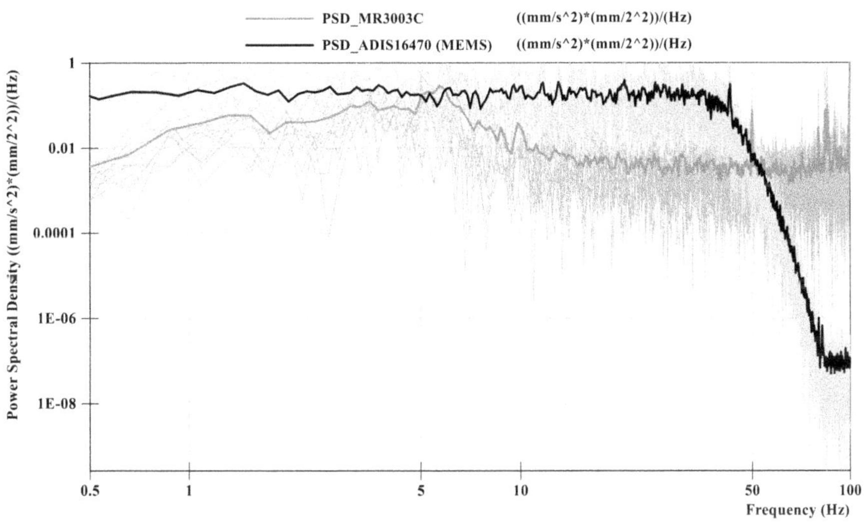

Figure 9. Comparison of PSD curves for noise recorded by the MEMS–IMU ADIS16470 accelerometer (black) and the classical accelerometer (red). (Stolecki *et al.* 2024).

5 CONCLUSIONS

Advances in information technology make it possible to develop automated systems for monitoring mining impacts at the ground surface; ASMOW is an example of such a solution. In addition to continuous and fully automated operation, a principal advantage of the system is its capability to provide real time visualisation via a web portal of surface changes induced by seismic tremors. A persistent limitation in scaling this class of systems is the cost of instrumentation deployed for surface measurements. This constraint can be mitigated by the use of low-cost sensors. Work on the feasibility of integrating low-cost sensors into monitoring systems was undertaken in the IPMOS project. Within this study, comparative analyses were performed to validate records obtained using a low-cost MEMS-IMU accelerometer. The results showed that, despite a slightly higher level of self-noise, MEMS-IMU accelerometers deliver highly reliable measurements, and the recorded vibration time histories correlate well with those obtained from a classical, calibrated instrument. Implementing such solutions in routine monitoring would not only significantly reduce the costs of constructing measurement networks, but would also provide the basis for increasing the spatial resolution of observations by expanding the coverage of the monitored area with measurement devices.

ACKNOWLEDGMENTS

This work was carried out under the projects ASMOW (POIR.04.01.04-00-0056/17) and IPMOS (POIR.01.01.01-00-0753/21-00).

REFERENCES

Adams J. (2013). Ground motion prediction equations for application to the 2015 Canadian national seismic hazard maps GM Atkinson, *Canadian Journal of Civil Engineering*, cdnsciencepub.com

Cieślik, J. Burtan, Z. Chlebowski, D. and Zorychta, A. (2017). Geomechanical analysis of location and conditions for mining-induced tremors in LGOM copper mines, *Journal of Sustainable Mining*, 16(3), 94–103. doi: 10.1016/j.jsm.2017.10.002

Campbell K.W. and Bozorgnia Y. (2014). NGA-West2 Ground Motion Model for the Average Horizontal Components of PGA, PGV, and 5% Damped Linear Acceleration Response Spectra, Earthquake Spectra, vol. 30, no. 3, pp. 1087–1115, https://doi.org/10.1193/062913EQS175M

Chodacki J. (2016). New ground motion prediction equation for peak ground velocity and duration of ground motion for mining tremors in Upper Silesia, *Acta Geophysica*, vol. 64, no. 6, pp. 2449–2470, DOI: 10.1515/acgeo-2016-0109

Fuławka K. Pytel W. and Pałac-Walko B. (2020). Near-Field Measurement of Six Degrees of Freedom Mining-Induced Tremors in Lower Silesian Copper Basin, *Sensors*, 20(23), 6801. doi: 10.3390/s20236801

Fuławka K. Mertuszka P. Szumny M. Stolecki L. and Szczerbiński K. (2022). *Application of MEMS-Based Accelerometers for Near-Field Monitoring of Blasting-Induced Seismicity*, *Minerals*, 12(5), 533. doi: 10.3390/min12050533

Goulet Ch. A. Bozorgnia Y. Kuehn N. Al Atik L. Youngs R.R. Graves R.W. and Atkinson G.M. (2021). NGA-East Ground-Motion Characterization model part I: Summary of products and model development, *Earthquake Spectra*, Vol 37, Issue 1_suppl, https://doi.org/10.1177/87552930211018723

Hinkle D.E. Wiersma W. and Jurs S.G. (2003). *Applied Statistics for the Behavioral Sciences*. 5th ed. Boston: Houghton Mifflin

Jana N. Singh C. Biswas R. Grewal N. and Singh A. (2017). Seismic noise analysis of broadband stations in the Eastern Ghat Mobile Belt of India using power spectral density, *Geomatics, Natural Hazards and Risk*, 8(2), 1622–1630. doi: 10.1080/19475705.2017.1365777

Kozłowska M. Orlecka-Sikora B. Rudziński L. Cielesta Mutke G. (2016). A typical evaluation of seismicity patterns resulting from the coupled natural, human-induced and coseismic stresses in a longwall coal mining environment, *Int. J. Rock Mech. Mining Sci.* 86, 5–15. DOI: 10.1016/j.ijrmms.2016.03.024

Lasocki S., Site specific prediction equations for peak acceleration of ground motion due to earthquakes induced by underground mining in Legnica-Głogów Copper District in Poland, *Acta Geophys.* 61, 5, 1130–1155, DOI: 10.2478/s11600-013-0139-8, 2013

Li Z. (2021). Recent advances in earthquake monitoring I: Ongoing revolution of seismic instrumentation, *Earthquake Science*, 34(2), 177–188. doi: 10.29382/eqs-2021-0011

McNamara D.E. Hutt C.R. Gee L.S. Benz H.M. and Buland R.P. (2009). A Method to Establish Seismic Noise Baselines for Automated Station Assessment, *Seismological Research Letters*, 80(4), 628–637. doi: 10.1785/gssrl.80.4.628

Mutke G. and Siata R. (2004). Ocena oddziaływania wstrząsów sejsmicznych na powierzchnię w rejonie LWB, Bełchatów, Warsztaty 2004 z cyklu, *Zagrożenia naturalne w górnictwie*, s. 409–418

NI. (2024) Available online: https://zone.ni.com (accessed on 10 december 2023).

Osada E. (2014). *Geodezyjne pomiary szczegółowe*. (Rozdz. 11. Numeryczny model terenu). UxLan Wrocław

Paziewski J., Tomaszewski D., Rapiński J. and Stolecki L. (2021). Wyznaczenie dynamicznych przemieszczeń punktów z wykorzystaniem autorskiego automatycznego modułu opracowania wysokoczęstotliwościowych obserwacji GNSS, *Czasopismo N-T Górnictwa Rud*, Cuprum", nr 3-4, s. 27–38

Paziewski J. Śmieja M. Rapiński J. Stępniak K. Sieradzki R. Wielgosz P. Duchnowski R. Wyszkowska P. Tomaszewski D. Janowski A. Dawidowicz K. Krzan G. Banach J. Srokosz P. Baryła R. Milanowska B.

Popielarczyk D. Zieliński T. Stolecki L. (2023). IPMOS: the system of precise displacement monitoring based on the integration of low-cost GNSS and accelerometer sensors. The 16th International Science and Technology Conference CURRENT PROBLEMS IN ENGINEERING SURVEYING, APGI and SCEgeo, *Engineering geodesy for the development of a sustainable industry* Poznan-Baranowo, wrzesień 21-23

Stolecki L. (2011). Badania rozkładów parametrów drgań generowanych wstrząsami górniczymi w kopalniach LGOM, Czasopismo N-T Górnictwa Rud „*Cuprum*, nr 3 (60), s. 29-37

Stolecki L. and Jaśkiewicz-Proć I. (2021). Skala GSI-2004/18 – skuteczne narzędzie oceny wpływu wstrząsów sejsmicznych na zabudowę kubaturową i na ludzi w LGOM CUPRUM, Czasopismo N-T Górnictwa Rud „*Cuprum*, nr 1-2, s. 5–19

Stolecki L. Osada E. Jaśkiewicz-Proć I. Kurpiński G. and Seta M. (2021). Metoda wyznaczania parametrów drgań generowanych aktywnością sejsmiczną towarzyszącą podziemnej eksploatacji złóż, *Czasopismo N-T Górnictwa Rud „Cuprum"*, nr 3-4, s. 17–26

Stolecki L. Fuławka K. Fruhwirt T. Schimmel M. (2024). IN-Field verification of the applicability of MEMS-IMU accelerometers for monitoring of mining-induced seismic activity, *Civil and Environmental Engineering Reports*, 34(4):563–581 DOI:10.59440/ceer/196753

Tatara T. (2012), Odporność dynamiczna obiektów budowlanych w warunkach wstrząsów górniczych. *Wydawnictwo Politechniki Krakowskiej*, Kraków

Włodarczyk K. Skrzypiec T. and Gumiela D. (2021). Automatyczny System Monitorowania Wpływów Wysokoenergetycznych Wstrząsów Parasejsmicznych na powierzchnię terenu – ASMoW, *Czasopismo N-T Górnictwa Rud „Cuprum"*, nr 3-4, s. 5–16

Zembaty Z. Kokot S. Bozzoni F. Scandella L. Lai C.G. Kuś J. and Bobra P. (2015). A system to mitigate deep mine tremor effects in the design of civil infrastructure, *Int. J. Rock Mining Sci.* 74, 81–90, DOI: 10.1016/j.ijrmms.2015.01.0004

Mine Surveying and Mining Area Protection – Sokoła-Szewioła and Jarczyk (Eds)
© 2025 The Author(s), ISBN: 978-1-041-10531-2

Applications and selection criteria of fly ash from hard coal fired power plants for post-mining activities in Upper Silesia Coal Basin

G. Strozik
Silesian University of Technology, Gliwice, Poland

ABSTRACT: Hard coal mines in Poland have been significant recipients and users of by-products and waste materials from coal combustion in the power sector, particularly fly ash. Following the decommissioning of coal mines, a range of technical and environmental activities are undertaken within the scope of post-mining management. Technologies employed for the reclamation of post-mining areas, spoil heaps, and industrial waste landfills, as well as for the neutralization of acidified soils and acidic groundwater, the stabilization of heavy metals, and the improvement of soil quality, either utilize fly ash or can benefit from its application. Fly ash derived from various combustion and flue gas desulfurization technologies exhibits differing levels of suitability for these purposes. Moreover, the specific conditions of individual technologies and the scale of environmental and technical challenges in post-mining areas influence the quantitative demand for fly ash. This paper presents a characterization of technologies used to address technical and environmental issues in post-mining areas where fly ash is applied, emphasizing the material properties required for its effective use. It also outlines the main differences in the characteristics of various types of fly ash and presents the results of laboratory tests on selected fly ash–water mixtures, confirming their suitability for specific post-mining applications.

1 INTRODUCTION

The Upper Silesia Coal Basin (USCB) is one of the largest and oldest industrial regions in Europe, developed around a rich hard coal deposit that has been intensively exploited continuously since the mid-19th century. At the beginning of the mining restructuring process, which beg (Kotarba 2021) an with the political and economic transformation in 1990, 65 mines were operating in the USCB, producing a total of approximately 140 million tonnes of hard coal annually. Currently, the number of active mining facilities has decreased to 18 (Frejowski 2021) (some of which are combined, forming a total of 12 mining enterprises), with their production in 2024 amounting to approximately 36 million tonnes (Kwidziński and Dusiło 2025, Przedlacki 2025). Over the same period, the share of coal (hard coal and lignite combined) in electricity generation declined from about 96% to 57% (Jeszke 2018a), with most of the decrease in coal's share in Poland's energy mix attributable to lignite. In 1990, hard coal accounted for approximately 56% of electricity generation and lignite for 40%, whereas in 2022 these shares were 44% and 26%, respectively (Jeszke 2018b).

To meet the needs of the heavy industry, which developed intensively in the USCB near hard coal mines, a significant number of coal-fired power plants and combined heat and power (CHP) plants were established. During the period when a substantial number of hard coal mines were closed, the number of operating hard coal-fired power plants and CHP plants in the USCB area did not change significantly. Currently, 13 coal-fired power and CHP plants are operating within the USCB, with a total installed capacity of 6240 MW. Including energy

DOI: 10.1201/9781003655558-24

facilities located outside the USCB but still in close proximity to hard coal mines, the total number of plants reaches 22, with a combined installed capacity of 11,173 MW.

According to current plans, all remaining hard coal mines are expected to be closed by 2040 (Ministry of State Assets 2020). At the same time, a complete phase-out of hard coal-based power generation is not anticipated during this period, and forecasts indicate that there may still be a demand for hard coal for large-scale power generation at a level of approximately 35 million tonnes per year (Chrzanowski *et al.* 2019, Ministry of Climate 2021).

These data are presented to demonstrate that, despite the closure of hard coal mines in the USCB, coal-fired power plants and CHP plants will continue to operate. Coal power plants produce significant amounts of solid materials, which can be classified either as waste or as combustion by-products (CBPs), depending on their intended use. These include primarily furnace slags, fly ash, and products of flue gas desulphurization, which may occur either mixed with fly ash (from dry and semi-dry FGD methods) or as synthetic gypsum from wet FGD methods.

The quantities of waste and combustion by-products generated in power and CHP plants depend on numerous factors, including plant efficiency, capacity, electricity production, combustion technology, fuel parameters, and others. For example, the Rybnik Power Plant, with a capacity of 1350 MW, annual electricity production of 7260 GWh in 2023, and an efficiency of 34.2%, generates approximately 624,000 tonnes of fly ash. In contrast, the Jaworzno III Power Plant, with a similar capacity (1354 MW) and an efficiency of 32%, produced 4,838 GWh of electricity in 2023 and 366,000 tonnes of fly ash.

In the 1990s, hard coal mines in the USCB began using fly ash (along with small amounts of other energy wastes) for sealing goafs as part of fire prevention measures and for the closure of workings in both active and decommissioned mines. The growing popularity of mining technologies utilizing fly ash coincided with the gradual decline in the number of operating mines and the volume of hard coal production. At its peak, between 2005 and 2015, the hard coal mining sector used between 4.0 and 4.5 million tonnes of fly ash annually.

It is noteworthy that mines also utilize fly ash with poorer physical and chemical properties, which cannot be applied as raw materials in other industrial sectors (e.g., in building materials production—ceramics, concrete, cement, binders, road construction, etc.).

The decline in the amount of fly ash utilized in hard coal mining, resulting from mine closures and decreasing coal production, encourages the search for alternative avenues for its management and utilization.

The closure of mining facilities is associated with the expansion of post-mining activities, including the reclamation of post-mining and post-industrial areas (degraded lands, waste dumps), the adaptation and development of infrastructure, the construction of new industrial facilities, and the introduction of new forms of land use. The scale of these activities can be substantial. The area of active mining sites and post-mining lands in the USCB is approximately 2000 km^2, compared to the total area of the region within Poland, which is 5400 km^2. The most urbanized part of the region is the Upper Silesian Metropolitan Area (GZM), comprising 41 cities and municipalities, with a total area of 2500 km^2 (Gasidło and Wyrzykowska 2022). The total area of post-industrial land within the GZM is around 53 km^2, of which approximately 18 km^2 consists of post-mining land. Much larger areas are affected by underground mining operations. In the first quarter-century of the 21st century alone, surface subsidence was recorded over a total area of 520 km^2 (Przyłucka *et al.* 2022). An equal area is occupied by lands vulnerable to surface deformations resulting from previous mining activities. Post-mining lands include, among others, 138 old industrial waste dumps (primarily mining waste).

It is therefore reasonable to ask whether, from the perspective of energy waste management (including CBPs), there are alternative ways of utilizing fly ash, aside from applications in underground mining technologies, offered by post-mining activities.

In the following sections of the article, selected technologies classified as post-mining activities, in which fly ash can be applied, are presented. Subsequently, the suitability of selected types of fly ash is evaluated based on an analysis of their physical and chemical properties and their mixtures with water.

2 POST-MINING TECHNOLOGIES UTILIZING FLY ASH

The scope of technical activities within post-mining operations is extensive, including those related to the utilization of fly ash. It should be noted that in many applications, fly ash is employed not as a substitute material but as a "first-choice" material due to its favorable physical and chemical properties (Smith *et al.* 2017, Kowalski and Nowak 2020). These technologies can be classified into five principal categories:

2.1 *Reclamation of spoil heaps and industrial waste disposal sites*

Fly ash can be applied to seal industrial waste bodies. In the case of coal-bearing mining waste, the use of fly ash provides protection against spontaneous ignition and combustion of mine waste, reduces the infiltration of precipitation water, and limits the leaching of chlorides and sulfates from the waste, as well as contributes to slope stabilization. Fly ash decreases the permeability of waste materials and enhances their compressive and shear strength (Narula *et al.* 2023).

In certain areas of the Upper Silesian Coal Basin (USCB), there are waste disposal sites containing metallurgical residues from zinc and lead ore mining, which exhibit elevated concentrations of lead, zinc, cadmium, copper, and other metals. Fly ash can be employed to immobilize heavy metals within these wastes, thereby reducing their mobility and mitigating associated environmental risks (Mandal *et al.* 2019).

In this way, fly ash helps stabilize the waste body, facilitates shaping the form of the landfill mass and its slopes, and supports the preparation of the surface for biological reclamation.

In the Upper Silesian Coal Basin (USCB), subgrade waste disposal sites are also present. These may consist of artificial or natural depressions (sink basins), or former open-pit excavations, filled with waste material. In such cases, fly ash can be used to impart the necessary geotechnical properties to the substrate for construction purposes (Zieliński and Mazur 2016).

2.2 *Filling of shallow underground voids and surface deformations caused by mining activities*

Fly ash–water mixtures are commonly used for filling shallow underground voids, sinkholes, fissures, and surface cracks, which frequently occur as a consequence of shallow hard coal mining. Such phenomena may sometimes manifest many years after the cessation of mining operations and pose significant hazards to the safety of people and surface structures. In areas of historical exploitation of shallow hard coal seams and zinc–lead ores, extensive systems of underground voids may be present, the remediation of which requires the use of thousands of cubic meters of fly ash. Very often, the presence of shallow underground voids manifests itself in the occurrence of zones of loosened soils with reduced bearing capacity. Fly ash can be used for their reinforcement and sealing. In such applications, fly ash–water mixtures serve as grouts for high-pressure or low-pressure injection. They can also be used as *flowable fill*, that is, controlled low-strength material (Dalal *et al.* 2019).

2.3 *Soils restoration and improvement*

Fly ash can be used for the remediation of soils containing contaminants and heavy metals. It helps restore the proper pH in acidic soils, improves groundwater retention in the soil, enhances soil texture and bulk density, and positively influences vegetation cover and land usability (Ciołek *et al.* 2021, Heidrich *et al.* 2013)

2.4 *Permeable reactive barriers for acid mine drainage (AMD) remediation*

Fly ash, due to its alkaline nature, can serve as a reactive medium in permeable reactive barriers (PRBs) to neutralize acidic mine drainage and immobilize dissolved contaminants, facilitating passive groundwater remediation on post-mining sites. The broader concept of PRBs is described as cost-effective in-situ barriers that allow contaminated groundwater to pass through reactive media which immobilize or transform pollutants. Moreover, literature on the beneficial reuse of coal combustion residues indicates that fly ash can neutralize AMD and act as a barrier to pyrite oxidation in abandoned mine workings and surface reclamation areas (Gupta and Chandra 2018).

2.5 *Remediation of abandoned underground mine workings*

Although post-mining activities generally occur after the closure of mines, in practice there is often a need to remediate remaining mine workings, particularly old shafts, adits, or other excavations that have direct contact with the ground surface. Such workings are primarily identified during the project planning phase. Such workings require remediation through tight filling, analogous to the backfilling of mine voids. The use of fly ash–water mixtures with favorable flow properties is advantageous in this context, as it allows the work to be conducted remotely, without the presence of personnel in areas at risk of roof collapse (Pavlik 2025).

3 KEY PROPERTIES OF FLY ASH RELEVANT TO POST-MINING APPLICATIONS

Fly ash, a by-product of coal combustion, possesses several key properties that make it suitable for various post-mining applications, including waste stabilization, soil remediation, and land reclamation. These properties include:

- Pozzolanic activity: Fly ash exhibits pozzolanic properties, meaning it can react with calcium hydroxide in the presence of water to form compounds possessing cementitious properties. This reactivity is influenced by the chemical composition and fineness of the fly ash particles. For instance, fly ashes with higher silica (SiO_2) and alumina (Al_2O_3) content demonstrate enhanced pozzolanic activity (Plewa *et al.* 2006),
- Particle size distribution: The particle size of fly ash affects its reactivity and suitability for various applications. Finer particles have a larger surface area, which can lead to higher reactivity. Studies have shown that fine fly ash particles, particularly those with a median diameter (d_{50}) of approximately 5 μm, exhibit increased pozzolanic reactivity (Plewa and Mysłek 2001, Schackow *et al.* 2025),
- Heavy metal content: Fly ash contains trace amounts of heavy metals such as lead (Pb), cadmium (Cd), zinc (Zn), and copper (Cu). While these metals can pose environmental risks, fly ash can also be utilized to immobilize these contaminants in soils and mining wastes, reducing their mobility and mitigating associated environmental risks (Chen *et al.* 2021; Majchrzak *et al.* 2022).

Physical and chemical properties of fly ash can vary widely depending on the combustion technology and flue gas desulphurisation method (FGD).

3.1 *Variability in fly ash properties based on combustion and FGD method*

The properties of fly ash are significantly influenced by the combustion process employed, with fluidized bed combustion (FBC) and pulverized coal combustion (PCC) producing ashes with distinct characteristics. Moreover, the desulfurization method employed, such as semi-dry (spray-dry) and dry (fluidized-bed combustion or dry sorbent injection) processes. critically affect the chemical composition, particle morphology, and reactivity of fly ash, which in turn determine its suitability for post-mining applications (Minh 2023).

Pulverized Coal Combustion (PCC) operates at high temperatures (1300–1700°C), producing predominantly spherical, glassy particles rich in silica (SiO_2) and alumina (Al_2O_3), which exhibit strong pozzolanic activity. This type of fly ash is highly suitable for applications requiring structural strength, such as soil stabilization and construction on reclaimed mining land.

Fluidized Bed Combustion operates at lower temperatures (800–900°C) and produces more irregular, often coarser particles. FBC fly ash tends to contain higher calcium content and unburned carbon, resulting in lower pozzolanic reactivity but enhanced lime-based stabilization potential. These characteristics make FBC fly ash particularly useful in neutralizing acidic soils or stabilizing heavy-metal-contaminated mining wastes (Rajczyk 2012).

In Polish coal-fired power plants, many boilers operate with PCC technology and employ dry or semi-dry flue gas desulphurization methods (PCC-SD). Presence of FGD products from these methods in the fly ash has a significant impact on the physical and chemical properties of the final combustion by-products. The main effects of the semi-dry desulphurization (spray-dry FGD) include (Popczyk 2016):

- Generating a partially hydrated sorbent mixed with ash resulted from injecting an alkaline slurry (typically lime or limestone) into flue gas
- Results in fly ash with increased calcium sulphate content, slightly higher moisture retention, and enhanced capacity to neutralize acidic soils or leachates,
- Particle morphology may include fine agglomerates, which improves packing density for filling voids or stabilizing waste bodies.

Dry Desulphurization (dry sorbent injection or FBC dry scrubbing – FBC-DD) involves injecting dry alkaline sorbents directly into flue gas that results in (Gawlicka and Wons 2023):

- Production of fly ash with lower moisture content, higher alkalinity, and higher proportion of unreacted lime,
- Enhanced potential for chemical immobilization of heavy metals and the stabilization of mine waste and spoil heaps (Styszko *et al.* 2018).

The combination of combustion type and desulfurization method influences key properties relevant to post-mining use as shown in Table 1.

After presenting the general conditions for the utilization of fly ashes from hard coal combustion in post-mining technologies, the results of laboratory tests on fly ashes and their mixtures with water, originating from power plants in the USCB area, are presented. These are the same types of fly ashes that are used in underground mining for filling of voids and workings, as well as grouting of goafs in longwalls.

Table 1. Key properties of fly ash from different combustion / FGD methods for post-mining applications.

Property / Process	PCC	FBC+DD	PCC + Dry FGD	PCC + semi-dry FGD	Relevance to Post-Mining Applications
Particle size	Fine	Coarse-medium	Fine	Fine-medium	Affects flowability, compaction, and sealing of voids
Pozzolanic reactivity	High	Low	Moderate	Low-moderate	Strength development in soil stabilization and backfill
Calcium content	High	Low-moderate	Moderate-high	High	Neutralization of acidic soils; heavy metal immobilization
Sulphur content	High	Low	Moderate-high	High	Influences handling, pumping, and placement in voids
Heavy metal immobilization potential	Moderate	Moderate-High	High	High	Reduces environmental risks in contaminated mining wastes

4 MATERIALS AND METHODS

For the assessment of their suitability for post-mining applications, four types of fly ash produced in power plants across the USCB were selected, originating from boilers employing different combustion and flue gas desulphurization technologies:

- Conventional fly ash (CFA) from PCC – without DFG – power plant „V",
- Fly ash with FDG by-products from PCC with semi-dry FDG method – power plant „X",
- Fly ash with FDG by-products from PCC with dry FDG method – power plant „Y",
- Fly ash from FBC boiler – power plant „Z"

Laboratory tests of fly ash and its mixtures with water were carried out in accordance with the standard PN-G 11011:1998, "Materials for solidified backfilling and sealing of goafs. Requirements and testing".

This standard includes a set of measurements that allow for a precise description of the setting process of ash–water mixtures and the mechanical properties of the resulting backfill. It is the only standard currently used in mining that relates specifically to the filling of underground voids. Therefore, it is appropriate for applications involving ash–water mixtures for filling various types of voids both within the rock mass and near the surface. The scope of testing defined by the standard includes measurements of viscosity, flowability and density of the mixture, the amount of excess water, setting time, compressive strength (up to 28 days of curing), and disintegration, i.e., the resistance of the solidified material to subsequent water action. The standard also specifies the maximum content of selected toxic contaminants in the aqueous eluate (Pierzyna 2022).

All measurement results represent average values obtained from tests on at least five specimens. Measurements requiring specimen conditioning (setting time, compressive strength) were performed on specimens stored under moist conditions at air humidity of 100% and temperature of 20 °C. Strength loss in water was measured for samples cured for 28 days under moist conditions and then immersed in water for 48 hours.

To assess the variability in properties of fly ashes originating from different boilers operating under identical combustion and flue gas desulfurization technologies, additional fly ash samples were collected from two other conventional pulverized coal combustion (PCC) boilers without FGD and from two fluidized bed boilers.

5 RESULTS AND DISCUSSION

All investigated types of fly ashes were subjected to chemical composition analysis (Table 2). Leachability tests of ions and heavy metals from water eluates were also performed (Table 3).

Table 2. Chemical characteristics of the fly ashes examined.

Compound	CFA	PCC-DD	PCC-SD	FBC-DD
SiO_2	51.68	50.69	39.82	39.4
Al_2O_3	29.06	18.80	20.18	20.27
Fe_2O_3	7.12	6.29	4.78	7.33
CaO	2.95	13.54	26.12	10.41
MgO	2.07	1.72	1.89	2.61
Na_2O	0.57	0.39	0.83	0.93
K_2O	3.00	2.22	2.19	1.93
SO_3	0.79	2.91	4.99	6.61
TiO_2	1.22	1.48	0.84	0.73
P_2O_5	0.45	0.66	0.36	0.26
BaO	0.21	0.18	0.12	0.16
LOI	2.19	4.71	5.64	8.89
Free CaO [% mass]	0.21	2.45	7.02	1.15

Table 3. Assessment of leachability of selected fly ashes [mg/dm^3].

Ion/element	Max. concentration (PN-G/11011)	CFA	PCC-SD	FBC-DD
SO_4^{2-}	500	404.2	1634.9	1569.3
S^{2-}	0,2	<0.2	<0.2	<0.02
Cl^-	1000	<3.5	187.8	71.19
CN^-	0.1	<0.025	<0.025	<0.025
Na^+	–	16.2	18.1	30.1
K^+	–	12.7	4.5	11.3
N_{NH4}	–	0.90	0.76	1.57
As	0.2	<0.05	<0.05	<0.05
Cd	0.1	<0.005	<0.005	<0.005
Cr	0.5	0.14	0.07	<0.02
Cu	0.5	<0.02	<0.02	<0.02
Fe	–	<0.02	0.03	0.11
Hg	0.02	<0.005	<0.005	<0.005
Mn	<0.004	<0.02	<0.02	<0.02
Ni	<0.004	<0.05	<0.05	<0.05
Pb	0.5	<0.05	<0.05	<0.05
Zn	–	0.04	<0.02	<0.02
pH	6.0 – 9.0	11.85	12.88	12.73
COD	100	<10	10.89	<10

The chemical composition of water extracts is an important indicator of the potential environmental applicability of fly ashes. According to the conducted tests, the maximum concentrations of selected ions and elements do not exceed the permissible values for substances introduced into groundwater.

Fly ashes from flue gas desulphurization processes are characterized by elevated sulfate anion (SO_4^{2-}) content. In general, fly ash eluates typically exhibit high pH values, in the range of 11–12, indicating the strong alkalinity of the fly ashes.

Fly ashes contain small amounts of heavy metals, which generally do not overshadow their beneficial properties as a material for stabilizing heavy metals in waste landfills or contaminated soils. However, the use of fly ashes to limit heavy metal migration must be preceded by specific analyses concerning the particular elements and toxic compounds involved.

According to the PN-EN ISO 14688 standard, 'Geotechnical investigation – Identification and classification of soils,' based on Eurocode 7, fly ashes contain a fine-grained fraction (clay and silt) and small amounts of coarse-grained fraction, mainly fine sand (Figure 1). Compared to PCC boiler fly ashes, those from fluidized bed boilers are characterized by

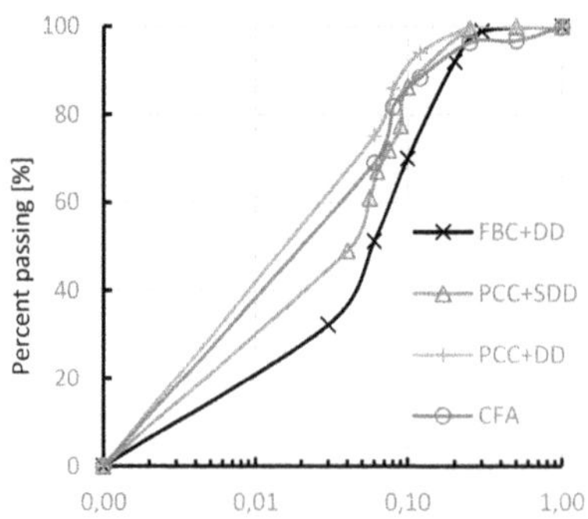

Figure 1. Particle size distribution curves of fly ash from different combustion processes and FGD metohodsto.

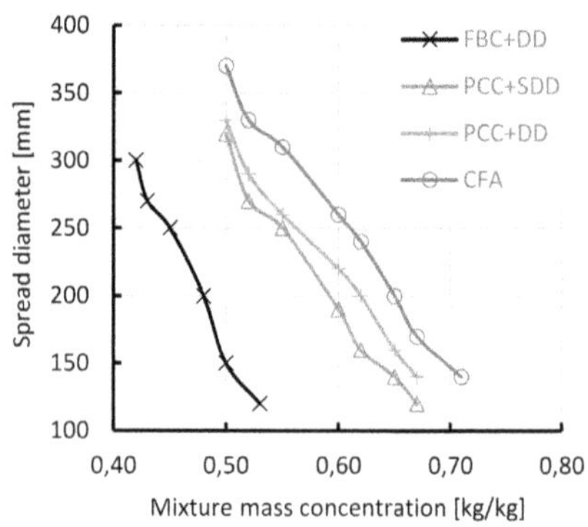

Figure 2. Spread diameter of mixtures prepared with different fly ashes versus mixture mass concentration.

slightly coarser particle sizes. The particle size distribution of fly ashes practically excludes their introduction into the environment in a dry state.

One of the main parameters determining the behaviour of ash–water mixtures is the concentration of solids in the mixture. Alternatively, the water-to-solids ratio, referred to as the W/S ratio, can be used (Figure 2). In most applications, the prepared ash–water mixture must be hydraulically transported to the place of use, where it undergoes setting and hardening. Table spread is a conventional parameter qualitatively describing the flowability of the mixture. As the mixture concentration increases, its density rises, leading to higher viscosity and a decrease in table spread (Figure 2). At low flowability, the mixture has limited ability to spread freely, which hampers hydraulic transport. On the other hand, mixtures with a relatively low water content achieve significantly better strength properties after setting and also exhibit much shorter setting times.

The curves of table spread versus mixture concentration shown in Figure 2 highlight two important issues. First, it should be noted that fly ashes from different sources exhibit highly variable physical and chemical properties, requiring different solids contents in the ash–water mixture to achieve similar flowability values. The second issue of practical significance is the steep relationship between concentration and table spread. This means that small changes in the mixture proportions can lead to significant deviations from the expected transport properties of the mixture.

It should be noted that table spread is only an indicative measure of the flowability of mixtures. Accurate rheological characteristics and predicted flow parameters in transport installations should be determined based on viscometric tests.

Figures 3 to 6 present a comparison of the variability of four basic parameters measured under laboratory conditions, depending on the type of fly ash, for two selected table spread values.

At a table spread value of 300 mm, the mixture can be easily transported through long pipelines. At a table spread of 200 mm, the mixture is noticeably thicker and may, for example, experience difficulties in penetrating small voids within the rock or soil. As shown by the presented graphs, in most cases mixtures with a table spread of 300 mm exhibit slightly inferior performance properties compared to mixtures with a table spread of 200 mm. An exception is the setting time for the mixture containing CFA (Figure 4), which is approximately 50% longer for the 300 mm table spread mixture than for the 200 mm one. This example demonstrates that selecting the appropriate proportions of mixture components can have a substantial impact on the mixture's properties.

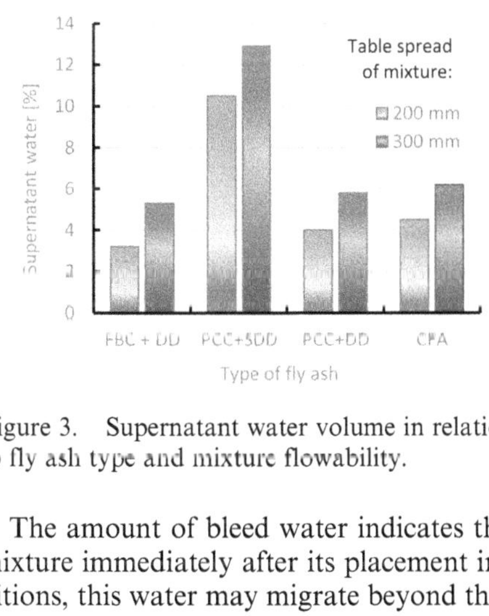

Figure 3. Supernatant water volume in relation to fly ash type and mixture flowability.

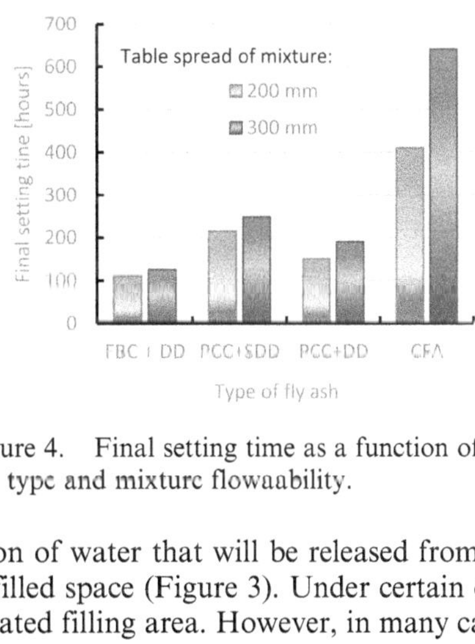

Figure 4. Final setting time as a function of fly ash type and mixture flowaability.

The amount of bleed water indicates the portion of water that will be released from the mixture immediately after its placement into the filled space (Figure 3). Under certain conditions, this water may migrate beyond the designated filling area. However, in many cases, if the bleed water remains in place, it will be absorbed during the setting process of the mixture. The outflow of bleed water can lead to a water deficit, resulting in reduced strength parameters of the hardened material.

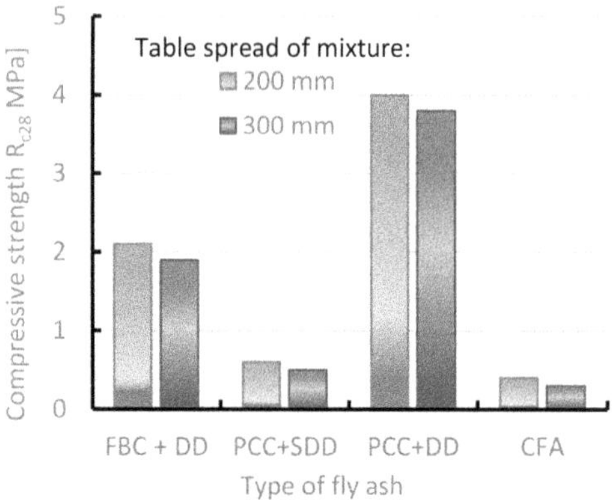

Figure 5. 28-Day unconfined compressive strength in relation to fly ash type and mixture flowability.

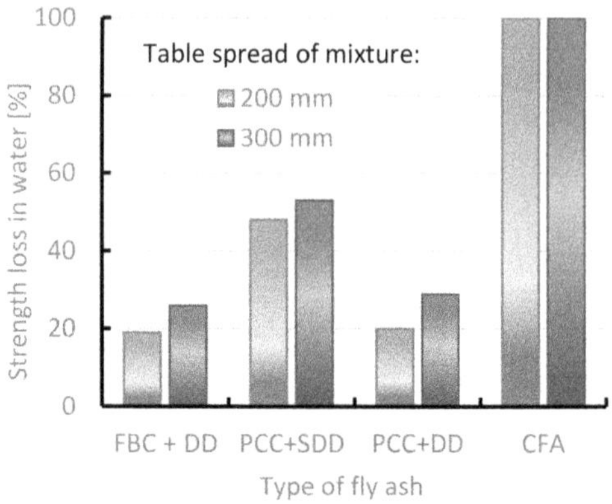

Figure 6. Relative compressive strength loss after 48 hours moist curing as function of mixture flowability and fly ash type.

Figure 5 illustrates the significant influence of fly ash type on the compressive strength of the hardened mixture. In cases where the utilization technology aims to achieve filling with relatively high strength properties, the best results are obtained using PCC boiler fly ash with products from dry flue gas desulfurization (Rc_{28} of approximately 4 MPa for a mixture with a table spread of 200 mm). Lower strength properties are exhibited by fly ashes from fluidized bed boilers with dry flue gas desulfurization (Rc_{28} of approximately 2 MPa). Mixtures prepared from fly ash without desulfurization products and with products from semi-dry desulfurization show very limited strength properties (Rc_{28} of approximately 0.5 MPa and 0.4 MPa, respectively).

This relationship is even more evident in the results of the strength loss in water test (Figure 6). For mixtures prepared from fly ashes with products of dry desulfurization (both PCC and FBC boilers), the reduction in strength after 48 hours of water immersion does not exceed 20% for mixtures with a table spread of 200 mm and 30% for those with a table spread of 300 mm. In contrast, the strength loss of samples made with PCC boiler fly ash containing semi-dry desulfurization products is approximately 50%, while samples of mixtures prepared with conventional fly ash underwent complete disintegration

As indicated by the presented relationships, for filling shallow underground voids or surface cavities, as well as for strengthening the subsoil and engineering structures (embankments, flood protection levees), it is advantageous to use fly ashes containing products of dry flue gas desulfurization. For creating impermeable sealing layers, particularly in environments where rock mass movements occur, it may be beneficial to use fly ashes whose mixtures with water retain a certain degree of plasticity (conventional fly ashes)

It should also be noted that, as shown by long-term studies, hardened fly ash–water mixtures exhibit an increase in strength parameters after several to over a dozen months of curing.

Finally, another set of setting time test results for fly ash–water mixtures is presented, comparing mixtures of fly ashes representing the same combustion technology and the same (or absence of) flue gas desulfurization method (Figure 7). As shown in the chart, mixtures prepared from fluidized bed boiler fly ashes practically complete setting before mixtures based on conventional fly ashes even begin to set. The setting process is also several times shorter for fluidized bed ashes, which is of considerable practical importance, as it allows for faster implementation of technological processes. The data in Figure 7 also indicate that fly ashes from a specific source may exhibit characteristic parameter values that deviate significantly from the average—as in the case of mixture CFA 3 compared to CFA 1 and CFA 2. Mixtures prepared from FBC 2 and FBC 3 ashes, in turn, set more slowly than the mixture with FBC 1 ash. It can also be observed that varying the table spread results in more regular and predictable changes in the physical properties of fly ash–water mixtures.

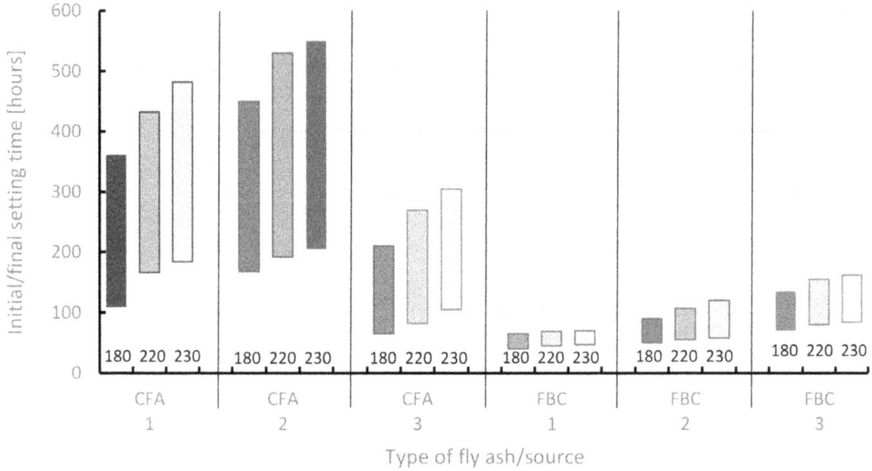

Figure 7. Initial and final setting times profiles of conventional (CFA) and fluidized bed (FBC) fly ashes from various power generation sources.

6 CONCLUSIONS

The variability of fly ash properties is thus a function of both combustion technology and flue gas desulfurization method. Understanding these factors is critical when selecting fly ash for post-mining applications, including:

- Stabilization of reclaimed soils and spoil heaps,
- Neutralization of acidic mine drainage,
- Filling shallow underground voids and depressions,
- Immobilization of heavy metals in mining waste.

By tailoring the choice of fly ash to the specific chemical and physical requirements of the post-mining site, environmental and engineering outcomes can be significantly improved.

Many types of fly ashes exhibit binding properties that allow them to harden without the addition of cement or other mineral binders, which beneficially reduces the cost of works involving their use. Fly ashes are also available in large quantities, which is important for post-mining applications, where filling underground voids can require thousands of cubic meters of mixtures

At present, conventional fly ashes without flue gas desulfurization products are increasingly rare on the market. Conversely, fly ashes from co-combustion of coal with biomass are becoming more common, exhibiting significantly more variable physical and chemical properties.

The variability of fly ash properties indicates the necessity of conducting detailed laboratory tests, taking into account the specifics of the given technology and using fly ashes from the particular, planned source of origin.

REFERENCES

Chen, W., Liu, H. and Wu, J. (2021). Characterization of fly ash from different combustion systems and its potential for mine reclamation. *Journal of Environmental Engineering* 147(3): 04021005

Chrzanowski Z., Baran B., Dudziak M. and Katzor R. (2019). Polska jako potencjalne źródło ubocznych produktów spalania (UPS) da rynków europejskich. XXVI Międzynarodowa Konferencja Popioły z energetyki.

Ciołek, M., Baran, A. and Antonkiewicz, J. (2021). Effect of coal combustion fly ash on soil properties and plant growth in degraded post-mining areas. *Environmental Science and Pollution Research*, 28(25): 31924–31937.

Dalal P.H., Patil M., Iyer K. and Dave T.N. (2023). Sustainable controlled low strength material from waste materials for infrastructure applications: State-of-the-art. *Journal of Environmental Management*, Vol. 342, 2023. DOI: 10.1016/j.jenvman.2023.118284

Frejowski A., Bondaruk J. and Dida A. (2021). Challenges and Opportunities for End-of-Life Coal Mine Sites: Black-to-Green Energy Approach. *Energies*, 10.3390/EN14051385.

Gasidło K. and Wyrzykowska A. (2022). Tereny pogórnicze w dokumentach strategicznych miast górnośląsko-zagłębiowskiej metropolii. *Builder*, vol. 305(12), 40–43.

Gawlicki M. and Wons W. (2023). Właściwości fizykochemiczne popiołów lotnych z kotłów fluidalnych i ich wpływ na wybrane cechy użytkowe mieszanek drogowych. *Prace Instytutu Ceramiki i Materiałów Budowlanych*, nr 12, 18–27

Gupta, A. and Chandra, R. (2018). Application of coal fly ash for neutralization of acid mine drainage and heavy metals stabilization. *Environmental Technology & Innovation* 10: 43–51

Heidrich, C., Feuerborn, H.J. and Weir, A. (2013). Coal combustion products: a global perspective. In F. Pinto (ed.), *Sustainable Use of Coal Combustion By-Products*: 3–30. Berlin: Springer.

Jeszke R. (red.) (2018). Klimat dla Polski – Polska dla Klimatu 1988–2018–2050. IOŚ PIB, Warszawa.

Jeszke R. (red.) (2018). Polska transformacja klimatyczno -energetyczna w liczbach IOŚ PIB, Warszawa.

Kotarba M. and Sechman H. (2021). Gas Hazard and Origin: Near-Surface Zone of the Upper and Lower Silesian Coal Basins. *NSG2021 27th European Meeting of Environmental and Engineering Geophysics.* 10.3997/2214-4609.202120008

Kowalski, Z. and Nowak, B. (2020). Utilization of coal combustion by-products in environmental technologies. In A. Maj and K. Wójcik (eds), *Sustainable Waste Management*: 103–117. Kraków: Politechnika Krakowska Press.

Kwidziński K. and Dusiło M. (2025), *Energy Transitom in Poland. Edition 2025*. Forum Energii, Warszawa

Majchrzak, M., Duda, P. and Szymański, T. (2022). Leaching behavior of fly ash in acidic environments: Implications for its use in reclamation. *International Journal of Coal Geology* 247: 103936

Mandal, S., Bhattacharya, A., Das, S. and Kumar, A. (2019). Toxic element leachability assessment from coal fly ash: Implications for environmental safety in mine land reclamation. *Science of the Total Environment* 651: 1357–1367

Minh V.T., Pham V.H., Tung V.H. and Phuong N.T. (2023). Firing-Associated Recycling of Coal Power Plant Fly Ash. *Journal od Analytical Methods in Chemistry*, Vol. 23(1). doi: 10.1155/2023/8597376.

Ministry of Climate, *Poland's Energy Policy until 2040*, Annex 2, "Findings from Prognostic Analyses for the Fuel-Energy Sector, 2021.

Ministry of State Assets, "Hard Coal Sector Restructuring Project", July 2020

Narula G., Sharma A. and Dogra V.K. (2023). Effect of Fly Ash on Geotechnical Properties of Soft Soil: A Critical Review. *Engineering Proceedings*, Vol 56(1). DOI: 10.3390/ASEC2023-16619.

Pavlik M., Gellendin M. and Rudolph T. Integrated Digital Twin and Geo-Fusion for Post-Mining Geomonitoring: Subsurface Modelling and Remote Sensing Application. *Civil and Environmental Engineering Reports*, 10.59440/ceer/205398

Pierzyna P.: Możliwość wykorzystania w górnictwie ubocznego produkty spalania pochodzącego z energetyki stosującej węgiel brunatny. *Rynek energii*, nr 2(159), 2022 s.54–60.

Plewa F. and Mysłek Z., (2001), Zagospodarowanie odpadów przemysłowych w podziemnych technologiach górniczych. Wydawnictwo Politechniki Śląskiej, Gliwice.

Plewa F., Popczyk M. and Mysłek Z. Możliwości wykorzystania w technologiach górniczych popiołów lotnych pochodzących z kotłów fluidalnych na przykładzie wybranych elektrowni. *Zeszyty Naukowe, Górnictwo, nr 1737*, Wydawnictwo Politechniki Śląskiej, Gliwice, 111–121.

Polski Komitet Normalizacyjny. (1998). PN-G 11011:1998 - Materiały do podsadzki zestalanej i doszczelniania zrobów. Wymagania I badania.

Popczyk M. (2016). Analiza właściwości fizyko-mechanicznych hydromieszanin wykonanych na bazie popiołów z kotłów fluidalnych pod kątem możliwości wykorzystania do likwidacji pustek i wyrobisk podziemnych, w *Górnictwo*, Gliwice: PA NOVA SA, 228–238.
Przedlacki W.: *Coal Brief, December 2024 and yearly summary*. www.energy.instrat.pl
Przyłucka M., Kowalski Z. and Perski Z. (2022). Twenty years of coal mining-induced subsidence in the Upper Silesia in Poland identified using InSAR. *Int J Coal Sci Technol.*, vol. 9(86). Doi: 10.1007/s40789-022-00541.
Rajczyk K. (2012). *Popioły lotne z kotłów fluidalnych i możliwości ich uszlachetniania*. Instytut Ceramiki i Materiałów Budowlanych, Opole.
Schackow A., Andrade Neto J., Hack L., Folgueras M.V., Kircchrim A.P., Martinez E.D.R. and de Matos P. R.: Coal bottom ash as supplementary cementitious material: Optimizing reactivity through particle grinding. *Sustainability*, Vol 17(11), doi: 10.3390/su17115031
Smith, R.D., Patel, S. and Howard, A. (2017). Pozzolanic materials in land restoration: A review. *Construction and Building Materials* 142: 50–64.
Styszko, A., Król, A., Drzymała, T. and Górecki, J. (2018). Leaching of heavy metals from coal combustion by-products and the potential use in mine reclamation. *Minerals* 8(10): 456–461.
Wang, J., Zhao, X., Li, H. and Shi, J. (2019). Influence of physical properties of fly ash on hydraulic backfill mixtures for mine reclamation. *Journal of Cleaner Production* 214: 740–749.
Zieliński, M. and Mazur, T. (2016). Environmental applications of fly ash in post-mining reclamation of hard coal spoil heaps. In T. Kowalczuk and M. Piekarski (eds), *Waste Management in Mining and Metallurgy*: 231–243. Wrocław: Instytut Górnictwa Politechniki Wrocławskiej.

Factors affecting structure damage in a post-mining area: Case study

P. Strzałkowski & L. Szojda
Silesian University of Technology, Gliwice, Poland

ABSTRACT: In the presented case, coal mining was carried out in the area surrounding the building in the 19th century at a depth of 30 meters below the surface and in the mid-20th century at a depth of 80 meters. For this reason, a potential sinkhole was analyzed using two methods, which concluded that under certain conditions, a sinkhole would not occur. The analyzed building is two-story and was constructed using traditional methods in two periods at the end of the 20th century, after mining operations ceased and surface deformations became apparent. In the first decade of the 21st century, damage occurred in the building. The tilt of the segments was slight, and the building structure remained intact. The damage was attributed to subsidence of the ground, which could have been caused by changes in soil and water conditions or the occurrence of the phenomenon of suffocation of soil material into the voids in the rock mass located under the building.

1 INTRODUCTION

The impact of underground mining on the surface has been the subject of many studies based on analyses of a large number of cases (Chudek 2010, Knothe 1984, Kowalski 2015, Strzałkowski 2010). This impact encompasses the atmosphere, hydrosphere, and lithosphere. The most burdensome impact on the lithosphere for mining communities is the impact on the lithosphere, manifesting itself in the occurrence of continuous and discontinuous deformations of the rock mass and rock mass tremors, which consequently lead to damage to structures on the surface (Szojda 2008, 2022, Szojda & Strzałkowski 2025). We are able to predict continuous deformations accompanying mining exploitation with sufficient accuracy for practical purposes (Chudek 2010, Knothe 1984, Kowalski 2015), which is largely possible thanks to professional computer software (Białek 2003, Ścigała 2008). The possibilities of forecasting discontinuous deformations are much smaller, both in terms of surface deformations (Chudek *et al.* 1988, Strzałkowski 2015, 2017) and discontinuous linear deformations (Orwat 2020, Ścigała & Szafulera 2019).

The impact of mining operations on the surface following mine closure includes some forms similar to those of active mines (Collective Work, ed. Kowalski 2000). For some time after mining operations cease, the process of subsidence continues (Blachowski *et al.* 2009, Cui *et al.* 2020, Strzałkowski 2021). A significant threat is posed by closed shafts due to the possibility of creating sinkholes (Kotyrba & Kortas 2016, Strzałkowski 2021) and mine gas emissions (Lagny 2014, Wrona *et al.* 2016).

In many of the mines operating in the Upper Silesian Coal Basin since the 19th century, mining operations were conducted at shallow depths, posing a serious risk of sinkholes. Voids left in the rock mass associated with mining operations generate sinkholes due to collapse or suffocation, which is difficult to predict. Damage to buildings in post-mining areas is often caused by these factors, although it should be noted that non-mining factors

DOI: 10.1201/9781003655558-25

also contribute to damage. This article highlights the above factors in relation to the analyzed case study.

2 ANALYSIS OF GEOLOGICAL AND MINING CONDITIONS IN THE ANALYZED AREA

2.1 *Lithology and stratigraphy*

Based on the drill holes, the rock mass in the area under consideration is composed of Quaternary overburden layers and productive Carboniferous rocks – Figure 1. The 17-meter-thick overburden is formed by a fill and layers of sand and clay. Beneath this are Carboniferous rocks in the form of 9-meter-thick sandstone, 1.3-meter-thick claystone, and 3-meter-thick hard coal – seam 207.

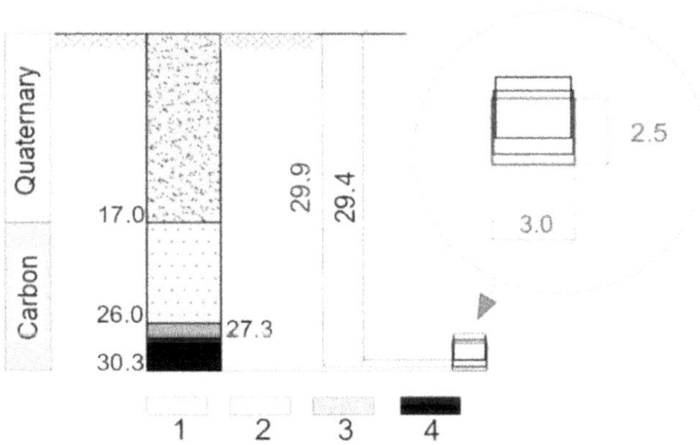

Figure 1. Geological structure of the rock mass in the considered region and voids: primary and secondary. 1 – overburden of loose Quaternary rocks, 2 – sandstone, 3 – claystone, 4 – coal.

2.2 *Tectonics*

The seam-coal maps of the property area do not show any tectonic disturbances in the form of faults.

2.3 *Completed mining exploitation*

Based on the geological and mining documentation held by the Archive of Survey and Geological Documentation of the State Mining Authority, it can be stated that the following mining works were carried out in the area of the property:

- Mining operations were conducted in seam 207, north of the building, between 1808 and 1852, with the roof caved in to a height of 4.2 m. The exploitation boundary practically adjoins the plot on which the analyzed building is located. Directly beneath the center of the property runs a gallery connecting the Aleksandra, Goepel, and Aleksander shafts. This extraction was most likely extracted in seam 207. The depth of the seam in the area of the property was approximately 30 m. A section of the map of seam 207 is shown in Figure 2.
- Then, mining was carried out using the longwall system in seam 208. The seam was mined in two layers with hydraulic backfill at a depth of approximately 80 m. The lower layer (mined first, hence layer 1) had a thickness of 2.3 ÷ 2.5 m. The seam was mined in this layer in 1955. The upper layer was mined in the years 1956 ÷ 1958 to a height of 2.1 ÷ 2.3 m.

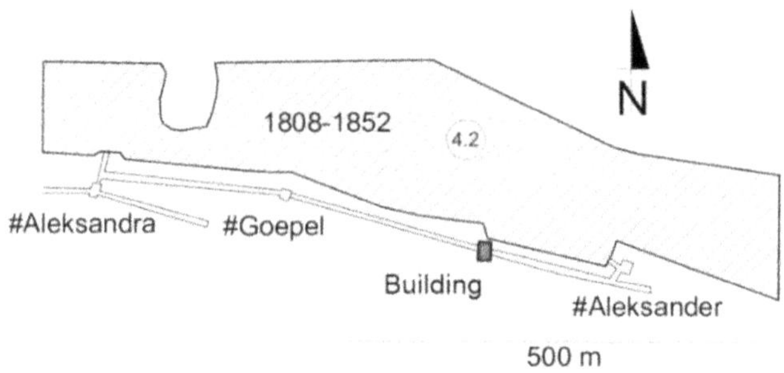

Figure 2. Section of the coal seam map 207.

3 ANALYSIS OF THE POSSIBILITY OF SINKHOLE CREATION

It was assumed that the risk of a sinkhole is associated with voids in seam 207. This assumption results from the depth of the seam and the fact that mining in seam 208 was carried out with hydraulic backfill. To analyze the possibility of a sinkhole occurrence, the Chudek-Olaszowski method (Chudek *et al.* 1988) and a method developed based on the pressure vault theory by A. Sałustowicz (Strzałkowski 2015) were used. The Chudek-Olaszowski method assumes the formation of caving and fracture zones above the void in each case – Figure 3. It was applied in a simplified version, limited to calculating the value of the Z index and the corresponding value of the probability of sinkhole formation – relationship (1).

$$Z = \frac{H - h}{g} \tag{1}$$

where: H =depth of the void ceiling, m; h =thickness of loose overburden, m; g =void height, m.

The probability of sinkhole formation can be calculated using the formula:

$$P(Z) = 0.00019 \cdot Z^2 - 0.036 \cdot Z + 1.34 \tag{2}$$

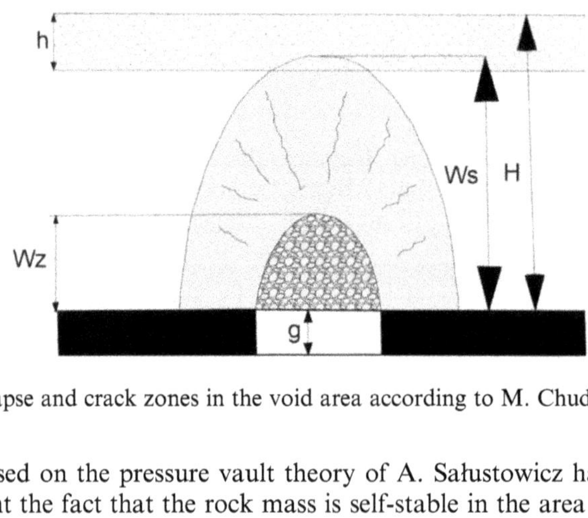

Figure 3. The collapse and crack zones in the void area according to M. Chudek – W. Olaszowski.

The method based on the pressure vault theory of A. Sałustowicz has the advantage of taking into account the fact that the rock mass is self-stable in the area of the void (extraction). The occurrence of a caving zone occurs when the stresses exceed the tensile strength of the rocks in the roof. The area of the caving zone (Figure 4) is expressed by the equation:

$$P_e = S_1 - \frac{gl}{2} + 2 \cdot S_2 = \frac{\pi a2b}{8} - \frac{wgl}{2} + \frac{2a2bk - 2gl}{9} = \frac{a2b(\pi + 2k) - 6gl}{8} \tag{3}$$

where: S_1 = surface area of the upper half of the fracture ellipse; a, b = ellipse axes; g, l = height and width of the extraction; k = arc measure of an angle α, therefore $k=\pi\alpha/180°$, $k = \frac{\pi}{180°} \cdot \arctan\frac{g}{l}$.

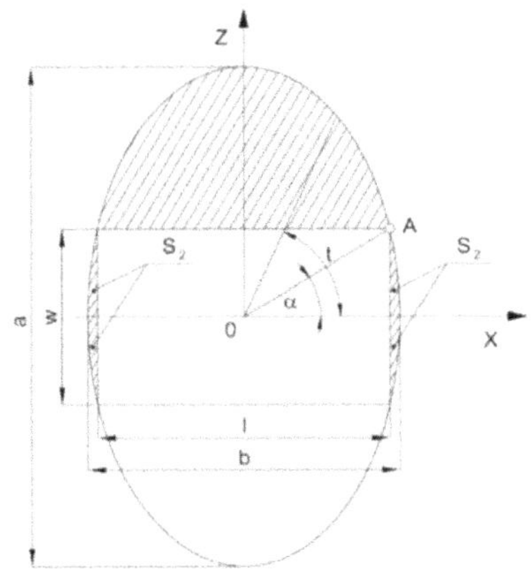

Figure 4. Relaxed zone around a void (extraction) in the rock mass (Strzałkowski 2015).

In the case of a passage through a cave-in, the rocks contained in the separate cavity are displaced by the void movement, filling the extraction. Let us denote:

$$P_1 = P_e \cdot k_r \tag{4}$$

$$P_2 = P_e + g \cdot l \tag{5}$$

where: k_r = rock loosening coefficient

If P_1=P_2, the void will self-propel, and the rocks contained in the destressed zone will completely fill the extraction (according to the Chudek-Olaszowski method). Then, there will no longer be a void in the rock mass. If $P_1 < P_2$, a secondary void will form in the upper part of the caving zone, with a volume resulting from the difference in areas P_2 - P_1. A cave-in occurs when the caving zone associated with the primary or secondary void reaches the loose overburden rocks.

3.1 *Calculation results according to the Chudek-Olaszowski method*

Using the method described above, it was decided to analyze the probability values of the sinkhole formation at different heights of the void related to the gallery in seam 207. The following assumptions were made:

- Depth of the void ceiling H = 27,3 m
- Thickness of loose overburden (including weathered) h = 17 m
- Void height g = 0,2 m – 2,0 m.

The calculation results are presented graphically in Figure 5.

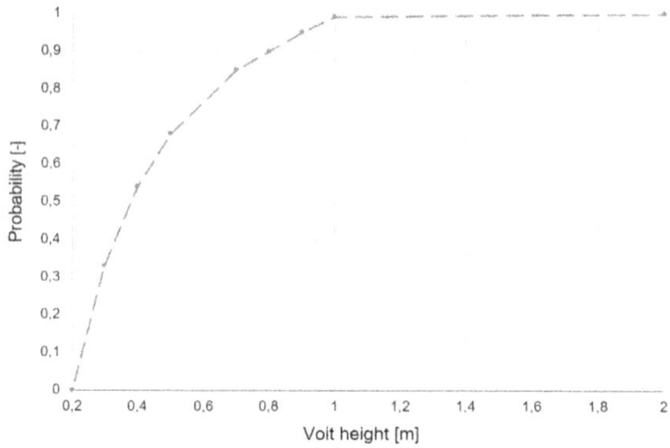

Figure 5. The relationship between the probability of sinkhole formation and the height of the void in seam 207 in the light of the calculation results according to M. Chudek – W. Olaszowski.

As can be seen from Figure 5, in the case of the void associated with the gallery, already at its height of 1 m, the probability of a sinkhole formation was 1.

3.2 *Calculation results using the pressure vault theory method*

In order to facilitate the calculations, a computer program was developed in C++ (author J. Strzałkowski, Strzałkowski 2017). Based on the lithological profile (Figure 1), the following data were adopted for calculations (Chudek 2010), summarized in Table 1. It was assumed that the bottom of the void occurs at a depth of H_{sp} = 30.3 m, the width of the void l = 2.5 m, and the height of the void g = 3 m. Based on the calculation results, the following can be stated:

There were two secondary voids:

- the first one with dimensions: 2.4 m × 2.9 m at the floor depth 29.9 m
- the second one with dimensions: 2.3 m × 2.7 m at a floor depth of 29.4 m

There was no stress-relieved zone above the second void due to the presence of sandstone at its roof, with an assumed strength of 3.5 MPa. Therefore, the rock mass within the void is stable. The locations of the secondary voids are shown in Figure 1.

Table 1. Summary of data for calculations.

Layer	Description	γ, MPa/m	R_r, MPa	k_r	h_i, m
1.	Quaternary	0.026	0.00	1.00	17.0
2.	sandstone	0.025	3.50	1.25	9.0
3.	claystone	0.025	0.02	1.20	1.3
4.	coal	0.015	0.40	1.20	3.0

It is worth adding that the results of gravimetric tests allowed us to determine the presence of a void in the rock mass in a place corresponding to the location of the gallery in seam 207.

4 CHARACTERIZATION OF THE OBJECT AND THE DAMAGES OCCURRING IN IT

4.1 *Characteristics of the structure of the facility*

The building consists of three separated sections: the entrance area with a staircase (A), the main building (B), and the garage, first floor, and partially usable attic (C). The ground floor plan is shown in Figure 6.

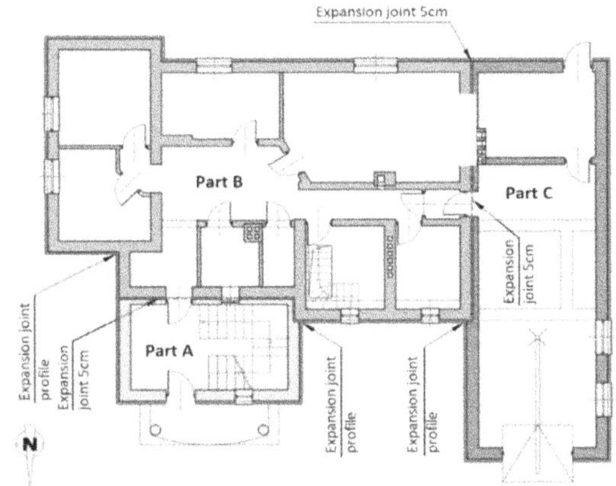

Figure 6. Ground floor plan of the building with marking of independent structural parts.

The oldest section of the building, part B, was built in the 1970s and is a two-story structure with a full basement and a multi-pitched roof. The structure features traditional concrete foundations; basement and ground floor walls are made of cinder blocks and perforated brick; first-floor walls are made of AAC and perforated brick; and partition walls are made of brick and AAC. The building's ceilings and inter-story stairs are made of reinforced concrete slabs. The roof structure is wooden, rafter-purlin, and covered with flat asbestos cement. This section has a partial basement.

The expansion of the building, including sections A and C, was carried out in the late 1990s. The enlargement included the construction of a new staircase outside the building on the north side and a garage on the west side. The structure of the expanded section consists of reinforced concrete foundations measuring 40 x 60 cm and 40 x 50 cm, the external walls are made of 29 cm thick MAX ceramic blocks and insulated with 10 cm of polystyrene, and the internal load-bearing walls are made of 25 cm thick solid brick. The inter-story ceilings and three-flight staircases are made of reinforced concrete slabs. Reinforced concrete lintels are installed above the window and door openings. The roof is a multi-pitched wooden structure, functionally connected to the previously constructed section. The roof is covered with metal roof tiles. After the expansion, the building's external dimensions are 16.9 x 14.2 m. Neither section has a basement.

The newly constructed sections were functionally connected to the existing section during the expansion, but a 5 cm expansion joint was left between them. According to the design, the foundations of both sections were constructed independently.

The facility was not protected against the impact of mining deformations in accordance with the decision of the relevant Director of the Mining Authority and the information that the facility is located outside the range of mining impacts.

4.2 *Damage to the facility*

During the damage inventory, two groups of damage were identified: scratches on the object caused by the independent operation of the old and new parts, and scratches on the object's finishing elements.

4.2.1 *Damage caused by independent operation of the structure*

Because the building structure consists of three separate volumes (Figure 6), the individual parts in the textured layer of the external plaster of the façade were secured with expansion joint profiles connected with flexible tape, which protects the joint from moisture and allows for movement of the joint. The location of the expansion joints, flexible profiles, and the deformation of these profiles in the expansion joints between volumes A and B, and B and C in the northern façade are shown in Figures 7-8.

Figure 7. Deformed expansion joint profile in the corner between parts A and B on the north side.

Figure 8. Deformed expansion joint profile in the corner between parts B and C on the north side.

On the eastern side of the building, the expansion joint profile does not align with the expansion joint between sections A and B (see Figure 6). No damage to the façade was found in this location.

However, damage related to the independent operation of the structure is visible at the functional connection point of parts A and B – the passage between the staircase and the living quarters. Because expansion joints were not used in these areas, and the internal surfaces of the passages were plastered and ceramic tiles were laid on the floors, damage was revealed in the form of scratched tiles and cracked walls. The extent of this damage increases with height and amounts to:

- scratch at the ground level on the floor, approximately 0.7 mm – Figure 9;
- crack in the lintel and walls of the opening, approximately 8 mm;
- crack in the first-floor level in the walls of the opening, approx. 13 mm – Figure 10;
- crack in the second-floor level in the walls of the opening, approx. 20 mm.

Figure 9. Damage to the ground floor in the transition between sections A and B.

Figure 10. Crack in the plaster of the jamb on the first floor in the transition between parts A and B.

According to the expansion design, the original roof of section B was extended and a single roof over section A was also added – these sections were not expanded. This resulted in cracks appearing in the second-story ceiling where the two sections joined (Figure 11).

The independent operation of parts B and C does not show significant deformation. No expansion joint profiles were used on the south elevation, and no plaster damage was observed. No cracks were found at the transition between the two parts on the ground floor. Crack damage occurred on the first floor, affecting the walls with a reveal width of up to 1.2 mm and the parquet flooring (slat spacing) of up to approximately 5 mm (Figure 12). In this case, no expansion joint covers were used to allow these parts to operate independently.

Figure 11. Crack under the roof slope in part A of the building.

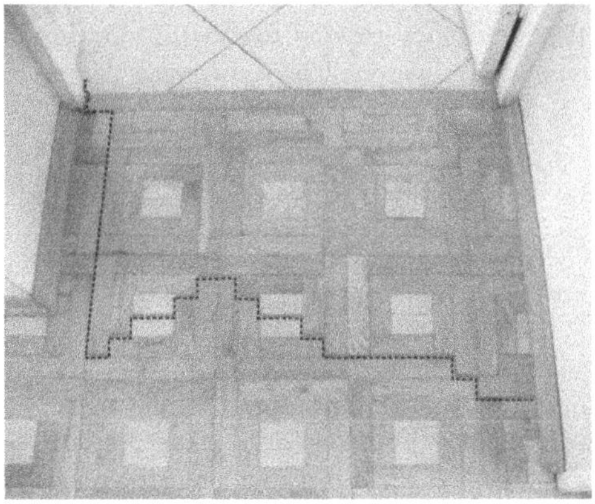

Figure 12. Crack in the parquet floor in the passage between parts B and C on the first floor.

4.2.2 *Damage to finishing elements*

Damage to the building's finishing elements is concentrated in sections B and C and primarily involves scratches on the ceramic floor tiles. The scratches were found in the ground and basement floors, which were laid on the ground, and they are irregular in shape. This indicates the possibility of ground movement between the foundations of section B of the building.

5 CAUSES OF OBJECT DAMAGE

5.1 *Causes of cracks when connecting parts of a building*

The main damage to the structure occurs at the junction of individual sections of the building. Deformation of the expansion joint profiles and internal damage to the plaster and flooring between sections A and B resulted from the subsidence of the ground beneath section A. The damage increases with the height of the structure, indicating that section A is deflecting from section B. Considering the crack widths on individual floors, at the junction between section A (staircase) and section B, section A deflection from the vertical of approximately 0.2° was observed. This translates to a floor slope of approximately 2.7‰ (significantly less than 10‰) and is imperceptible to users. The deflection from the vertical in section C relative to section B is much smaller, but still present.

In the case under consideration, the effect of the structure's settlement after many years of operation may be caused by changes in the soil and water conditions beneath the foundations, which reduce the soil's load-bearing capacity, potentially leading to further settlement. Damage may also be exacerbated by the emergence of subsoil deformation resulting from the activation of post-mining voids or suffocation, as mentioned above.

It should be noted that the cracks appearing at the expansion joints between segments do not constitute structural damage. Instead, they reduce the aesthetic and functional value of the building. The use of rigid, non-deformable finishing materials causes damage and prevents the building from operating independently, which is the basic assumption behind the use of expansion joints.

5.2 *Causes of damage to finishing elements*

Damage to building finishing elements, such as scratches on ceramic tiles on the basement floors and parquet floors on the ground floor, is primarily caused by the side effects of building subsidence or changes in ground and water conditions. Regardless of the cause, all these scratches are not structural in nature and do not threaten the building's load-bearing capacity. However, they do reduce the aesthetic and functional value of the premises.

6 SUMMARY AND CONCLUSIONS

The analyses carried out as part of the work allow for the following comments and conclusions to be presented:

Mining operations were conducted at a depth of approximately 30 meters in the area surrounding the property, posing a significant risk of a sinkhole. However, no such deformation occurred, and a gravimetric method revealed the presence of voids in the rock mass. Using a method based on the pressure vault theory (Strzałkowski 2015) allowed for obtaining results consistent with the results of geophysical surveys.

The damage to the facility was influenced by factors indicated early on in the form of scratches in the flooring and stiff door frame finishes resulting from the deflection of parts A and C from the vertical relative to part B. The deflection was caused by subsidence due to changes in ground and water conditions within the foundations of the new parts of the building. It is also possible that the suffocation phenomenon, on small areas beneath the building, contributed to some damage to the building. Damage to the ceramic tile finishes in the ground floor and basement could also have been caused by changes in ground and water conditions of the floors laid on the ground.

The presented case study indicates that, despite the high risk of sinkholes, the building's damage was caused by changes in groundwater levels, resulting in changes in soil and water conditions, or by suffocation. Both of these phenomena can lead to a loss of subsoil load-bearing capacity, and consequently, to its compaction by the foundation, while simultaneously tilting the building. Therefore, the mere presence of voids at shallow depths does not always lead to unfortunate consequences.

ACKNOWLEDGEMENT

This work was carried out with the support of the Silesian University of Technology, Grant No. BK-250/RB6/2025.

REFERENCES

Białek J. (2003). Algorytmy i programy komputerowe do prognozowania deformacji terenu górniczego. *Wydawnictwo Politechniki Śląskiej*, Gliwice.

Blachowski J., Cacoń S., and Milczarek W. (2009). Analysis of post-mining ground deformations caused by underground coal extraction in complicated geological conditions. *Acta Geodyn. Geomater.*, vol. 6, no. 3 (155), 351–357.

Chudek M., Janusz W. and Zych J. (1988). Studium dotyczące rozpoznania tworzenia się i prognozowania deformacji nieciągłych pod wpływem podziemnej eksploatacji złóż. *Zeszyty Naukowe Politechniki Śląskiej*, seria Górnictwo, zeszyt nr 141, Gliwice.
Chudek M. (2010). Mechanika górotworu z podstawami zarządzania ochroną środowiska w obszarach górniczych i pogórniczych. *Wydawnictwo Politechniki Śląskiej*, Gliwice.
Cui X., Zhao Y., Wang G., Zhang B and Li C. (2020). Calculation of Residual Surface Subsidence Above Abandoned Longwall Coal Mining. *Sustainability* 12, 1528; doi:10.3390/su12041528
Knothe S. (1984). Prognozowanie wpływów eksploatacji górniczej. *Wydawnictwo Śląsk*, Katowice.
Kotyrba A. and Kortas Ł. (2016). Sinkhole hazard assessment in the area of abandoned mining shaft basing on microgravity survey and modelling — Case study from the Upper Silesia Coal Basin in Poland. *Journal of Applied Geophysics*. Vol. 130, str. 62–70 https://doi.org/10.1016/j.jappgeo.2016.04.007
Kowalski A (red.). (2000). Eksploatacja górnicza a ochrona powierzchni. Doświadczenia z wałbrzyskich kopalń. *Wydawnictwo GIG Katowice*.
Kowalski A. (2015). Deformacje powierzchni w Górnosląskim Zagłebiu Węglowym. *Wydawnictwo GIG Katowice*.
Lagny C. (2014). The emissions of gases from abandoned mines: role of atmospheric pressure changes and air temperature on the surface. *Environ Earth Sci* 71: 923. https://doi.org/10.1007/s12665-013-2495-8
Orwat J. (2020). Causes analysis of occurrence of the terrain surface discontinuous deformations of a linear type. *J. Phys., Conf. Ser.* 2020 vol. 1426, art. no. 012016, str. 1–9
Strzałkowski P. (2015). Mathematical model of forecasting the formation of sinkhole using Salustowicz's theory. *Archives of Min. Sci.* 1/2015. Str. 63–71
Strzałkowski P. (2017). The proposal of predicting formation of sinkholes with an exemplary application. *Journal of Mining Science* vol. 53 iss. 1/2017, str. 53–58
Strzałkowski P. (2021). Sinkhole hazard caused by inactive mining shafts as illustrated by a selected example. *Pure and Applied Geophysics*, 178, 1697–1707. https://doi.org/10.1007/s00024-021-02716-z
Strzałkowski P. (2020). Duration of the final phase of mining area deformation process in the conditions of Upper Silesia (Poland). *Bulletin of Engineering Geology and the Environment*, 80, 5769–5780. https://doi.org/10.1007/s10064-021-02272-9
Szojda L. (200). Analiza numeryczna wpływu nieciągłych deformacji podłoża na budynki ścianowe. *Wydawnictwo Politechniki Śląskiej*, Gliwice.
Szojda L. (2022). Aspekty konstrukcyjne zabezpieczania budynków na terenach górniczych. *Wydawnictwo Politechniki Śląskiej*, Gliwice.
Szojda L. and Strzałkowski P. (2025). Numerical analysis of building damage caused by the exposure of a terrain threshold as a result of underground mining-induced subsidence. *Scientific Reports, Nature Publishing Group*, vol. 15, nr 1, no: 12138, pp. 1–17, DOI:10.1038/s41598-025-96648-6
Ścigała R. (2008). Komputerowe wspomaganie prognozowania deformacji górotworu i powierzchni wywołanych podziemną eksploatacją górniczą. *Wydawnictwo Politechniki Śląskiej*, Gliwice.
Ścigała R. and Szafulera K. (2020). Linear discontinuous deformations created on the surface as an effect of underground mining and local geological conditions - case study. *Bulletin Engineering Geology and the Environment*, Vol. 79, pp. 2059–2068; DOI: https://doi.org/10.1007/s10064-019-01681-1
Wrona P., Różański Z. and Pach G. (2016). Closed coal mine shaft as a source of carbon dioxide emissions. *Environ Earth Sci* 75, 1139. https://doi.org/10.1007/s12665-016-5977-7.

Mine Surveying and Mining Area Protection – Sokoła-Szewioła and Jarczyk (Eds)
© 2025 The Author(s), ISBN: 978-1-041-10531-2

Factors influencing the accuracy of recording resources extracted from a deposit exploited beneath the water table

A. Szafarczyk
Cracow University of Technology, Poland

ABSTRACT: The article presents a methodology for calculating the volume of mined mineral extracted from an open pit or quarry beneath the water table, with particular emphasis on the impact of the accuracy of bathymetric measurements on the calculated volume. The legal requirements regarding the record-keeping of deposit resources and the preparation of inventory reports are discussed. Environmental, technical, and operational factors influencing the precision of depth measurements using single-beam and multi-beam echo sounders are analyzed. Field studies conducted at the Wał–Ruda Borowce facility using three sondes, showed a measurement error in depth ranging from 0.1 to 0.16 m, dependent on the inclination of the measured bottom. GIS tools were used for the automated calculation of the volume of the excavation, employing Voronoy polygons. The results indicate that the accuracy of bathymetric measurements is crucial for the reliable estimation of the amount of mined raw material, and the methodology used allows for the control of measurement errors, which is of significant importance for maintaining resource records and managing exploitation.

1 INTRODUCTION

In Poland, supervision and control of mines is exercised by the State Mining Authority and its subordinate district mining offices. According to the State Mining Authority, as of December 31, 2024, 7,322 quarries extracting minerals from deposits covered by land ownership were subject to supervision and control by the mining offices. According to the Polish law (Ustawa 2011), minerals extracted from deposits covered by land ownership are shallow-lying industrial mineral and rock deposits, especially comprising natural aggregates, clay minerals, limestone, and others. Most of these quarries hold a license to extract minerals from deposits granted by the district head. Polish law obliges mining companies to maintain records of mineral deposit resources and regulates the preparation of a mineral deposit inventory report. The mineral deposit inventory takes into account changes in resources resulting from, among other things, more detailed exploration of the deposit, mining and resulting losses, changes in deposit boundaries or division, environmental protection and occupational health and safety requirements, and reclassification of resources and losses between geological and industrial categories. The mineral deposit inventory includes the annual preparation of a mineral deposit inventory report, completed by February 28th, as of December 31st of the previous year. For solid mineral deposits, the inventory report is prepared based on excavation surveys. In the case of activities conducted on the basis of a concession granted by the starosta, namely chairman of the executive body of the district, the measurement of excavations is carried out every three years, while in the annual cadastral reports the state of resources, the amount of extraction and losses are determined on the basis of estimates, leaving detailed arrangements until the measurement is carried out.

In Poland, natural aggregates are exclusively mined in quarries Sand and gravel are extracted primarily from beneath the water table (approximately 75% of extracted amount),

DOI: 10.1201/9781003655558-26

while rock aggregates are mined on the surface, in quarries at various depths (Kozioł *et al.* 2016). The most commonly used methods for measuring open pit mine workings include classic surveying using total stations, global positioning systems (GNSS), laser scanning (LIDAR), and aerial or drone photogrammetry based on the analysis of aerial photographs or those taken from unmanned aerial vehicles, enabling the creation of orthophotomaps and high-resolution digital terrain models.

When mining from beneath the groundwater table, measurements cannot be made using any of the above-mentioned standard measurement methods applied in mine surveying, and mapping of the open pit floor is performed applying bathymetric measurements (Choiński and Ptak 2014). This article presents a method for calculating the volume of minerals extracted annually in a mine operating from beneath the water surface. The accuracy of the depth determination using bathymetric measurement technology is considered.

2 FACTORS INFLUENCING THE ACCURACY OF BATHYMETRIC MEASUREMENTS

When conducting field measurements to determine the volume of overburden or extracted minerals, a combination of two measured values is most often used. The first determines the thickness, and the second determines the location of points. In open-pit mines and quarries not flooded with water, many different geodetic measurement technologies can be applied, based on the use of electromagnetic waves, especially radio waves and microwaves (total stations, GNSS receivers, scanners, drones, InSAR measurements). For example, GNSS (Global Navigation Satellite System) systems emit radio waves that are received on the Earth's surface and enable precise location determination of points (Nguyen *et al.* 2024). Techniques such as interferometric synthetic aperture radar (InSAR), use microwaves, which allow for the creation of terrain deformation maps and elevation models with high precision (Tang *et al.* 2020). These waves travel through the atmosphere and their measurement incorporates corrections related to the influence of the ionosphere and troposphere. Geodetic instruments using electromagnetic waves, such as advanced electronic total stations or GNSS systems, achieve very high measurement accuracy, down to the millimeter range, depending on the technology used and measurement conditions. Unfavorable environmental factors, such as strong electromagnetic fields, can reduce the accuracy of such measurements.

In the case of a flooded open-pit mine, measurements of the bottom profile are performed using bathymetry, which employs acoustic and ultrasonic waves. The most commonly used are echosounders, which send sound pulses towards the bottom and measure their return time after reflection from the water bed (Grządziel *et al.* 2015). This allows for determining the depth at the measurement location. More advanced bathymetric techniques utilize multibeam echosounders, which simultaneously emit several hundred narrow acoustic wave beams, allowing for precise mapping of the reservoir bottom over a wide area. Information regarding the bottom elevation is typically obtained at selected measurement points arranged in a regular grid. The existing bathymetric LiDAR method using a green laser (Szafarczyk and Toś 2023), as a remote technology, is very convenient but due to the limitations resulting from its measurement range (up to 3 Sechi depths depending on the model) as well as the very high costs of the measuring device, it is not used on a large scale.

The depth accuracy of bathymetric instruments using acoustic and ultrasonic waves, such as echosounders and sonars, depends on many factors. Precisely determining the velocity of sound in water is crucial, as it varies with temperature, density, and pressure, directly impacting the quality of the results. Under typical conditions, modern echosounders can provide vertical accuracy of several centimeters, although in more challenging conditions, this accuracy can drop to around 10–20 cm (Gołuch *et al.* 2010). The accuracy of bathymetric measurements can also vary depending on the sonar technology used — multibeam systems allow for more detailed and accurate data on the reservoir bottom than standard single-beam systems.

Single-beam echosounders are widely used in shallow waters for coastal surveys and bathymetric mapping of small areas due to their low cost (Grządziel and Wąż 2015). An example of the use of single-beam echosounders is the study conducted at the Stobreč marina, during which their accuracy was verified by comparing the results with more precise tachymetric methods. Based on the study, it was concluded that, thanks to their relatively high accuracy, they meet the International Hydrographic Organization (IHO) requirements for high-accuracy measurements (Tuno *et al.* 2024). However, it should be noted that measurements with this type of echosounder are limited to point depths of the bottom, without detailed information about its topography (Bassett and Zeiden 2024).

One of the most important variables to consider in bathymetric measurements is environmental disturbances (temperature, salinity, water currents), which significantly affect echosounding accuracy (Bassett and Zeiden 2024, Tuno *et al.* 2024). Temperature affects the speed of sound in water. This speed increases by approximately 4.3 m/s with a 1°C increase in temperature (Tuno *et al.* 2024). Salinity also affects the speed of sound (the higher the salinity, the faster the speed), but it can also affect signal distortion (Bassett and Zeiden 2024). Water currents cause transducer movement underwater or cause secondary reflections. Another factor affecting measurement accuracy is the characteristics of the water bottom. Parameters such as the type of bottom or the hardness of its surface, resulting from its structure (compact or loose such as rocks or sand), generate clear, strongly reflected signals, as opposed to a soft bottom (mud, organic sediments), which causes signal absorption and much weaker reflections (Bassett and Zeiden 2024). The evenness of the bottom surface causes further interference; a flat bottom will be easier to interpret due to its "straight reflections", as opposed to a bottom that is wavy or covered with obstacles, depending on the type of reservoir, which may be debris, wrecks, or vegetation, making them more difficult to read (Szafarczyk and Gawałkiewicz 2023).

3 MEASUREMENT OF AN OPEN PIT IN AN EXEMPLARY MINING PLANT OPERATING FROM BENEATH THE WATER TABLE

An example test facility was the reservoir of the "Wał-Ruda Borowce" open-pit mining plant, covering an area of 0.08 km^2, located in Radłów, Lesser Poland Voivodeship (Poland). This quarry, which extracts natural aggregate, ceased mining operations in 2024. Bathymetric measurements of the bottom were taken from a boat powered by an electric motor, at a constant speed of 5 km/h. There were no significant environmental disturbances during the measurements, and the day was windless. The measurements were taken in fresh, still water, free of currents, and at a constant temperature, which is optimal for ensuring the highest measurement accuracy. The bottom of the quarry was hard and free of vegetation, and due to the method of extracting the aggregate from beneath the water surface (using a dredger), it was relatively flat. The maximum depth of the reservoir was 11 m. For a selected water body, which is part of an open-pit mine for common minerals, test measurements were conducted using a Humminbird HELIX 7 CHIRP MEGA SI GPS G4 echosounder, the readings of which were verified with a handheld probe and a Humminbird 718 Sonar. The characteristics of the devices used for depth measurement are presented in Table 1.

Table 1. Characteristics of the measuring devices used to measure depth.

Device	Humminbird HELIX 7	Humminbird 718 Sonar	Hand probe
Principle of depth measurement	Measurement of wave propagation time	Measurement of wave propagation time	Direct measurement
Nominal reading accuracy	0.1 m	0.1 m	0.01 m

Measurements were planned at 86 selected points, evenly distributed throughout the water reservoir. The locations of both probes were determined based on real-time location information obtained on the screen of the GNSS receiver (Hi-Target H32, serial number: 11602722). Based on the X, Y coordinate measurement report in the PUWG 2000 coordinate system, zone 7, the locations of the depth measurement points were visualized in QGIS (Figure 1).

Figure 1. Location of points, where bathymetric measurements were made using the devices listed in Table 1.

At 30 points, sufficient samples for statistical calculations, a hand-held probe was used. This is significantly more time-consuming than echosounder measurements and requires the involvement of a human operator (inserting and retrieving the probe from the water). In the shallower areas of the reservoir, the Helix 7 probe was compared with the hand-held probe; in the central and eastern parts, where depths range from 8 to 11 meters, the Helix 7 probe was compared with another probe (718 sonar).

The first comparison was made between the Helix 7 probe and the hand-held probe for thirty points located at shallower depths of the reservoir (numbered 1 to 30 in Figure 1), ranging from

0 to 8 meters, in coastal zones with a steeply sloping bottom of the excavation. The error calculated from pairs of observations, after eliminating gross errors (for points numbered 9 and 18) exceeding three times the mean error, was 0.36 meters. When using the hand-held probe, the vertical positioning of the linear ruler is an absolute requirement for obtaining accurate depth readings. At greater depths, this is very difficult to achieve. Therefore, the hand-held probe was discontinued for reservoir depths exceeding 8 meters.

A second comparison was made between the Helix 7 and the 718 Sonar probes for sixty four points located at greater depths (from 3 to 11 meters) in the reservoir (numbered 23 to 86 in Figure 1), covering the entire central part of the excavation, with a bottom with slight height differences resulting from the dredger-based extraction technology. The error calculated from pairs of observations, after rejecting gross errors (for points numbered 38, 39, 54, and 79) exceeding three times the mean error, was 0.16 meters.

The results of all depth measurements, after corrections were made (removing observations with gross errors), were divided into two groups using the tank bottom slope at the measurement location as the criterion. Slopes of 0-6.5% were assumed to be flat, while slopes of 6.5-13% were assumed to be sloping. The mean depth measurement error of the Helix 7 echosounder (after removing gross errors) was 0.10 m for a flat bottom and 0.16 m for a sloping bottom.

Based on the calculated correlation coefficient, no relationship was found between the error values and the depth of individual measurement points.

4 CALCULATION OF THE VOLUME OF MINERAL EXTRACTED FROM BENEATH THE WATER TABLE

The basis for calculating the volume of the extracted minerals using the polygon method is the assumption that the information obtained at a given point also applies to its surroundings. The studied area can therefore be divided into smaller, simpler geometric shapes, such as triangles, rectangles, or polygons. Polygons are constructed by connecting the nearest measurement points. The resulting segments are used to determine the perpendicular bisectors, which define the polygon contours. Each figure is the base of a prism, whose height is the thickness at the point where it is located. The solids are analyzed separately, and then the results are added to obtain the final volume.

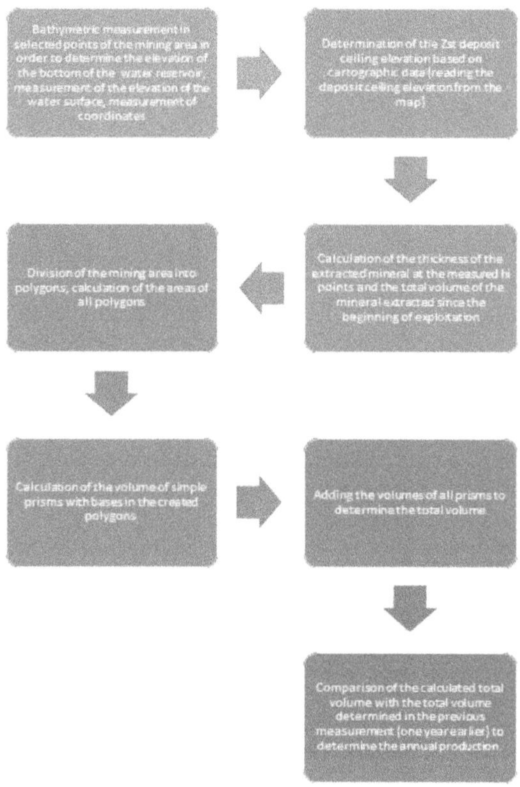

Figure 2. Stages of an automated procedure for calculating the volume of mineral extracted from beneath the water surface using GIS tools.

The annual volume of extracted mineral was calculated using GIS tools based on the automated procedure presented by Szafarczyk and Siwek (2025). Since the reference level in this method is the upper edge of the deposit slope, calculating the excavation volume in 2023 using this method yields the volume of mineral extracted from the beginning of mining to the measurement. Similarly, data from 2024 are calculated by calculating the volume of mineral extracted from the beginning of mining to the measurement. Therefore, subtracting the volume calculated in 2023 from the total volume calculated in 2024 yields, the volume of material extracted between the indicated dates is yielded. The calculation procedure is presented in Figure 2. The first step is to divide the entire study area into polygons built around the measurement points. In QGIS, this is achieved using the Voronoy Polygons function. This function creates polygons, each of which gathers all the points on the surface that lie closer to the base point of that polygon than other base points. Objects constructed this way also inherit the attributes of the base point. In this case, the outer polygons had to be limited. This was achieved by clipping the polygon to the reservoir boundary. The boundary was obtained from a vectorized scan of the situational-elevation map of the mine in question. The Voronoy polygons created for the 2024 data are shown in Figure 3.

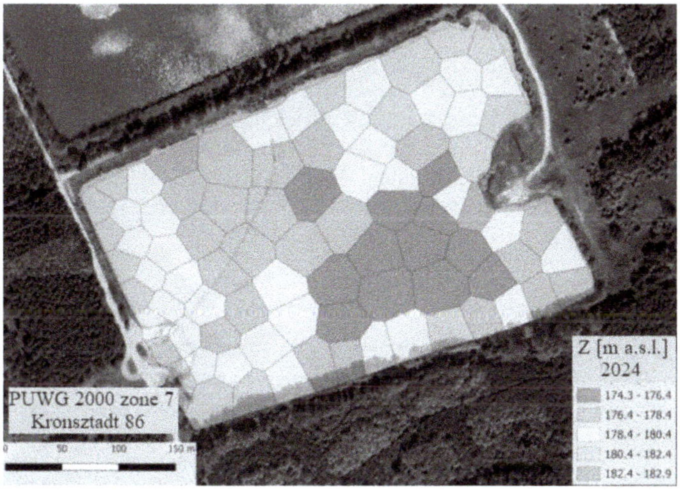

Figure 3. Voronoy polygons with their bottom elevation from 2024.

The second step was to calculate the volume for both measurement sessions. This was performed using the field calculator in both layers. The algorithm calculated the volume for each polygon by multiplying the difference between the average upper boundary of the deposit (186.6 m above sea level) and the seabed elevation in the polygon by its area. To sum and compare these values, a statistical analysis tool was used. This tool allows for obtaining simple statistical indicators for a selected attribute of all objects in the vector layer. The sum of the volumes of all polygons in 2024 and 2023 was compared. This difference determines the amount of mineral extracted in 2024, which is 6354.4 m^3.

5 SUMMARY AND CONCLUSIONS

Given the same surface area of an open pit being mined, the percentage error in volume calculation will be greater for a shallow pit and smaller for a deep pit. As mining continues in subsequent years of production records, the error in estimating the amount of mined mineral will decrease.

In the case under consideration, the total volume of mineral extracted in the considered excavation of the open-pit mine since the beginning of operation is 650,880 m^3. Assuming a depth measurement error of 0.16 m during bathymetric measurements and assuming an area of all polygons of 81,360 m^2, the resulting volume is 13,017 m^3, which represents less than 2%

of the total volume of mineral extracted from the deposit during the period of mining in the mine in question. Assuming a depth measurement error determined during bathymetric measurements of 0.10 m, the percentage error in the total volume of mineral extracted is 1.2%.

In summary, the precision of depth measurements performed using bathymetric devices is determined by a complex set of factors that can be divided into three basic categories: environmental, technical, and operational. In the context of environmental factors, variable physicochemical properties of water, such as temperature, salinity, pressure, and density, play key roles. These factors determine the propagation speed of acoustic waves and thus, influence the accuracy of depth measurements. Furthermore, the presence of pollutants, suspended materials, and plankton causes attenuation and scattering of the acoustic signal, which can reduce the quality of bathymetric data. The morphological characteristics of the reservoir bottom—including terrain, substrate diversity, and the presence of bottom vegetation—significantly affect acoustic wave reflection and complicate the interpretation of obtained signals. Technical aspects concern the parameters of the measurement instruments, such as the echosounder type (single- or multi-beam), its calibration, operating frequency, and the stability of the measurement platform, which translates into the reliability and repeatability of measurements. Operational factors also influence the measurement system, including data collection methodology, measurement density, and the effectiveness of protection against external interference. The final quality of the measurements also depends on data processing, which includes sound velocity correction, noise suppression, and signal filtering. Comprehensive consideration and proper management of these factors is essential to minimize measurement errors and obtain reliable, precise seabed topographic data and depth values, which are crucial for surveying and operational purposes.

REFERENCES

Bassett, C. and Zeiden, K. (2024). Calibration and Processing of Nortek Signature1000 Echosounders.

Choiński, A. and Ptak, M. (2014). Najnowsze sondowania wybranych jezior Pojezierza Wielkopolsko Kujawskiego. Badania Fizjograficzne nad Polską Zachodnią. *Seria A: Geografia Fizyczna* 65: 55–63.

Gołuch, P., Dombek, A. and Kapłon J. (2010). Ocena dokładności danych uzyskanych z pomiaru batymetrycznego wykonanego echosondą Lowrance LMS-527C DF iGPS, *Archiwum Fotogrametrii, Kartografii i Teledetekcji*, 21:109–118.

Grządziel, A. and Wąż, M. (2015). System echosondy wielowiązkowej w pomiarach batymetrycznych planowanych tras żeglugowych. *Logistyka* 3:4250–4257.

Grządziel, A., Wąż, M., Naus, K. and Felski, A. (2015). Niepewność pomiaru głębokości w pasie pomiarowym swath echosondy wielowiązkowej. *Journal of Hydrographic Surveying* 12: 2575–2584.

Kozioł, W., Machniak, Ł., Borcz, A., Baic, I. (2016). Górnictwo kruszyw w Polsce – szanse i zagrożenia. *Inżynieria Mineralna* 2:175–182.

Nguyen, H.V., Pham, K.C., Nguyen, D.B. and Nguyen, L.Q. (2024). Zastosowanie metod GNSS w monitorowaniu przemieszczeń powierzchni terenów pórniczych: przegląd systemowy. *Inżynieria Mineralna 3*(2). https://doi.org/10.29227/IM-2024-02-96

Szafarczyk, A. and Gawałkiewicz, R. (2023). An inventory of opencast mining excavations recultivated in the form of water reservoirs as an example of activities increasing the retention potential of the natural environment: a case study from Poland. *Geology, Geophysics & Environment* 49(4): 401–418.

Szafarczyk, A. and Siwek, M. (2025). GIS-Based Process Automation of Calculating the Volume of Mineral Extracted from a Deposit. *Geosciences* 15:315. https://doi.org/10.3390/geosciences15080315

Szafarczyk, A, and Toś, C. (2023). The Use of Green Laser in LiDAR Bathymetry: State of the Art and Recent Advancements. *Sensors* 23(1):292. https://doi.org/10.3390/s23010292

Tang, W., Motagh, M. and Zhan, W. (2020). Monitoring active open-pit mine stability in the Rhenish coalfields of Germany using a coherence-based SBAS method. *International Journal of Applied Earth Observation and Geoinformation* 93: 102217, https://doi.org/10.1016/j.jag.2020.102217

Tuno, N., Kulo, N., Perić, D., Đidelija, M., Hamzić, A., Topoljak, J., Mulahusić, A., and Kogoj, D. (2024). Analyzing Depth Uncertainty of Near-Shore Bathymetric Survey Conducted by Single Beam Echo Sounder. *Technical Journal* 18(si1): 84–91.

Ustawa z dnia 9 czerwca 2011 r. Prawo geologiczne i górnicze Dz.U.2023.0.633.

Mine Surveying and Mining Area Protection – Sokoła-Szewioła and Jarczyk (Eds)
© 2025 The Author(s), ISBN: 978-1-041-10531-2

Assessment of drainage condition of structures using UAVs- case studies

M. Wróblewska & M. Grygierek
Silesian University of Technology, Gliwice, Poland

ABSTRACT: Monitoring structures requires a series of activities to check their technical condition, including geometry. Geodetic measurements are the basic tool here. Technological advances have made it possible to perform very detailed measurements using dense point clouds. One example is the use of low-altitude photogrammetry with unmanned aerial vehicles (UAVs). A huge advantage is the remote mapping of large areas without compromising safety in hazardous and often inaccessible areas. The results of UAV measurements provide a basis for verifying the condition of structures and ultimately constitute a source of data necessary to determine corrective measures. The study shows an example of measurements using UAVs, including in the area of a road embankment affected by underground mining. In the cases discussed, the geometry of the structures was analyzed in the context of the slopes necessary for proper water drainage. The observations carried out revealed changes such as: a significant reduction in the slope of the culvert under the road embankment from 1.9% to 0.07%. Alignment with the national coordinate system was achieved using ground control points. The georeferencing accuracies were up to 0.86 cm for XY (RMSE XY) and up to 0.53 cm for Z (RMSE Z).

1 INTRODUCTION

Inspecting the technical condition of engineering structures is one of the key tasks of the managers of these structures, primarily due to the safety of their use. The impact of external environmental factors causes changes to the originally designed characteristics of the structure, including its geometry. This exposes the structure to a number of negative phenomena, including those related to the impact of water (Politechnika Śląska and Tamaka 2014), (Łupieżowiec 2024), (Pilecka and Manterys 2017). Improper drainage can be a key factor in failures such as a reduction in the bearing capacity of the ground (Wróblewska *et al.* 2023). In the case of linear structures such as roads, excess rainwater will cause flooding and at the same time have a destructive effect on the surface (Huang, 2015), (Grygierek and Wróblewska 2023). These situations threaten traffic safety and pavement durability. It is therefore necessary to monitor the condition of structures so that, in the event of environmental hazards, they can respond to emerging threats, prevent damage, and ultimately provide the data necessary to determine corrective actions.

The basis for verifying the technical condition of structures is geodetic measurements to determine their geometry. The obtained x, y, and z coordinates enable observations across the entire range. In the case of proper drainage of embankment structures, correctly shaped slopes will be important. Currently used measurement techniques enable results to be obtained in the form of a dense point cloud, allowing for very detailed observation of the condition of objects. Laser (Dreier *et al.* 2022), photogrammetric (Cho *et al.* 2022), and satellite (Nam *et al.* 2021) technologies can be distinguished here. In the case of structures that are difficult to access or exposed to constant vehicle traffic (as in the case of roads),

DOI: 10.1201/9781003655558-27

remote techniques that do not require direct access to the observation area are used. This has a significant impact on the safety of the measurements performed (Howard *et al.* 2018). An example of this is low-altitude photogrammetric measurements using unmanned aerial vehicles (commonly known as drones) (Tatum and Liu 2017), (Nowobilski 2020). The application of this technology in the area of road embankments and post-mining waste embankments (heaps) is presented in this article.

2 METHODOLOGY OF MEASUREMENTS USING LOW-ALTITUDE PHOTOGRAMMETRY

The study used a multi-rotor aircraft equipped with a 1-inch 20 MPix camera. A series of photographs was taken, which allowed the study areas to be mapped and then processed using specialized software. The following flight parameters were used:

- double flight trajectory (double grid),
- flight altitude of 50 m, corresponding to a terrain pixel of 1.38 cm/px,
- speed of 4 m/s,
- transverse and longitudinal coverage of 80%.

Photogrammetric processing began with loading a series of photographs and then aligning them. Alignment with the national coordinate system was achieved using 4 ground control points marked in the field and tracked using GNSS technology. A dense point cloud was obtained, where each measurement point was assigned x, y, and z coordinates. The georeferencing accuracies were up to 0.86 cm for XY (RMSE XY) and up to 0.53 cm for Z (RMSE Z). As a result, 3D models of the study areas, i.e., the road embankment (Figure 1) and the post-mining waste heap (Figure 2), were obtained. The analysis also used digital elevation models (DEM) obtained with a resolution of 4.34 cm/px.

Figure 1. 3D model of the road embankment area.

Figure 2. 3D model of the post-mining dumps 3D.

3 OBSERVATION OF BUILDING DRAINAGE CONDITION USING UAV

3.1 *Observations of road drainage in the mining area*

The surface drainage condition was determined for a provincial road located in a mining area. The predicted impact of underground mining was classified as category IV mining area ($T \leq 15$ mm/m). During construction works, in accordance with the design documentation, an HDPE pipe culvert with a diameter of 1500 mm was constructed to drain rainwater from the north to the south (Figure 3) with a slope of 1.9%.

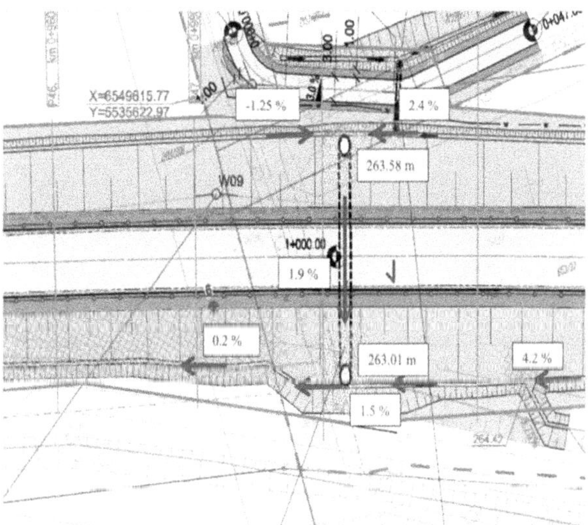

Figure 3. View of the designed state of surface drainage of the right-of-way along with the specified direction of water runoff and the slope value.

In 2021, a floodplain was found on the north side, reaching nearby agricultural plots (Figure 4). Photogrammetric measurements using an unmanned aerial vehicle, which were started at that time, showed a floodplain area of over 2700 m^2 (Figure 4). Processing of photogrammetric images from the aerial survey made it possible to compare the designed slopes with the existing ones for the culvert and ditches on the left and right sides of the road (Figure 5). The terrain elevations were additionally checked by measurement using GNSS satellite technology.

Figure 4. Photograph of the floodplain (own source).

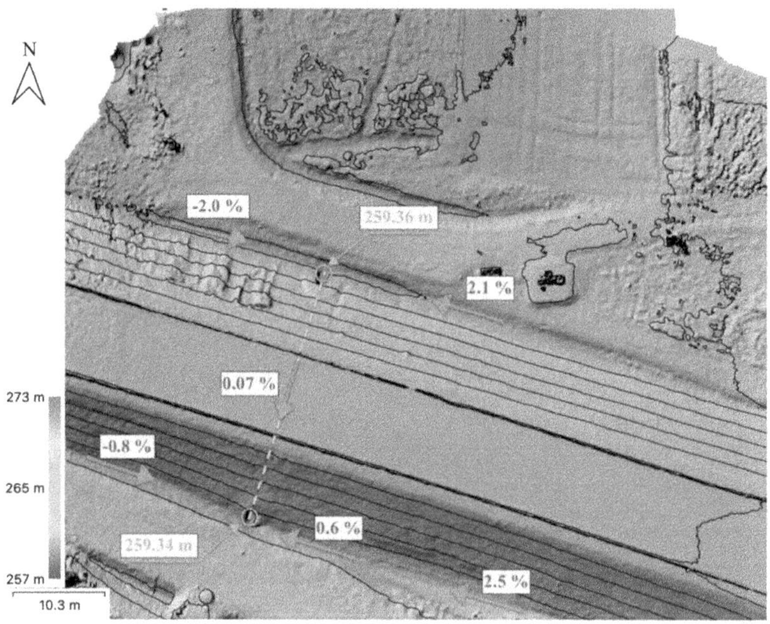

Figure 5. DEM model (Digital Elevation Model) with the specified value and direction of slopes for the culvert under the right-of-way and ditches on the left and right side of the road.

When analyzing the results obtained, a significant reduction in the slope of the culvert under the road embankment can be observed, from 1.9% to 0.07%. Similar changes, significant in terms of drainage efficiency, can be observed for the ditch on the south side of the road, where the slope values have been reduced by half. Even on a certain section of the ditch, a counter-slope is noticeable, from the designed value of 0.2% to -0.8%. Disturbances in the slopes from before the deformation, both for the culvert and the drainage ditches, cause a deterioration in the efficiency of gravitational drainage of rainwater in this area, resulting in flooding of local plots and the culvert itself.

3.2 *Observations of post-mining waste dumps*

Another object of research is a spoil tip covering over 8 hectares of land, which is a remnant of a coal mine that ceased operations in Upper Silesia over 20 years ago. The analyzed section of the embankment is 8 m high and runs along a busy street for a length of approx. 600 m.

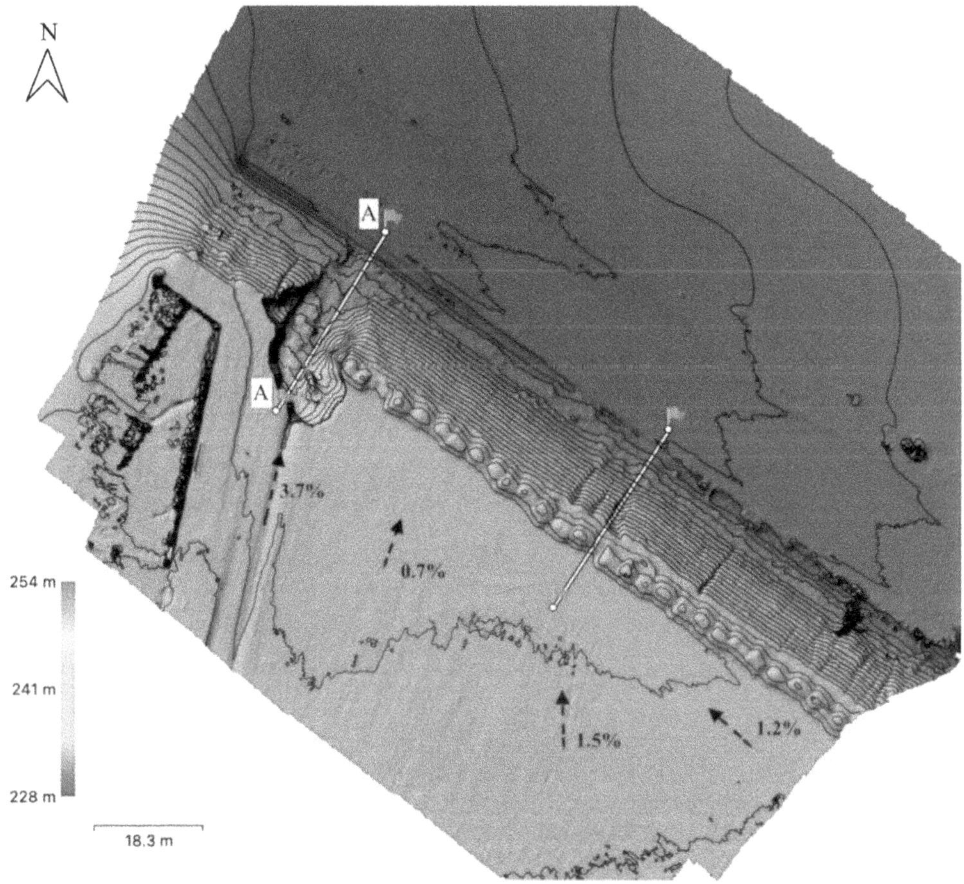

Figure 6. DEM model (Digital Elevation Model) with the specified value and direction of slopes with section AA.

The first observations took place in 2022, shortly after heavy rainfall caused part of the material to slide onto the nearest lane of the road. This resulted in a significant gap, the cross-section of which is shown in Figure 7. Using the generated DEM models, it was possible to calculate the volume of the loss, which amounted to approximately 1309.4 m^3. It is worth noting that the main decline in the upper surface of the spoil heap, including the largest (3.7%), was directed towards the site of the damage, which must have caused a constant flow of water to the analyzed site after heavy rainfall and was the most likely cause of instability.

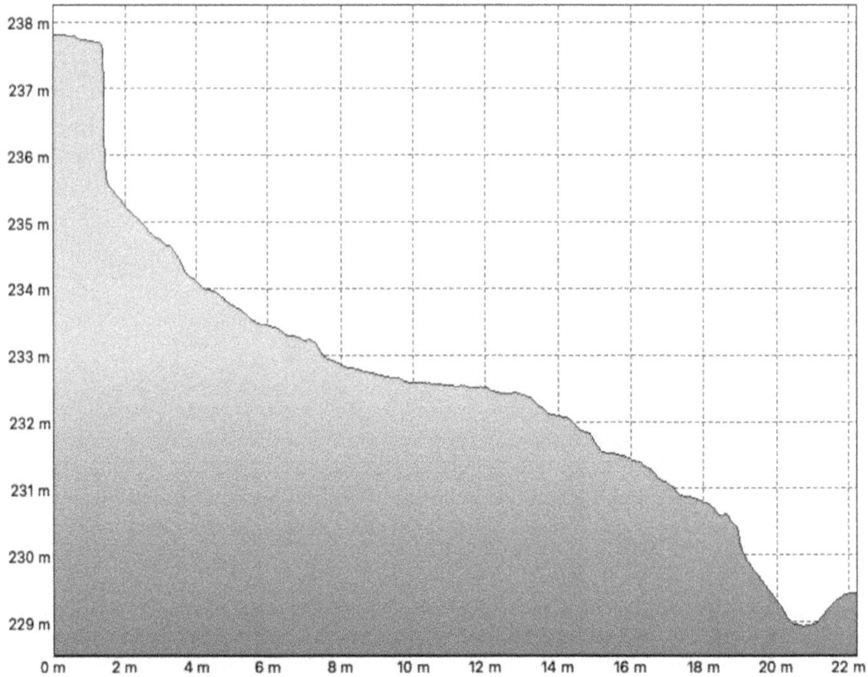

Figure 7. A-A cross-section of the slope fragment.

After the failure, the embankment was reinforced by installing openwork concrete slabs and constructing a drainage channel. Less than six months later, the slope collapsed again, including in the vicinity of the newly installed reinforcements (Figure 8).

Figure 8. 3D model of the post-mining dumps- after repairs.

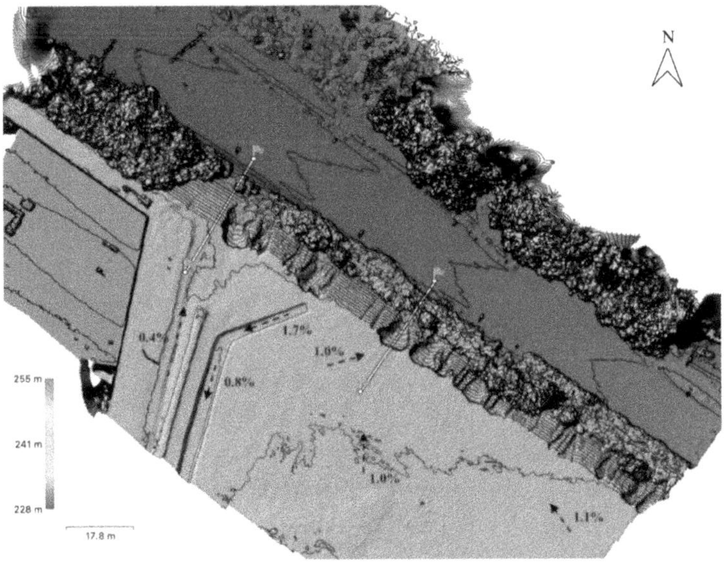

Figure 9. DEM model (Digital Elevation Model) with the specified value and direction of slopes.

The number of damages in the form of material slippage was significantly greater than that observed initially. As can be seen in Figure 9, the upper surface of the spoil heap slopes towards the road. In addition, the flattening of the upper part of the spoil heap by removing the crest of the slope (the material was probably used for backfilling) facilitated the access of rainwater to the slope. The measurement results indicated a significant problem related to the incorrect slope of the upper surface of the spoil heap as well as the slope of the embankment itself, which, combined with heavy rainfall, results in a risk of destabilization.

4 CONCLUSIONS

Measurements controlling the condition of engineering structures are a key task in the process of their management. Today, thanks to modern remote techniques, including low-altitude photogrammetry, it is possible to check their technical condition in a much safer way than using classic measurements. At the same time, the measurement data obtained provides a spatial representation of the object under investigation.

In the examples discussed in this study, the condition of embankments was analyzed in the context of their geometry, including the slopes necessary for proper rainwater drainage. In the context of a road embankment located in a mining area, the functioning of the road culvert and drainage ditches was checked. A reduction in the gradients was observed, which could contribute to improper drainage of rainwater. In the case of a post-mining waste disposal site, UAV measurements also helped to observe the causes of the failures. In addition, it was observed that the remedial measures taken by the owner of the facility (covering the slope with concrete slabs and constructing a single drainage channel) proved to be insufficient.

REFERENCES

Cho, J., Lee, J., and Lee, B. (2022). Application of UAV Photogrammetry to Slope-Displacement Measurement. *KSCE Journal of Civil Engineering*, 26(4), 1904 1913. https://doi.org/10.1007/s12205-021-1374-1

Dreier, A., Kuhlmann, H., and Klingbeil, L. (2022, June 20). The potential of UAV-based Laser Scanning for Deformation Monitoring – Case Study on a Water Dam. Proceedings of the 5th Joint International Symposium on Deformation Monitoring - JISDM 2022. *5th Joint International Symposium on Deformation Monitoring*. https://doi.org/10.4995/JISDM2022.2022.13833

Grygierek, M., and Wróblewska, M. (2023). Drogi kołowe na terenach górniczych w świetle obserwacji terenowych. In J. Pizoń, M. Gołaszewska, B. Orlik-Kożdoń, and Ł. Drobiec (Eds), Wyzwania budownictwa na terenach górniczych, pogórniczych i zdegradowanych: Monografia problemowa: Praca zbiorowa. *Krynicka Konferencja Naukowa*, Gliwice. Wydawnictwo Politechniki Śląskiej.

Howard, J., Murashov, V., and Branche, C.M. (2018). Unmanned aerial vehicles in construction and worker safety. *American Journal of Industrial Medicine*, 61(1), 3–10. https://doi.org/10.1002/ajim.22782

Huang, S.-C. (Ed.). (2015). *Advances in asphalt materials: Road and pavement construction*. Woodhead Publ.

Łupieżowiec, M. 2024. Wpływ ciśnienia spływowego wody na stateczność skarp i zboczy. *Materiały Budowlane*, 1(8), 36–48. https://doi.org/10.15199/33.2024.08.06

Nam, B.X., Van Anh, T., Bui, L.K., Long, N.Q., Le Thu Ha, T., and Goyal, R. (2021). Mining-Induced Land Subsidence Detection by Persistent Scatterer InSAR and Sentinel-1: Application to Phugiao Quarries, Vietnam. In D. Tien Bui, H.T. Tran, and X.-N. Bui (Eds), *Proceedings of the International Conference on Innovations for Sustainable and Responsible Mining* (Vol. 108, pp. 18–38). Springer International Publishing. https://doi.org/10.1007/978-3-030-60269-7_2

Nowobilski, T. (2020). Bezzałogowe statki powietrznew kontroli obiektów budowlanych. *BUILDER*, 271(2), 18–20. https://doi.org/10.5604/01.3001.0013.7500

Pilecka, E., and Manterys, T. (2017). Wpływ nawodnienia gruntu na stateczność nasypu drogowego na osuwisku. Autobusy. *Technika, Eksploatacja, Systemy Transportowe*, 18(12), 376–380.

Politechnika Śląska, and Tamaka, S. (2014). The influence of permeability of soil on the stability of railway embankment. *Journal of Civil Engineering, Environment and Architecture*, XXXI(61 (3/I/14)), 347–359. https://doi.org/10.7862/rb.2014.65

Tatum, M.C., and Liu, J. (2017). Unmanned Aircraft System Applications in Construction. *Procedia Engineering*, 196, 167–175. https://doi.org/10.1016/j.proeng.2017.07.187

Wróblewska, M., Kowalska, M., and Łupieżowiec, M. (2023). Slope stability analysis of post-mining dumps with the use of photogrammetric geometry measurements – a case study. *Archives of Civil Engineering*. https://doi.org/10.24425/ace.2023.147654

Mine Surveying and Mining Area Protection – Sokoła-Szewioła and Jarczyk (Eds)
© 2025 The Author(s), ISBN: 978-1-041-10531-2

Analysis of the impact of Earth's curvature on the transformation of boundary point coordinates using the Method of Independent Models

M. Żogała, V. Sokoła-Szewioła & M. Poniewiera
Silesian University of Technology, Gliwice, Poland

ABSTRACT: The coordinate transformation methods used so far between local reference systems used in mining enterprises are characterized by limited accuracy, resulting in inconsistencies in the coordinates of corresponding boundary points, despite their precise location in the field. This problem, which is significant from the point of view of ensuring the integrity of geospatial data, has been thoroughly analyzed in a research study conducted at the Silesian University of Technology. In response to the identified topological inconsistencies, a new transformation procedure based on the Method of Independent Models (MIM) has been developed. This method enables precise mapping of the spatial relationships between points and effective elimination of errors arising from the simplified assumptions of classical transformation methods. The article presents the algorithm of the authors' transformation procedure, which accounts for the Earth's curvature, including the ellipsoid and cartographic projections and the results of a transformation task applied to 14 mining areas located in southern Poland, within the Upper Silesian Coal Basin. The results obtained were compared with the coordinates derived from transformations performed separately for each mining area and those from the MIM method without considering Earth's curvature. The positional differences of corresponding boundary points in the analyzed areas after applying the MIM method (in two variants) ranged from just 0.002 to 0.066 meters.

1 INTRODUCTION

The legal regulations governing mining activities in Poland, in particular the Geological and Mining Law Act (Act 2011) and the Regulation on Mine Surveying and Geological Documentation (Regulation 2015), clearly define the responsibilities of the mining entrepreneur in relation to maintaining and updating the required documentation. The use of local coordinate systems in documents is permitted only if they can be transformed into a geodetic reference system, which is an integral part of the national spatial reference system, in accordance with the provisions of the Geodetic and Cartographic Law (Act 1989) and theRegulation on the State System of Spatial References (Regulation 2012).

The issue of coordinate transformation between different reference systems has been widely addressed in scientific studies by both Polish and international authors. Notable examples include the works of Kadaj (2002), Maciaszek (2010), Even–Tzur (2020), Banković *et al.* (2024), Janssen (2009), Sisman (2014), and James (2012), highlighting the significance and complexity of this issue in practice.

The coordinate transformation methods currently used between local reference systems in Poland's mining sector exhibit limited precision, leading to inconsistencies in the map location of identical boundary points in comparison to their real location. This issue is of critical importance for maintaining the quality and consistency of geospatial data used in mining operations, as well as for the safe management of mining and post-mining areas.

DOI: 10.1201/9781003655558-28

This issue was analyzed within a research project conducted at the Silesian University of Technology (Sokoła *et al.* 2017), whose main objective was to verify the accuracy of selected coordinate transformation methods in the context of mine surveying applications. The results revealed significant limitations in traditional approaches, justifying the need to develop new, more advanced transformation methods.

In response to the identified topological discrepancies, an innovative coordinate transformation procedure based on the use of local independent models was developed (Żogała 2022). This method allows for more accurate representation of spatial relationships between points and minimizes errors resulting from the simplified assumptions typical of classical transformation techniques. The proposed solution marks a significant step forward in improving the accuracy and reliability of coordinate transformations when working with different reference systems in mining and geological documentation related to exploited deposits. Given the need to perform transformation tasks both for large and small spatial areas, the procedure was developed in two variants: a basic version that does not account for Earth's curvature (including the ellipsoid and projection), and an extended version that does. The first variant is discussed in detail in (Żogała *et al.* 2025).

This article presents the algorithm of the authors' transformation procedure for use in large areas, taking into account Earth's curvature, including the ellipsoid and cartographic projection.

2 MATERIALS

The study was conducted in the southern part of Poland, in the area of the Upper Silesian Coal Basin. Data were collected for 14 adjacent mining areas located within this region. The dataset included the coordinates of 248 boundary points in a local reference system (Sucha Góra 1901, 1926 – Bessel ellipsoid) and the coordinates of 52 shafts (in Sucha Góra 1901, 1926 systems and in the national PL–2000 system – GRS 80 ellipsoid). These data were provided by mining companies, the Higher Mining Authority, and the Archive of the Higher Mining Authority.

The coordinates of boundary points and shafts in local reference systems obtained from mine resources were compared with data obtained from the Higher Mining Authority and the Authority's Archives. Additionally, in order to confirm the accuracy of the data obtained for one of the mining areas (Szczygłowice), a field measurement was performed using GNSS (Global Navigation Satellite System) technology at points located in the immediate vicinity of shafts I and II. The study also utilized results of a transformation task carried out using the algorithm developed for the basic variant, as well as results from transformations performed separately for each mining area, as discussed in (Żogała *et al.* 2025).

Figure 1 presents the locations of the aforementioned mining areas and mines, including boundary (tie) points and shafts (adjustment points).

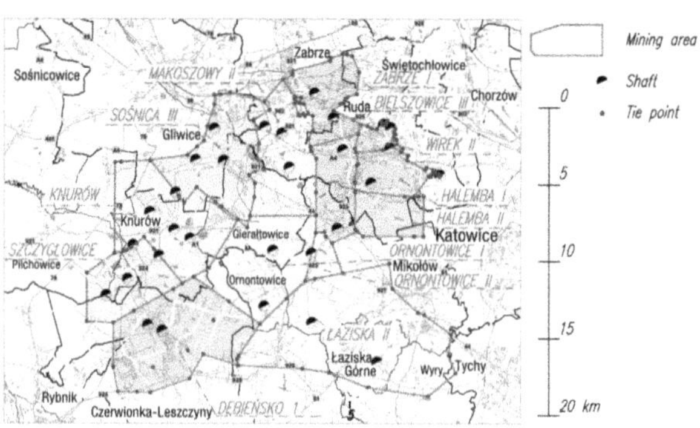

Figure 1. Location of mining areas and mines with boundary points and shafts.

3 ALGORITHM OF THE AUTHORIAL TRANSFORMATION PROCEDURE BASED ON THE METHOD OF INDEPENDENT MODELS FOR LARGE AREAS

To perform the transformation task for boundary points of adjacent mining areas, collectively covering a large spatial area, it is necessary to account for Earth's curvature in the transformation process, including the ellipsoid and cartographic projection. Therefore, the basic transformation procedure using the method of independent models, presented in (Żogała *et al.* 2025), was modified by the authors through the introduction of an additional stage involving the conversion of adjustment point coordinates from the PL–2000 system to geocentric Cartesian XYZ coordinates of the GRS 80 ellipsoid (Geodetic Reference System '80).

In accordance with the assumptions made in this study, identical boundary points found in adjacent mining areas were treated as tie points in the transformation process, while shafts served as adjustment points. The algorithm of the procedure described is presented in Figure 2.

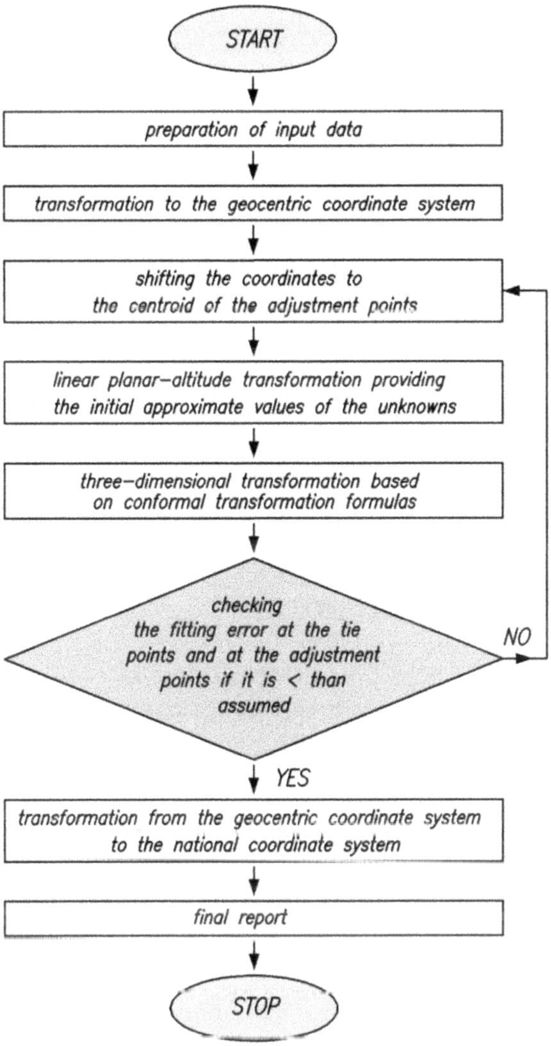

Figure 2. Authorial transformation procedure using the MIM method, incorporating Earth's curvature by transitioning through a geocentric reference system.

The transformation task that takes into account Earth's curvature (ellipsoid and projection) should be carried out according to the following stages:

(I) Preparation of input data is performed analogously to Stage I of the transformation task without including curvature (obtaining the coordinates of the boundary points of mining areas and shafts located in these areas; analyzing the obtained data for completeness and quality; if necessary, conducting field measurements to determine the coordinates of the shafts; preparing the appropriate input files) – (Żogała *et al.* 2025).
(II) Transformation to the geocentric coordinate system.

To eliminate the effect of Earth's curvature, the system of equations of the method of independent models should be solved using Cartesian XYZ coordinates, as described by Formula 1.

$$X = \lambda \cdot A \cdot x + X_0 \tag{1}$$

Where λ – scale change coefficient; A – rotation matrix; X – matrix of coordinates in the secondary system; x – matrix of coordinates in the primary system; X_0 – translation matrix.

Transformation from PL-2000 system (Gauss–Krüger projection) to XYZ system may consist of the following stages (Kadaj 2002):

(1) Conformal transformation of the Gauss–Krüger projection plane to the Mercator projection plane.

$$\begin{aligned} &(\varphi, \Delta\lambda) -> (B, \Delta L) \\ &B = \varphi + c_2 \cdot \sin(2 \cdot \varphi) + c_4 \cdot \sin(4 \cdot \varphi) + c_6 \cdot \sin(6 \cdot \varphi) + \ldots \\ &(\Delta L) -> (\Delta\lambda) \\ &\Delta L = L - L_0 \end{aligned} \tag{2}$$

Where ϕ, λ – geographic coordinates corresponding to a point on Lagrange's sphere (with $\lambda = L$); B, L – geodetic coordinates of the point; L_0 – central meridian longitude (longitude of the meridian tangent to the projection cylinder); c_2, c_4, c_6 – trigonometric series coefficients.

(2) Conformal cylindrical transverse projection of the plane onto the sphere (transverse Mercator projection).

$$\begin{aligned} &(x_{MERC}, y_{MERC}) -> (\varphi, \Delta\lambda) \\ &\alpha = \frac{x_{MERC}}{R_0}, \beta = y_{MERC}/R_0 \\ &w = 2 \cdot arctg[\exp(\beta)] - \pi/2 \\ &\varphi = \arcsin[\cos(w)] \cdot \sin(\alpha)] \\ &\Delta\lambda = arctg[tg(w)/\cos(\alpha)] \end{aligned} \tag{3}$$

where x_{MERC}, y_{MERC} – Mercator projection coordinates; α, β – Mercator coordinates for the unit sphere; w – geographic latitude on the sphere; R_0 – radius of Lagrange's sphere.

(3) Conformal projection of the sphere onto the ellipsoid (Lagrange projection).

$$\begin{aligned} &(x_{GK}, y_{GK}) -> (x_{MERC}, y_{MERC}) \\ &z = [(x_{GK} - a_0) \cdot s, y_{GK} \cdot s] \\ &z_{MERC} = b_0 + z(b_1 + z(b_2 + z(b_3 + z(b_4 + z(b_5 \cdot b_6))))) \end{aligned} \tag{4}$$

where x_{GK}, y_{GK} – Gauss–Krüger projection coordinates; $z_{MERC} = (x_{MERC}, y_{MERC})$ – Mercator projection coordinates in complex form; s – normalization parameter, a_0 – centering parameter; b_0, b_1, b_2, b_3, b_4, b_5, b_6 – polynomial coefficients. Others as above.

In the calculations, an approximate ellipsoidal height *H* of the transformed point was assumed, appended to the *BL* geodetic coordinates. This introduces a point error of about 0.000024 × H[m], which is negligible in standard cartographic processing (Kadaj 2002).

(4) Conversion from geodetic to cartesian coordinates (*BLH*) ⇒ (*XYZ*):

Geodetic coordinates *B*, *L*, *H* are converted to Cartesian coordinates *XYZ* using the following formulas:

$$
\begin{aligned}
x &= (N + H) \cdot \cos B \cdot \cos L \\
y &= (N + H) \cdot \cos B \cdot \sin L \\
z &= \left(\tau^{-1} N + H\right) \cdot \sin B
\end{aligned}
\tag{5}
$$

Where τ and N were calculated from Equations (6 and 7).

$$\tau^{-1} = 1 - e^2 \tag{6}$$

$$N = \frac{a}{\sqrt{\tau^{-1} \sin^2 B}} \tag{7}$$

where *B, L, H* geodetic coordinates; *N* – radius of curvature in the prime vertical; *a* – semi-major axis (equatorial radius); e^2 – squared first eccentricity of the ellipsoid.

Geometric scheme of the Gauss–Krüger projection realization was presented in Figure 3.

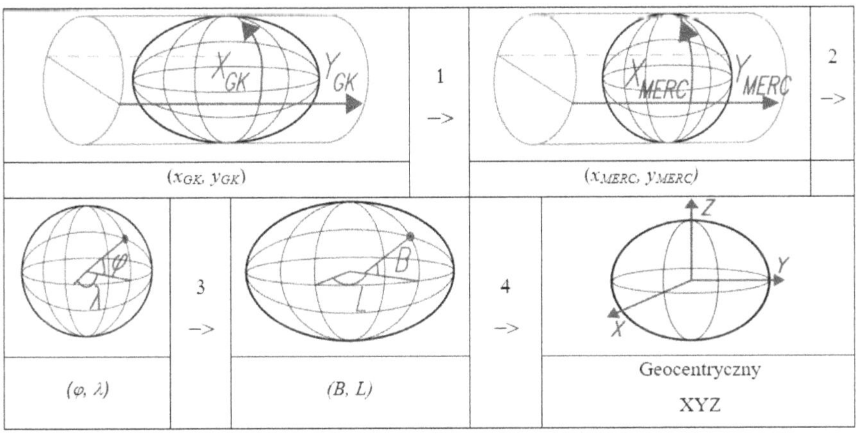

Figure 3. Geometric scheme of the Gauss–Krüger projection realization.

STAGES III–VI - these are executed analogously to stages II–V described in (Żogała *et al.* 2025).

III. Shifting coordinates to the centroid of the adjustment points (shafts)

IV. Plan-height linear transformation – providing initial approximations.

V. 3D transformation based on conformal transformation formulas.

VI. Adjustment error check on tie points; if error exceeds threshold, return to step III.

VII. Final Stage: Transformation from geocentric to secondary system — inverse of step II (Kadaj 2002).

VIII. Final report — Generated if the adjustment error is within acceptable limits.

4 RESULTS

To perform the required calculations, three specialized software programs were used. The transformation using the separate method was carried out with the GEOLISP - version 25 (Geolisp 2025) and GEONET – Geonet 2006 (Geonet 2025) software. Numerical computations related to the transformation task based on the MIM approach (Method of Independent Models), according to the authorial procedure, were performed using the GEONET Dimensional Control package – version 2020 (Geonet 2025).

Tables 1 and 2 present the coordinates of selected corresponding boundary points obtained using the separate method and MIM (variant I and variant II), respectively. Tables 3 and 4 show excerpts of the results concerning differences in coordinates *x*, *y*, and *xy* for both methods (separate and MIM), respectively for variants I and II.

Figure 4 presents the quantitative distribution of the number of identical boundary points in relation to the differences in their location after the transformation in twelve class intervals.

Table 1. Coordinates of selected corresponding boundary points after transformations (Separate method vs. MIM – Variant I).

		"Separate" method		MIM - Variant I	
Point no.	Bordering Areas	*X* [m]	*Y* [m]	*X* [m]	*Y* [m]
8185	Łaziska II – Dębieńsko 1	5555980.639	6552623.952	5555979.690	6552623.730
8184	Łaziska II – Dębieńsko 1	5556375.944	6552565.313	5556374.430	6552565.440
8183	Łaziska II – Dębieńsko 1	5556381.990	6552569.242	5556380.970	6552569.250
8181	Łaziska II – Dębieńsko 1	5558684.534	6554067.191	5558683.970	6554066.960
9461	Łaziska II – Ornontowice I	5560362.640	6555782.467	5560362.060	6555782.180
8181	Dębieńsko 1 – Ornontowice I	5558683.674	6554066.974	5558683.970	6554066.960

Table 2. Coordinates of selected corresponding boundary points after transformations (Separate method vs. MIM – Variant II).

		"Separate" method		MIM -Variant II	
Point no.	Bordering Areas	*X* [m]	*Y* [m]	X [m]	Y [m]
8185	Łaziska II – Dębieńsko 1	5555980.639	6552623.952	5555979.691	6552623.726
8184	Łaziska II – Dębieńsko 1	5556375.944	6552565.313	5556374.432	6552565.443
8183	Łaziska II – Dębieńsko 1	5556381.990	6552569.242	5556380.968	6552569.248
8181	Łaziska II – Dębieńsko 1	5558684.534	6554067.191	5558683.965	6554066.957
9461	Łaziska II – Ornontowice I	5560362.640	6555782.467	5560362.064	6555782.184
8181	Dębieńsko 1 – Ornontowice I	5558683.674	6554066.974	5558683.965	6554066.957

Table 3. Coordinate differences for selected identical boundary points after transformation (Separate method vs. MIM – Variant I).

Point no.	Bordering Areas	x_{I} [m]	y_{I} [m]	xy_{I} [m]
8185	Łaziska II – Dębieńsko 1	0.949	0.222	0.975
8184	Łaziska II – Dębieńsko 1	1.524	0.137	1.530
8183	Łaziska II – Dębieńsko 1	1.030	0.008	1.030
8181	Łaziska II – Dębieńsko 1	0.574	0.221	0.615
9461	Łaziska II – Ornontowice I	0.570	0.277	0.634
8181	Dębieńsko 1 – Ornontowice I	0.286	0.004	0.286

Table 4. Coordinate differences for selected identical boundary points (Separate method vs. MIM with inclusion of Earth curvature – Variant II).

Point no.	Bordering Areas	x_{II} [m]	y_{II} [m]	xy_{II} [m]
8185	Łaziska II – Dębieńsko 1	0.948	0.226	0.974
8184	Łaziska II – Dębieńsko 1	1.512	0.130	1.517
8183	Łaziska II – Dębieńsko 1	1.022	0.006	1.022
8181	Łaziska II – Dębieńsko 1	0.569	0.234	0.615
9461	Łaziska II – Ornontowice I	0.576	0.283	0.641
8181	Dębieńsko 1 – Ornontowice I	0.291	0.017	0.291

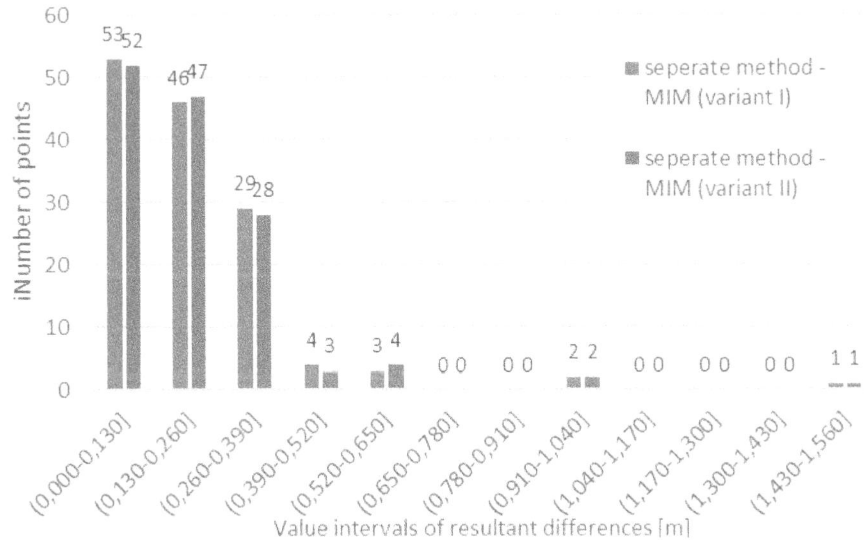

Figure 4. The quantitative distribution of the number of identical boundary points in relation to the differences in their location (variant I and II).

The complete dataset was then analyzed, and the results are presented in Chapter 5.

5 DISCUSSION

The analysis of the results of the coordinate transformation of boundary points performed using the algorithm presented in this article was based on a comparison of the coordinate differences obtained using the aforementioned variant and the basic variant that does not take into account the curvature of the Earth, as well as on a comparison of the coordinate difference results after transformation by the above methods with the results obtained from a separate transformation.

To evaluate the obtained results, key statistical measures were calculated: $\mathrm{Min}_{\Delta X}$ (minimum absolute difference in x-coordinates), $\mathrm{Max}_{\Delta X}$ (maximum absolute difference in x-coordinates), $\sigma_{\Delta X}$ (standard deviation of x-coordinate differences), $\mathrm{Min}_{\Delta Y}$ (minimum absolute difference in y-coordinates), $\mathrm{Max}_{\Delta Y}$ (maximum absolute difference in y-coordinates), $\sigma_{\Delta Y}$ (standard deviation of y-coordinate differences), $\mathrm{Min}_{\Delta XY}$ (minimum resultant difference), $\mathrm{Max}_{\Delta XY}$ (maximum resultant difference), $\sigma_{\Delta XY}$ (standard deviation of resultant differences). The results are presented in tables containing values calculated for coordinates after transformation using the separate method and the MIM-based method – Variant I (Table 5), the separate method and the MIM-based method – Variant II (Table 6), and between the MIM-based method – Variant I and Variant II (Table 7).

Table 5. Statistical summary (Separate method vs. MIM – Variant I).

$Min_{\Delta x}$ [mm]	$Max_{\Delta x}$ [mm]	$\sigma_{\Delta X}$ [mm]	$Min_{\Delta y}$ [mm]	$Max_{\Delta y}$ [mm]	$\sigma_{\Delta Y}$ [mm]	$Min_{\Delta xy}$ [mm]	$Max_{\Delta xy}$ [mm]	$\sigma_{\Delta XY}$ [mm]
13	1527	418	47	404	82	16	1544	349

Table 6. Statistical summary (Separate method vs. MIM – Variant II).

$Min_{\Delta x}$ [mm]	$Max_{\Delta x}$ [mm]	$\sigma_{\Delta X}$ [mm]	$Min_{\Delta y}$ [mm]	$Max_{\Delta y}$ [mm]	$\sigma_{\Delta Y}$ [mm]	$Min_{\Delta xy}$ [mm]	$Max_{\Delta xy}$ [mm]	$\sigma_{\Delta XY}$ [mm]
15	1512	197	42	421	70	19	1529	196

Table 7. The coordinate differences of identical boundary points after performing both transformations, "MIM Variant I and Variant II".

$Min_{\Delta x}$ [mm]	$Max_{\Delta x}$ [mm]	$\sigma_{\Delta X}$ [mm]	$Min_{\Delta y}$ [mm]	$Max_{\Delta y}$ [mm]	$\sigma_{\Delta Y}$ [mm]	$Min_{\Delta xy}$ [mm]	$Max_{\Delta xy}$ [mm]	$\sigma_{\Delta XY}$ [mm]
2	66	10	0	19	6	2	66	10

A comparison of the basic statistical indicators presented in Tables 5 and 6 indicates that, in the case of using the MIM-based algorithm in Variant II for the transformation task, significantly lower values of the key statistical measures were obtained. The position differences of identical boundary points in the analyzed areas, after performing the transformation using the separate method and MIM (Variant II), ranged from 0.019 m to 1.529 m, whereas in the case of the transformation using the separate method and MIM (Variant I), they ranged from 0.016 m to 1.544 m. The coordinate difference Δx in the analysis for Variant II reached a maximum value of approximately 1.512 m, while in Variant I it was 1.527 m. The coordinate differences Δy reached 0.421 m in Variant II and 0.404 m in Variant I, respectively.

The position differences of identical boundary points in the analyzed areas, after performing the transformation using the MIM method (in both variants), ranged from 0.002 m to 0.066 m. The maximum Δx coordinate difference reached 0.066 m, and the Δy coordinate difference reached 0.019 m (Table 7).

The above indicates more favorable statistical measures in the case of using the algorithm variant that accounts for the curvature of the Earth, including the ellipsoid and map projection. However, the comparison of coordinate differences resulting from the transformation using Variant I and Variant II in the study area shows that these values do not significantly affect the final result (Figure 4), as they fall within the acceptable error margins for determining the coordinates of boundary points and do not exceed 0.100 m.

6 SUMMARY

This publication addresses the issue of discrepancies in the coordinates of identical boundary points in adjacent areas, which arise as a result of coordinate transformations between different coordinate systems. To ensure that boundary points have identical coordinates after transformation, the authors proposed an algorithm based on MIM (Method of Independent Models). The algorithm was developed in two variants: Variant I, which does not directly account for the Earth's curvature, and Variant II, which incorporates the Earth's curvature, including the ellipsoid and map projection.

This paper presents the algorithm of the author's transformation procedure based on MIM in Variant II, intended for application over large areas, along with the results of a transformation task conducted using this approach. The procedure was tested on 14 mining areas located in the southern part of Poland, in the region of the Upper Silesian Coal Basin. The developed transformation procedure can be applied to very large areas, but due to the data obtained, it was limited only to the indicated mining areas. The results were compared with coordinates obtained through separate transformations performed individually for each mining area, as well as with those derived from the algorithm in Variant I.

In the studied region, the coordinates obtained using the authors' procedure with curvature correction are close to those obtained with Variant I. The differences are minor, averaging approximately 0.004 m. The additional adjustment to the ellipsoid had no significant impact on the final result. The total area of the analyzed mining regions in this case amounted to 343 km^2.

The reason for such small differences lies in the fact that in Variant I, the coordinates of the secondary system (PL–2000) already incorporate the GRS 80 ellipsoid. The MIM algorithm iteratively fits and deforms the entire aeroblock in order to minimize errors at the target points on the ellipsoid. Nevertheless, it is important to be aware of the presence of a certain systematic error, the magnitude of which depends on the size of the transformed area and the type of coordinate system used. Therefore, it is always advisable to use the variant proposed in this article, which directly accounts for the Earth's curvature, including the ellipsoid and the cartographic projection.

REFERENCES

Act of June 9, Geological and Mining Law/Ustawa z Dnia 9 Czerwca 2011 r. Prawo Geologiczne i Górnicze. 2011. *Dz. U.* 163.981. https://isap.sejm.gov.pl/isap.nsf/DocDetails.xsp?id=wdu20111630981.

Act of May 17, 1989 Geodetic and cartographic law/Ustawa z dnia 17 maja 1989 r. Prawo geodezyjne i kartograficzne. 1989. *Dz. U.* 30.163.

Banković, T., Dubrović, M., Banko, A. and Pavasović, M. (2024). Detail Explanation of Coordinate Transformation Procedures Used in Geodesy and Geoinformatics on the Example of the Territory of the Republic of Croatia, *Appl. Sci.* 14, 1067. https://doi.org/10.3390/app14031067.

Even–Tzur, G. (2020). Coordinate transformation with variable number of parameters. *Survey Review*: Volume 52, Issue 370. https://doi.org/10.1080/00396265.2018.1517477.

Geolisp – Mine Digital Map System/Geolisp – System kopalnianych map numerycznych. 2025.
www.geolisp.pl (accessed 1 June 2025).

Geonet – Algorithms and IT solutions for geodesy, cartography, and GNSS systems/ Geonet – Algorytmy i rozwiązania informatyczne dla geodezji, kartografii, systemów GNSS. www.geonet.net.pl (accessed 1 June 2025).

James, B. and Steed, B. (2012). A practical approach to transformation between commonly used reference systems. *Journal Australian Surveyor*: 35, 248–264. https://doi.org/10.1080/00050326.1990.10438680.

Janssen, V. (2009). Understanding coordinate systems, datums and transformations in Australia. *International Journal of Geoinformatics*: 5(4), 697–715. https://www.researchgate.net/publication/228649896.

Kadaj, R. (2002). Coordinate Systems in Poland: Transformation Methods, Algorithms, and Programs/ Polskie układy współrzędnych. Formuły transformacyjne, algorytmy i programy. Rzeszów.

Maciaszek, J. (2010). Archival Maps and Mining Fields Information System for Spatial Development Planning/System informacji o archiwalnych mapach i polach górniczych na potrzeby zagospodarowania przestrzennego. *Wydawnictwo AGH*. Kraków.

Regulation of the Minister of the Environment of October 28, 2015 on Surveying and Geological Documentation/Rozporządzenie Ministra Środowiska z dnia 28 Października 2015 r. w Sprawie

Dokumentacji Mierniczo-Geologicznej. 2015. *Dz. U.* 2015, poz. 1941. https://isap.sejm.gov.pl/isap.nsf/DocDetails.xsp?id=WDU20150001941.

Regulation of the Council of Ministers of October 15, 2012 on the State System of Spatial References/Rozporządzenie Rady Ministrów z Dnia 15 Października 2012 r. w Sprawie Państwowego Systemu Odniesień Przestrzennych. 2012. *Dz. U.* 2012, poz. 1247. https://isap.sejm.gov.pl/isap.nsf/DocDetails.xsp?id=WDU20120001247.

Sisman, Y. (2014). Coordinate transformation of cadastral maps using different adjustment methods. *Journal of the Chinese Institute of Engineers*: 37, 869–882. https://doi.org/10.1080/02533839.2014.888800.

Sokoła–Szewioła, V. *et al.* (2017). Statutory research project BK No. 06/050/BK17/0042. Evaluation of accuracy and development of transformation parameters between local rectangular plane coordinate systems used in mining enterprises and the national coordinate system using GIS-class software/Badania statutowe BK nr 06/050/BK17/0042. Ocena dokładności i opracowanie parametrów trans–formacji lokalnych układów współrzędnych prostokątnych płaskich stosowanych w przedsiębiorstwach górniczych do układu państwowego z oprogramowaniem w systemie klasy GIS, *Politechnika Śląska*. Gliwice.

Żogała, M. (2022). Application of the independent models method for the transformation of local coordinate systems used in hard coal mines/Zastosowanie metody niezależnych modeli do transformacji lokalnych układów współrzędnych stosowanych w kopalniach węgla kamiennego. *Praca doktorska*, Politechnika Śląska, Gliwice.

Żogała, M., Sokoła–Szewioła, V., Poniewiera, M. (2025). Transformation of Coordinates of Boundary Points of Neighboring Mining Areas Using an Authorial Procedure Based on the Method of Independent Models – A Case Study. *Appl. Sci.*:15(11), 5812. https://doi.org/10.3390/app15115812.

© 2025 The Editor(s), ISBN: 978-1-041-10531-2

Author index

Apanowicz, B. 1, 12

Bętkowski, P. 20
Błotnicki, J. 69
Balcarczyk, L. 187
Blachowski, J. 30
Bonczyk, M. 12
Bryt-Nitarska, I. 40, 50

Chomacki, L. 197

Dai, H.-y. 111
Dyduch, G. 101

Florkowska, L. 50
Fuławka, K. 217

Gromysz, K. 60
Gruchlik, P. 12
Gruszczyński, M.F. 69
Grygierek, M. 255

Hejmanowski, R. 111

Jaśkowski, W. 127
Jarczyk, P. 79, 206

Karsznia, K. 89
Kowalski, A. 12
Kruczkowski, M. 101
Książek, K. 89
Kura, K. 12
Kwinta, A. 111

Łabojko, S. 136
Lipecki, T. 127

Malinowska, A. 111, 136
Mazur, J. 155
Menéndez-García, L.A. 181
Menéndez Fernández, M. 181
Mierzejowska, A. 145
Modlińska, P. 30

Niedbalska, K. 12
Niewiadomski, A.P. 187

Pałęcki, A. 69
Pach, G. 187
Paszek, J. 155
Pawliński, J. 187
Pierzyna, P. 164
Pomykoł, M. 40, 164
Poniewiera, M. 164, 173, 263
Popczyk, M. 69

Różański, Z. 187
Rodríguez-Galindo Díez, R. 181
Rusek, J. 197

Słowik, L. 197
Sokoła-Szewioła, V. 181, 206, 263
Sowa, W. 69
Stasch, K. 69
Stolecki, L. 217
Strozik, G. 226
Strzałkowski, P. 238
Stypiński, B. 136
Szafarczyk, A. 248
Szafulera-Bałecka, K. 40, 79
Szojda, L. 238
Szostak-Chrzanowski, A. 1

Wala, A. 187
Walerysiak, N. 30
Wróblewska, M. 255
Wrona, P. 187
Wysocka, M. 12

Żogała, M. 263

For Product Safety Concerns and Information please contact our EU
representative GPSR@taylorandfrancis.com
Taylor & Francis Verlag GmbH, Kaufingerstraße 24, 80331 München, Germany

www.ingramcontent.com/pod-product-compliance
Ingram Content Group UK Ltd.
Pitfield, Milton Keynes, MK11 3LW, UK
UKHW052213190926
475702UK00006B/992

* 9 7 8 1 0 4 1 1 0 5 3 1 2 *